Student Solutions Manual
for
Tipler and Mosca's
Physics for Scientists and Engineers
Fifth Edition
Volume 2

DAVID MILLS
Professor Emeritus
College of the Redwoods

with

CHARLES L. ADLER
Saint Mary's College of Maryland

EDWARD A. WHITTAKER
Professor of Physics
Stevens Institute of Technology

GEORGE ZOBER
Yough Senior High School

PATRICIA ZOBER
Ringgold High School

W. H. Freeman and Company
New York

Copyright © 2004 by W. H. Freeman and Company

All rights reserved.

Printed in the United States of America

ISBN: 0-7167-8334-7

First printing 2003

W. H. Freeman and Company
41 Madison Avenue
New York, NY 10010
Houndmills, Basingstoke RG21 6XS, England

Contents

To the Student, v

Acknowledgments, vii

About the Authors, ix

Chapter 21   The Electric Field I: Discrete Charge Distributions, 1

Chapter 22   The Electric Field II: Continuous Charge Distributions, 29

Chapter 23   Electric Potential, 51

Chapter 24   Electrostatic Energy and Capacitance, 71

Chapter 25   Electric Current and Direct-Current Circuits, 97

Chapter 26   The Magnetic Field, 135

Chapter 27   Sources of the Magnetic Field, 159

Chapter 28   Magnetic Induction, 189

Chapter 29   Alternating-Current Circuits, 219

Chapter 30   Maxwell's Equations and Electromagnetic Waves, 253

Chapter 31   Properties of Light, 275

Chapter 32   Optical Images, 291

Chapter 33   Interference and Diffraction, 315

Chapter 34   Wave-Particle Duality and Quantum Physics, 337

Chapter 35   Applications of the Schrödinger Equation, 353

Chapter 36   Atoms, 363

Chapter 37   Molecules, 379

Chapter 38   Solids and the Theory of Conduction, 393

Chapter 39   Relativity, 407

Chapter 40   Nuclear Physics, 429

Chapter 41   Elementary Particles and the Beginning of the Universe, 447

To the Student

This solution manual accompanies *Physics for Scientists and Engineers, 5e*, by Paul Tipler and Gene Mosca. Following the structure of the solutions to the Worked Examples in the text, we begin the solutions to the back-of-the-chapter numerical problems with a brief discussion of the physics of the problem, represent the problem pictorially whenever appropriate, express the physics of the solution in the form of a mathematical model, fill in any intermediate steps as needed, make the appropriate substitutions and algebraic simplifications, and complete the solution with the substitution of numerical values (including their units) and the evaluation of whatever physical quantity is called for in the problem. This is the problem-solving strategy used by experienced learners of physics, and it is our hope that you will see the value in such an approach to problem solving and learn to use it consistently.

Believing that it will maximize your learning of physics, we encourage you to create your own solution before referring to the solutions in this manual. You may find that, by following this approach, you will find different, but equally valid, solutions to some of the problems. In any event, studying the solutions contained herein without having first attempted the problems will do little to help you learn physics.

You'll find that nearly all problems with numerical answers have their answers given to three significant figures. Most of the exceptions to this rule are in the solutions to the problems on Significant Figures and Order of Magnitude and the problems dealing with nuclear physics. When the nature of the problem makes it desirable to do so, we keep more than three significant figures in the answers to intermediate steps and then round to three significant figures for the final answer. Some of the Estimation and Approximation Problems have answers to fewer than three significant figures.

Physics for Scientists and Engineers, 5e includes numerous spreadsheet problems. Most of them call for the plotting of one or more graphs. The solutions to these problems were generated using Microsoft Excel and its "paste special" feature, so that you can easily make changes to the graphical parts of the solutions.

Acknowledgments

Charles L. Adler (Saint Mary's College of Maryland), Ed Whittaker (Stevens Institute of Technology, George Zober (Yough Senior High School) and Patricia Zober (Ringgold High School) are the authors of the new problems appearing in the Fifth Edition. Chuck, Ed, George, and Patricia saved me (dm) many hours of work by providing rough-draft solutions to these new problems, and I thank them for their help. Gene Mosca (United States Naval Academy and the co-author of the Fifth Edition) helped me tremendously by reviewing my work, helping me clarify many of my solutions, and providing solutions when I was unsure how best to proceed. It was a pleasure to collaborate with Gene in the creation of this solutions manual. All of us who were involved in the creation of this solutions manual hope that you will find the solutions useful in learning physics.

We want to thank Lay Nam Chang (Virginia Polytechnic Institute), Brent A. Corbin (UCLA), Alan Cresswell (Shippensburg University), Ricardo S. Decca (Indiana University–Purdue University), Michael Dubson (The University of Colorado at Boulder), David Faust (Mount Hood Community College), Philip Fraundorf (The University of Missouri–Saint Louis), Clint Harper (Moorpark College), Kristi R. G. Hendrickson (University of Puget Sound), Michael Hildreth (The University of Notre Dame), David Ingram (Ohio University), James J. Kolata (The University of Notre Dame), Eric Lane (The University of Tennessee–Chattanooga), Jerome Licini (Lehigh University), Laura McCullough (The University of Wisconsin–Stout), Carl Mungan (United States Naval Academy), Jeffrey S. Olafsen (University of Kansas), Robert Pompi (The State University of New York at Binghamton), R. J. Rollefson (Wesleyan University), Andrew Scherbakov (Georgia Institute of Technology), Bruce A. Schumm (University of Chicago), Dan Styer (Oberlin College), Daniel Marlow (Princeton University), Jeffrey Sundquist (Palm Beach Community College–South), Cyrus Taylor (Case Western Reserve University), and Fulin Zuo (University of Miami), for their reviews of the problems and their solutions.

Jerome Licini (Lehigh University), Michael Crivello (San Diego Mesa College), Paul Quinn (University of Kansas), and Daniel Lucas (University of Wisconsin–Madison) error-checked the solutions. Without their thorough and critical work, many errors would have remained to be discovered by the users of this solutions manual. Their assistance is greatly appreciated. In spite of their best efforts, there may still be errors in some of the solutions, and for those I (dm) assume full responsibility. Should you find errors or think of alternative solutions that you would like to call to my attention, I would appreciate it if you would communicate them to me by sending them to asktipler@whfreeman.com.

It was a pleasure to work with Brian Donnellan, Media and Supplements Editor for Physics, who guided us through the creation of this solution manual. Our thanks to Amanda McCorquodale and Eileen McGinnis for organizing the reviewing and error-checking process.

September 2003

David Mills
Professor Emeritus
College of the Redwoods

Charles L. Adler
Saint Mary's College of Maryland

Edward A. Whittaker
Professor of Physics
Stevens Institute of Technology

George Zober
Yough Senior High School

Patricia Zober
Ringgold High School

About the Authors

David Mills, Professor Emeritus, College of the Redwoods, retired in May of 2000 after a teaching career of 42 years. He earned his bachelor's degree at Humboldt State College, his master's degree at California State University–Hayward, and his doctoral degree at the University of Northern Colorado. His teaching career included experience with the Physical Science Study Committee materials, the Harvard Project curriculum, the Personalized System of Instruction, Microcomputer-Based Laboratory instruction, and the interactive-engagement movement in physics education. A 1996 NSF.ILI grant allowed him to transform instruction in physics at the College of the Redwoods from a traditional lecture-laboratory delivery system to one that was microcomputer based, eliminate the distinction between lecture and laboratory, and utilize interactive-engagement teaching and learning strategies. He authored the Test Bank to accompany *Physics for Scientists and Engineers, 3e* and *4e*. He now lives in Henderson, NV and is an Adjunct Professor at the Community College of Southern Nevada.

Charles L. Adler is a professor of physics at St. Mary's College of Maryland. He received his undergraduate, masters, and doctoral degrees in physics from Brown University before doing his postdoctoral work at the Naval Research Laboratory in Washington, D.C. His research covers a wide variety of fields, including nonlinear optics, electrooptics, acoustics, cavity quantum electrodynamics, and pure mathematics. His current interests concern problems in light scattering, inverse scattering, and atmospheric optics. Dr. Adler is the author of over 30 publications.

Edward A Whittaker has been a professor of physics at Stevens Institute of Technology since 1984. His research interests include laser spectroscopy, quantum optics, and optical communications. In 2003 he was named an American Institute of Physics State Department Science Fellow.

George Zober is a teacher of Advanced Placement Physics at Yough High School in western Pennsylvania. He serves as a Physics Consultant with the College Board and teaches Advanced Placement Physics Workshops at Wilkes University and Manhattan College during his summers.

Patricia J. Zober teaches Advanced Placement Physics at Ringgold High School in Monongahela, Pennsylvania and is a Physics Consultant for the College Board. Patricia presents Advanced Placement Workshops in Physics and during her summers is a faculty physics project leader with the Governor's School for the Sciences at Carnegie Mellon University.

Chapter 21
The Electric Field 1: Discrete Charge Distributions

Conceptual Problems

*1 •• Discuss the similarities and differences in the properties of electric charge and gravitational mass.

Similarities:	Differences:
The force between charges and masses varies as $1/r^2$.	There are positive and negative charges but only positive masses.
The force is directly proportional to the product of the charges or masses.	Like charges repel; like masses attract.
	The gravitational constant G is many orders of magnitude smaller than the Coulomb constant k.

*5 •• Two uncharged conducting spheres with their conducting surfaces in contact are supported on a large wooden table by insulated stands. A positively charged rod is brought up close to the surface of one of the spheres on the side opposite its point of contact with the other sphere. (*a*) Describe the induced charges on the two conducting spheres, and sketch the charge distributions on them. (*b*) The two spheres are separated far apart and the charged rod is removed. Sketch the charge distributions on the separated spheres.

Determine the Concept Because the spheres are conductors, there are free electrons on them that will reposition themselves when the positively charged rod is brought nearby.

(*a*) On the sphere near the positively charged rod, the induced charge is negative and near the rod. On the other sphere, the net charge is positive and on the side far from the rod. This is shown in the diagram.

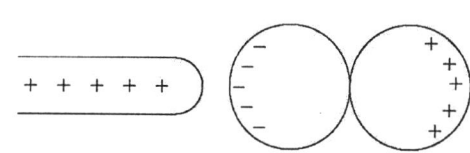

(*b*) When the spheres are separated and far apart and the rod has been removed, the induced charges are distributed uniformly over each sphere. The charge distributions are shown in the

*7 • A positive charge that is free to move but is at rest in an electric field $\vec{E}$ will
(a) accelerate in the direction perpendicular to $\vec{E}$.
(b) remain at rest.
(c) accelerate in the direction opposite to $\vec{E}$.
(d) accelerate in the same direction as $\vec{E}$.
(e) do none of the above.

Determine the Concept The acceleration of the positive charge is given by $\vec{a} = \dfrac{\vec{F}}{m} = \dfrac{q_0}{m}\vec{E}$. Because q_0 and m are both positive, the acceleration is in the same direction as the electric field. $\boxed{(d) \text{ is correct.}}$

*8 • If four charges are placed at the corners of a square as shown in Figure 21-33, the field $\vec{E}$ is zero at

(a) all points along the sides of the square midway between two charges.
(b) the midpoint of the square.
(c) midway between the top two charges and midway between the bottom two charges.
(d) none of the above.

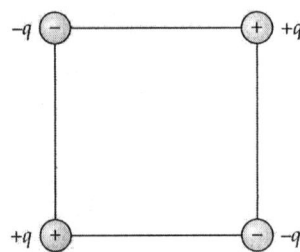

Figure 21-33 Problem 8

Determine the Concept $\vec{E}$ is zero wherever the net force acting on a test charge is zero. At the center of the square the two positive charges alone would produce a net electric field of zero, and the two negative charges alone would also produce a net electric field of zero. Thus, the net force acting on a test charge at the midpoint of the square will be zero. $\boxed{(b) \text{ is correct.}}$

*11 • Two charges $+q$ and $-3q$ are separated by a small distance. Draw the electric field lines for this system.

Determine the Concept We can use the rules for drawing electric field lines to draw the electric field lines for this system. In the field-line sketch to the right we've assigned 2 field lines to each charge q.

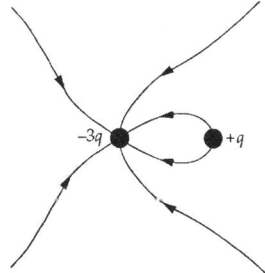

*12 • Three equal positive point charges are situated at the corners of an equilateral triangle. Sketch the electric field lines in the plane of the triangle.

Determine the Concept We can use the rules for drawing electric field lines to draw the electric field lines for this system. In the field-line sketch to the right we've assigned 7 field lines to each charge q.

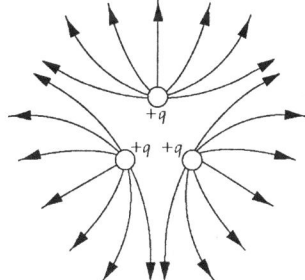

*14 • The electric field lines around an electrical dipole are best represented by which, if any, of the diagrams in Figure 21-34?

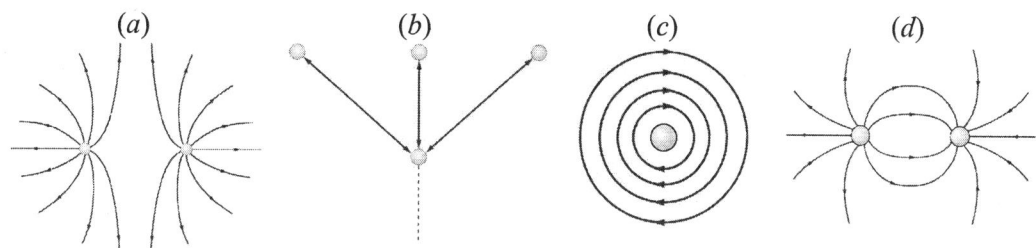

Figure 21-34 Problem 14

Determine the Concept Electric field lines around an electric dipole originate at the positive charge and terminate at the negative charge. Only the lines shown in (d) satisfy this requirement. $\boxed{(d) \text{ is correct.}}$

*15 •• A molecule with electric dipole moment $\vec{p}$ is oriented so that $\vec{p}$ makes an angle θ with a uniform electric field $\vec{E}$ that is in the direction of increasing x. The dipole is free to move in response to the force from the field. Describe the motion of the dipole. Suppose the electric field is nonuniform and is larger in the x direction. How will the motion be changed?

Determine the Concept Because $\theta \neq 0$, a dipole in a uniform electric field will experience a restoring torque whose magnitude is $pE_x \sin\theta$. Hence it will oscillate about its equilibrium orientation, $\theta = 0$. If $\theta \ll 1$, $\sin\theta \approx \theta$, and the motion will be simple harmonic motion. Because the field is nonuniform and is larger in the x direction, the force acting on the positive charge of the dipole (in the direction of increasing x) will be greater than the force acting on the negative charge of the dipole (in the direction of decreasing x) and thus there will be a net electric force on the dipole in the direction of increasing x. Hence, the dipole will accelerate in the x direction as it oscillates about $\theta = 0$.

*18 ••   A metal ball is positively charged. Is it possible for it to attract another positively charged ball? Explain.

Determine the Concept Yes. A positively charged ball will induce a dipole on the metal ball, and if the two are in close proximity, the net force can be attractive.

*19 ••   A simple demonstration of electrostatic attraction can be done simply by tying a small ball of tinfoil on a hanging string, and bringing a charged wand near it. Initially, the ball will be attracted to the wand, but once they touch, the ball will be repelled violently from it. Explain this behavior.

Determine the Concept Assume that the wand has a negative charge. When the charged wand is brought near the tinfoil, the side nearer the wand becomes positively charged by induction, and so it swings toward the wand. When it touches the wand, some of the negative charge is transferred to the foil, which, as a result, acquires a net negative charge and is now repelled by the wand.

Estimation and Approximation

*23 ••   A popular classroom demonstration consists of rubbing a "magic wand" made of plastic with fur to charge it, and then placing it near an empty soda can on its side (Figure 21-35). The can will roll toward the wand, as it acquires a charge on the side nearest the wand by induction. Typically, if the wand is held about 10 cm away from the can, the can will have an initial acceleration of about 1 m/s². If the mass of the can is 0.018 kg, estimate the charge on the rod.

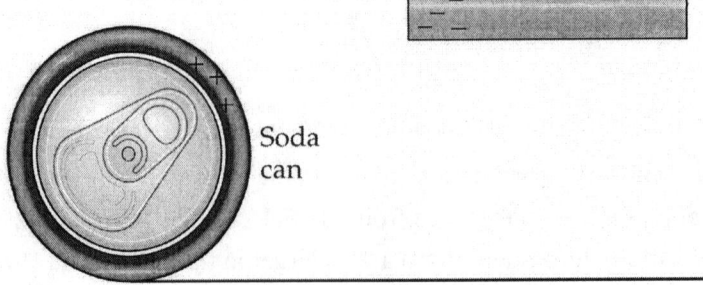

Figure 21-35 Problem 23

Picture the Problem We can use Coulomb's law to express the charge on the rod in terms of the force exerted on it by the soda can and its distance from the can. We can apply Newton's 2nd law in rotational form to the can to relate its acceleration to the electric force exerted on it by the rod. Combining these equations will yield an expression for Q as a function of the mass of the can, its distance from the rod, and its acceleration.

Use Coulomb's law to relate the force on the rod to its charge Q and distance r from the soda can:

$$F = \frac{kQ^2}{r^2}$$

Solve for Q to obtain:

$$Q = \sqrt{\frac{r^2 F}{k}} \quad (1)$$

Apply $\sum \tau_{\text{center of mass}} = I\alpha$ to the can:

$$FR = I\alpha$$

Because the can rolls without slipping, we know that its linear acceleration a and angular acceleration α are related according to:

$$\alpha = \frac{a}{R}$$

where R is the radius of the soda can.

Because the empty can is a hollow cylinder:

$$I = MR^2$$
where M is the mass of the can.

Substitute for I and α and solve for F to obtain:

$$F = \frac{MR^2 a}{R^2} = Ma$$

Substitute for F in equation (1):

$$Q = \sqrt{\frac{r^2 Ma}{k}}$$

Substitute numerical values and evaluate Q:

$$Q = \sqrt{\frac{(0.1\,\text{m})^2 (0.018\,\text{kg})(1\,\text{m/s}^2)}{8.99 \times 10^9 \,\text{N} \cdot \text{m}^2 / \text{C}^2}}$$

$$= \boxed{141\,\text{nC}}$$

Electric Charge

*27 • How many coulombs of positive charge are there in 1 kg of carbon? Twelve grams of carbon contain Avogadro's number of atoms, with each atom having six protons and six electrons.

Picture the Problem We can find the number of coulombs of positive charge there are in 1 kg of carbon from $Q = 6n_C e$, where n_C is the number of atoms in 1 kg of carbon and the factor of 6 is present to account for the presence of 6 protons in

6 Chapter 21

each atom. We can find the number of atoms in 1kg of carbon by setting up a proportion relating Avogadro's number, the mass of carbon, and the molecular mass of carbon to n_C.

Express the positive charge in terms of the electronic charge, the number of protons per atom, and the number of atoms in 1 kg of carbon:

$$Q = 6n_C e$$

Using a proportion, relate the number of atoms in 1 kg of carbon n_C, to Avogadro's number and the molecular mass M of carbon:

$$\frac{n_C}{N_A} = \frac{m_C}{M} \Rightarrow n_C = \frac{N_A m_C}{M}$$

Substitute to obtain:

$$Q = \frac{6 N_A m_C e}{M}$$

Substitute numerical values and evaluate Q:

$$Q = \frac{6(6.02 \times 10^{23} \text{ atoms/mol})(1 \text{ kg})(1.6 \times 10^{-19} \text{ C})}{0.012 \text{ kg/mol}} = \boxed{4.82 \times 10^7 \text{ C}}$$

Coulomb's Law

*32 •• A point charge of -2.5 μC is located at the origin. A second point charge of 6 μC is at $x = 1$ m, $y = 0.5$ m. Find the x and y coordinates of the position at which an electron would be in equilibrium.

Picture the Problem The positions of the charges are shown in the diagram. It is apparent that the electron must be located along the line joining the two charges. Moreover, because it is negatively charged, it must be closer to the -2.5 μC than to the 6.0 μC charge, as is indicated in the figure. We can find the x and y coordinates of the electron's position by equating the two electrostatic forces acting on it and solving for its distance from the origin.

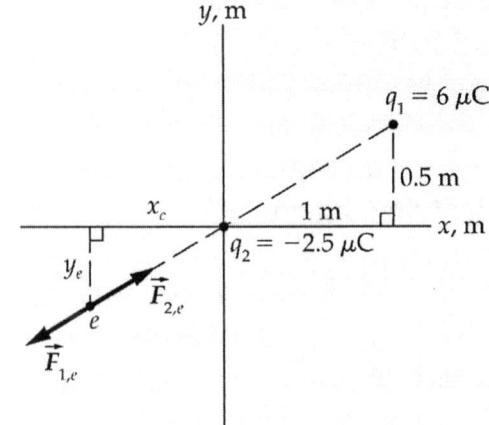

We can use similar triangles to express this radial distance in terms of the x and y coordinates of the electron.

Express the condition that must be satisfied if the electron is to be in equilibrium:	$F_{1,e} = F_{2,e}$
Express the magnitude of the force that q_1 exerts on the electron:	$F_{1,e} = \dfrac{kq_1 e}{\left(r + \sqrt{1.25\,\text{m}}\right)^2}$
Express the magnitude of the force that q_2 exerts on the electron:	$F_{2,e} = \dfrac{k|q_2|e}{r^2}$
Substitute and simplify to obtain:	$\dfrac{q_1}{\left(r + \sqrt{1.25\,\text{m}}\right)^2} = \dfrac{|q_2|}{r^2}$
Substitute for q_1 and q_2 and simplify:	$\left(-1.4\,\text{m}^{-2}\right)r^2 + \left(2.2361\,\text{m}^{-1}\right)r + 1.25\,\text{m} = 0$
Solve for r to obtain:	$r = 2.036\,\text{m}$ and $r = -0.4386\,\text{m}$ Because $r < 0$ is unphysical, we'll consider only the positive root.
Use the similar triangles in the diagram to establish the proportion involving the y coordinate of the electron:	$\dfrac{y_e}{0.5\,\text{m}} = \dfrac{2.036\,\text{m}}{1.12\,\text{m}}$
Solve for y_e:	$y_e = 0.909\,\text{m}$
Use the similar triangles in the diagram to establish the proportion involving the x coordinate of the electron:	$\dfrac{x_e}{1\,\text{m}} = \dfrac{2.036\,\text{m}}{1.12\,\text{m}}$
Solve for x_e:	$x_e = 1.82\,\text{m}$
The coordinates of the electron's position are:	$(x_e, y_e) = \boxed{(-1.82\,\text{m},\, -0.909\,\text{m})}$

*33 •• A charge of −1.0 µC is located at the origin; a second charge of 2.0 µC is located at $x = 0$, $y = 0.1$ m; and a third charge of 4.0 µC is located at $x = 0.2$ m, $y = 0$. Find the forces that act on each of the three charges.

Picture the Problem Let q_1 represent the charge at the origin, q_2 the charge at (0, 0.1 m), and q_3 the charge at (0.2 m, 0). The diagram shows the forces acting on each of the charges. Note the action-and-reaction pairs. We can apply Coulomb's law and the principle of superposition of forces to find the net force acting on each of the charges.

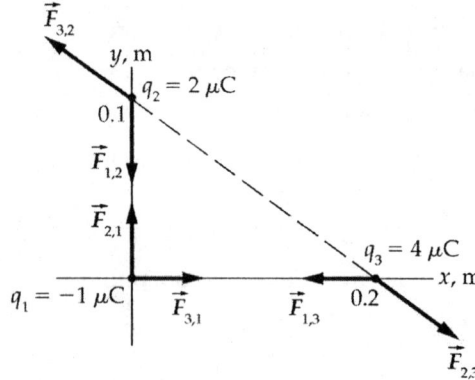

Express the net force acting on q_1:
$$\vec{F}_1 = \vec{F}_{2,1} + \vec{F}_{3,1}$$

Express the force that q_2 exerts on q_1:
$$\vec{F}_{2,1} = \frac{kq_2q_1}{r_{2,1}^2}\hat{r}_{2,1} = \frac{kq_2q_1}{r_{2,1}^2}\frac{\vec{r}_{2,1}}{r_{2,1}} = \frac{kq_2q_1}{r_{2,1}^3}\vec{r}_{2,1}$$

Substitute numerical values and evaluate $\vec{F}_{2,1}$:

$$\vec{F}_{2,1} = (8.99\times 10^9\ \text{N}\cdot\text{m}^2/\text{C}^2)(2\,\mu\text{C})\frac{(-1\,\mu\text{C})}{(0.1\,\text{m})^3}(-0.1\,\text{m})\hat{j} = (1.80\,\text{N})\hat{j}$$

Express the force that q_3 exerts on q_1:
$$\vec{F}_{3,1} = \frac{kq_3q_1}{r_{3,1}^3}\vec{r}_{3,1}$$

Substitute numerical values and evaluate $\vec{F}_{3,1}$:

$$\vec{F}_{3,1} = (8.99\times 10^9\ \text{N}\cdot\text{m}^2/\text{C}^2)(4\,\mu\text{C})\frac{(-1\,\mu\text{C})}{(0.2\,\text{m})^3}(-0.2\,\text{m})\hat{i} = (0.899\,\text{N})\hat{i}$$

Substitute to find $\vec{F}_1$:
$$\vec{F}_1 = \boxed{(0.899\,\text{N})\hat{i} + (1.80\,\text{N})\hat{j}}$$

Express the net force acting on q_2:
$$\vec{F}_2 = \vec{F}_{3,2} + \vec{F}_{1,2}$$
$$= \vec{F}_{3,2} - \vec{F}_{2,1}$$
$$= \vec{F}_{3,2} - (1.80\,\text{N})\hat{j}$$

because $\vec{F}_{1,2}$ and $\vec{F}_{2,1}$ are action-and-reaction forces.

Express the force that q_3 exerts on q_2:

$$\vec{F}_{3,2} = \frac{kq_3q_2}{r_{3,2}^3}\vec{r}_{3,2}$$

$$= \frac{kq_3q_2}{r_{3,2}^3}\left[(-0.2\,\text{m})\hat{i} + (0.1\,\text{m})\hat{j}\right]$$

Substitute numerical values and evaluate $\vec{F}_{3,2}$:

$$\vec{F}_{3,2} = (8.99\times 10^9\,\text{N}\cdot\text{m}^2/\text{C}^2)(4\,\mu\text{C})\frac{(2\,\mu\text{C})}{(0.224\,\text{m})^3}\left[(-0.2\,\text{m})\hat{i} + (0.1\,\text{m})\hat{j}\right]$$

$$= (-1.28\,\text{N})\hat{i} + (0.640\,\text{N})\hat{j}$$

Find the net force acting on q_2:

$$\vec{F}_2 = \vec{F}_{3,2} - (1.80\,\text{N})\hat{j} = (-1.28\,\text{N})\hat{i} + (0.640\,\text{N})\hat{j} - (1.80\,\text{N})\hat{j}$$

$$= \boxed{(-1.28\,\text{N})\hat{i} - (1.16\,\text{N})\hat{j}}$$

Noting that $\vec{F}_{1,3}$ and $\vec{F}_{3,1}$ are an action-and-reaction pair, as are $\vec{F}_{2,3}$ and $\vec{F}_{3,2}$, express the net force acting on q_3:

$$\vec{F}_3 = \vec{F}_{1,3} + \vec{F}_{2,3} = -\vec{F}_{3,1} - \vec{F}_{3,2} = -(0.899\,\text{N})\hat{i} - \left[(-1.28\,\text{N})\hat{i} + (0.640\,\text{N})\hat{j}\right]$$

$$= \boxed{(0.381\,\text{N})\hat{i} - (0.640\,\text{N})\hat{j}}$$

The Electric Field

*37 • A charge of 4.0 μC is at the origin. What is the magnitude and direction of the electric field on the x axis at (a) $x = 6$ m, and (b) $x = -10$ m? (c) Sketch the function E_x versus x for both positive and negative values of x. (Remember that E_x is negative when $\vec{E}$ points in the negative x direction.)

Picture the Problem Let q represent the charge at the origin and use Coulomb's law for $\vec{E}$ due to a point charge to find the electric field at $x = 6$ m and -10 m.

(a) Express the electric field at a point P located a distance x from a charge q:

$$\vec{E}(x) = \frac{kq}{x^2}\hat{r}_{P,0}$$

10 Chapter 21

Evaluate this expression for $x = 6$ m:

$$\vec{E}(6\,\text{m}) = \frac{(8.99 \times 10^9 \,\text{N} \cdot \text{m}^2/\text{C}^2)(4\,\mu\text{C})}{(6\,\text{m})^2}\hat{i}$$

$$= \boxed{(999\,\text{N/C})\hat{i}}$$

(b) Evaluate $\vec{E}$ at $x = -10$ m:

$$\vec{E}(-10\,\text{m}) = \frac{(8.99 \times 10^9 \,\text{N} \cdot \text{m}^2/\text{C}^2)(4\,\mu\text{C})}{(10\,\text{m})^2}(-\hat{i}) = \boxed{(-360\,\text{N/C})\hat{i}}$$

(c) The following graph was plotted using a spreadsheet program:

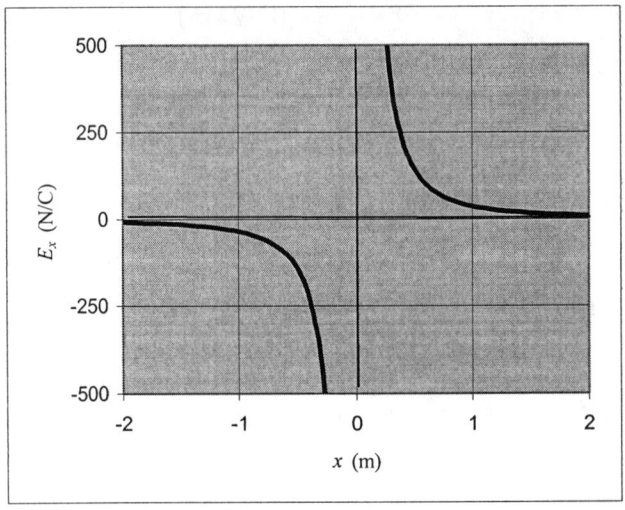

*38 •   Two charges, each $+4\,\mu\text{C}$, are on the x axis, one at the origin and the other at $x = 8$ m. Find the electric field on the x axis at (a) $x = -2$ m, (b) $x = 2$ m, (c) $x = 6$ m, and (d) $x = 10$ m. (e) At what point on the x axis is the electric field zero? (f) Sketch E_x versus x.

Picture the Problem Let q represent the charges of $+4\,\mu\text{C}$ and use Coulomb's law for $\vec{E}$ due to a point charge and the principle of superposition for fields to find the electric field at the locations specified.

Noting that $q_1 = q_2$, use Coulomb's law and the principle of superposition to express the electric field due to the given charges at a point P a distance x from the origin:

$$\vec{E}(x) = \vec{E}_{q_1}(x) + \vec{E}_{q_2}(x) = \frac{kq_1}{x^2}\hat{r}_{q_1,P} + \frac{kq_2}{(8\,\text{m}-x)^2}\hat{r}_{q_2,P} = kq_1\left(\frac{1}{x^2}\hat{r}_{q_1,P} + \frac{1}{(8\,\text{m}-x)^2}\hat{r}_{q_2,P}\right)$$

$$= (36\,\text{kN}\cdot\text{m}^2/\text{C})\left(\frac{1}{x^2}\hat{r}_{q_1,P} + \frac{1}{(8\,\text{m}-x)^2}\hat{r}_{q_2,P}\right)$$

(a) Apply this equation to the point at $x = -2$ m:

$$\vec{E}(-2\,\text{m}) = (36\,\text{kN}\cdot\text{m}^2/\text{C})\left[\frac{1}{(2\,\text{m})^2}(-\hat{i}) + \frac{1}{(10\,\text{m})^2}(-\hat{i})\right] = \boxed{(-9.36\,\text{kN/C})\hat{i}}$$

(b) Evaluate $\vec{E}$ at $x = 2$ m:

$$\vec{E}(2\,\text{m}) = (36\,\text{kN}\cdot\text{m}^2/\text{C})\left[\frac{1}{(2\,\text{m})^2}(\hat{i}) + \frac{1}{(6\,\text{m})^2}(-\hat{i})\right] = \boxed{(8.00\,\text{kN/C})\hat{i}}$$

(c) Evaluate $\vec{E}$ at $x = 6$ m:

$$\vec{E}(6\,\text{m}) = (36\,\text{kN}\cdot\text{m}^2/\text{C})\left[\frac{1}{(6\,\text{m})^2}(\hat{i}) + \frac{1}{(2\,\text{m})^2}(-\hat{i})\right] = \boxed{(-8.00\,\text{kN/C})\hat{i}}$$

(d) Evaluate $\vec{E}$ at $x = 10$ m:

$$\vec{E}(10\,\text{m}) = (36\,\text{kN}\cdot\text{m}^2/\text{C})\left[\frac{1}{(10\,\text{m})^2}(\hat{i}) + \frac{1}{(2\,\text{m})^2}(\hat{i})\right] = \boxed{(9.35\,\text{kN/C})\hat{i}}$$

(e) From symmetry considerations: $E(4\,\text{m}) = \boxed{0}$

(f) The following graph was plotted using a spreadsheet program:

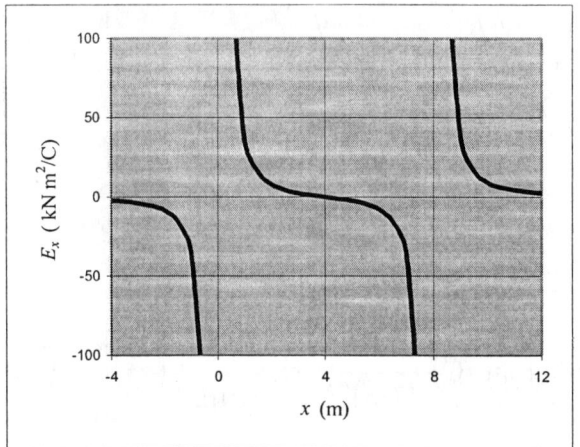

*42 •• A point charge of +5.0 µC is located at $x = -3.0$ cm, and a second point charge of −8.0 µC is located at $x = +4.0$ cm. Where should a third charge of +6.0 µC be placed so that the electric field at $x = 0$ is zero?

Picture the Problem If the electric field at $x = 0$ is zero, both its x and y components must be zero. The only way this condition can be satisfied with the point charges of +5.0 µC and −8.0 µC are on the x axis is if the point charge of +6.0 µC is also on the x axis. Let the subscripts 5, −8, and 6 identify the point charges and their fields. We can use Coulomb's law for $\vec{E}$ due to a point charge and the principle of superposition for fields to determine where the +6.0 µC charge should be located so that the electric field at $x = 0$ is zero.

Express the electric field at $x = 0$ in terms of the fields due to the charges of +5.0 µC, −8.0 µC, and +6.0 µC:

$$\vec{E}(0) = \vec{E}_{5\,\mu C} + \vec{E}_{-8\,\mu C} + \vec{E}_{6\,\mu C} = 0$$

Substitute for each of the fields to obtain:

$$\frac{kq_5}{r_5^2}\hat{r}_5 + \frac{kq_6}{r_6^2}\hat{r}_6 + \frac{kq_{-8}}{r_{-8}^2}\hat{r}_{-8} = 0$$

or

$$\frac{kq_5}{r_5^2}\hat{i} + \frac{kq_6}{r_6^2}(-\hat{i}) + \frac{kq_{-8}}{r_{-8}^2}(-\hat{i}) = 0$$

Divide out the unit vector $\hat{i}$ to obtain:

$$\frac{q_5}{r_5^2} - \frac{q_6}{r_6^2} - \frac{q_{-8}}{r_{-8}^2} = 0$$

Substitute numerical values to obtain:

$$\frac{5}{(3\,\text{cm})^2} - \frac{6}{r_6^2} - \frac{-8}{(4\,\text{cm})^2} = 0$$

Solve for r_6:

$$r_6 = \boxed{2.38\,\text{cm}}$$

*45 •• A 5-μC point charge is located at $x = 1$ m, $y = 3$ m; and a -4-μC point charge is located at $x = 2$ m, $y = -2$ m. (a) Find the magnitude and direction of the electric field at $x = -3$ m, $y = 1$ m. (b) Find the magnitude and direction of the force on a proton at $x = -3$ m, $y = 1$ m.

Picture the Problem The diagram shows the electric field vectors at the point of interest P due to the two charges. We can use Coulomb's law for $\vec{E}$ due to point charges and the superposition principle for electric fields to find $\vec{E}_P$. We can apply $\vec{F} = q\vec{E}$ to find the force on a proton at (−3 m, 1 m).

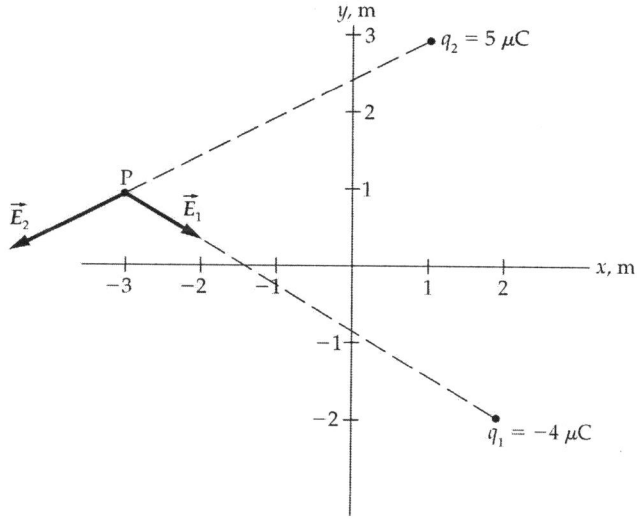

(a) Express the electric field at (−3 m, 1 m) due to the charges q_1 and q_2:

$$\vec{E}_P = \vec{E}_1 + \vec{E}_2$$

Evaluate $\vec{E}_1$:

$$\vec{E}_1 = \frac{kq_1}{r_{1,P}^2}\hat{r}_{1,P} = \frac{(8.99\times 10^9 \text{ N}\cdot\text{m}^2/\text{C}^2)(-4\,\mu\text{C})}{(5\text{m})^2 + (3\text{m})^2}\left(\frac{(-5\text{m})\hat{i} + (3\text{m})\hat{j}}{\sqrt{(5\text{m})^2 + (3\text{m})^2}}\right)$$

$$= (-1.06 \text{ kN/C})(-0.857\,\hat{i} + 0.514\,\hat{j}) = (0.908 \text{ kN/C})\hat{i} + (-0.544 \text{ kN/C})\hat{j}$$

Evaluate $\vec{E}_2$:

$$\vec{E}_2 = \frac{kq_2}{r_{2,P}^2}\hat{r}_{2,P} = \frac{(8.99\times 10^9 \text{ N}\cdot\text{m}^2/\text{C}^2)(5\,\mu\text{C})}{(4\text{m})^2 + (2\text{m})^2}\left(\frac{(-4\text{m})\hat{i} + (-2\text{m})\hat{j}}{\sqrt{(4\text{m})^2 + (2\text{m})^2}}\right)$$

$$= (2.25 \text{ kN/C})(-0.894\,\hat{i} - 0.447\,\hat{j}) = (-2.01 \text{ kN/C})\hat{i} + (-1.01 \text{ kN/C})\hat{j}$$

Substitute and simplify to find $\vec{E}_P$:

$$\vec{E}_P = (0.908\,\text{kN/C})\hat{i} + (-0.544\,\text{kN/C})\hat{j} + (-2.01\,\text{kN/C})\hat{i} + (-1.01\,\text{kN/C})\hat{j}$$
$$= (-1.10\,\text{kN/C})\hat{i} + (-1.55\,\text{kN/C})\hat{j}$$

The magnitude of $\vec{E}_P$ is:

$$E_P = \sqrt{(1.10\,\text{kN/C})^2 + (1.55\,\text{kN/C})^2}$$
$$= \boxed{1.90\,\text{kN/C}}$$

The direction of $\vec{E}_P$ is:

$$\theta_E = \tan^{-1}\left(\frac{-1.55\,\text{kN/C}}{-1.10\,\text{kN/C}}\right) = \boxed{235°}$$

Note that the angle returned by your calculator for $\tan^{-1}\left(\dfrac{-1.55\,\text{kN/C}}{-1.10\,\text{kN/C}}\right)$ is the reference angle and must be increased by 180° to yield θ_E.

(b) Express and evaluate the force on a proton at point P:

$$\vec{F} = q\vec{E}_P = (1.6\times10^{-19}\,\text{C})\left[(-1.10\,\text{kN/C})\hat{i} + (-1.55\,\text{kN/C})\hat{j}\right]$$
$$= (-1.76\times10^{-16}\,\text{N})\hat{i} + (-2.48\times10^{-16}\,\text{N})\hat{j}$$

The magnitude of $\vec{F}$ is:

$$F = \sqrt{(-1.76\times10^{-16}\,\text{N})^2 + (-2.48\times10^{-16}\,\text{N})^2} = \boxed{3.04\times10^{-16}\,\text{N}}$$

The direction of $\vec{F}$ is:

$$\theta_F = \tan^{-1}\left(\frac{-2.48\times10^{-16}\,\text{N}}{-1.76\times10^{-16}\,\text{N}}\right) = \boxed{235°}$$

where, as noted above, the angle returned by your calculator for $\tan^{-1}\left(\dfrac{-2.48\times10^{-16}\,\text{N}}{-1.76\times10^{-16}\,\text{N}}\right)$ is the reference angle and must be increased by 180° to yield θ_E.

*48 ••• Two positive point charges $+q$ are on the y axis at $y = +a$ and $y = -a$ as in Problem 44. A bead of mass m carrying a negative charge $-q$ slides without friction along a thread that runs along the x axis. (a) Show that for small displacements of $x \ll a$, the bead experiences a restoring force that is proportional to x and therefore undergoes simple harmonic motion. (b) Find the period of the motion.

Picture the Problem In Problem 44 it is shown that the electric field on the x axis, due to equal positive charges located at $(0, a)$ and $(0,-a)$, is given by $E_x = 2kqx(x^2 + a^2)^{-3/2}$. We can use $T = 2\pi\sqrt{m/k'}$ to express the period of the motion in terms of the restoring constant k'.

(a) Express the force acting on the on the bead when its displacement from the origin is x:

$$F_x = -qE_x = -\frac{2kq^2 x}{(x^2 + a^2)^{3/2}}$$

Factor a^2 from the denominator to obtain:

$$F_x = -\frac{2kq^2 x}{a^2\left(\dfrac{x^2}{a^2}+1\right)^{3/2}}$$

For $x \ll a$:

$$\boxed{F_x = -\frac{2kq^2}{a^3}x}$$

i.e., the bead experiences a linear restoring force.

(b) Express the period of a simple harmonic oscillator:

$$T = 2\pi\sqrt{\frac{m}{k'}}$$

Obtain k' from our result in part (a):

$$k' = \frac{2kq^2}{a^3}$$

Substitute to obtain:

$$T = 2\pi\sqrt{\frac{m}{\dfrac{2kq^2}{a^3}}} = \boxed{2\pi\sqrt{\frac{ma^3}{2kq^2}}}$$

Motion of Point Charges in Electric Fields

*50 • (a) Compute e/m for a proton, and find its acceleration in a uniform electric field with a magnitude of 100 N/C. (b) Find the time it takes for a proton initially at rest in such a field to reach a speed of $0.01c$ (where c is the speed of light).

Chapter 21

Picture the Problem We can use Newton's 2nd law of motion to find the acceleration of the proton in the uniform electric field and constant-acceleration equations to find the time required for it to reach a speed of $0.01c$ and the distance it travels while acquiring this speed.

(a) Use data found at the back of your text to compute e/m for an electron:

$$\frac{e}{m_p} = \frac{1.6 \times 10^{-19} \text{ C}}{1.67 \times 10^{-27} \text{ kg}}$$

$$= \boxed{9.58 \times 10^7 \text{ C/kg}}$$

Apply Newton's 2nd law to relate the acceleration of the electron to the electric field:

$$a = \frac{F_{net}}{m_p} = \frac{eE}{m_p}$$

Substitute numerical values and evaluate a:

$$a = \frac{(1.6 \times 10^{-19} \text{ C})(100 \text{ N/C})}{1.67 \times 10^{-27} \text{ kg}}$$

$$= \boxed{9.58 \times 10^9 \text{ m/s}^2}$$

> The direction of the acceleration of a proton is in the direction of the electric field.

(b) Using the definition of acceleration, relate the time required for an electron to reach $0.01c$ to its acceleration:

$$\Delta t = \frac{v}{a} = \frac{0.01c}{a}$$

Substitute numerical values and evaluate Δt:

$$\Delta t = \frac{0.01(3 \times 10^8 \text{ m/s})}{9.58 \times 10^9 \text{ m/s}^2} = \boxed{313 \, \mu s}$$

*54 •• A particle leaves the origin with a speed of 3×10^6 m/s at 35° to the x axis. It moves in a constant electric field $\vec{E} = E_y \hat{j}$. Find E_y such that the particle will cross the x axis at $x = 1.5$ cm if the particle is (a) an electron, and (b) a proton.

Picture the Problem We can use constant-acceleration equations to express the x and y coordinates of the particle in terms of the parameter t and Newton's 2nd law to express the constant acceleration in terms of the electric field. Eliminating the parameter will yield an equation for y as a function of x, q, and m that we can solve for E_y.

Express the x and y coordinates of the particle as functions of time:

$$x = (v \cos \theta) t$$

and
$$y = (v\sin\theta)t - \tfrac{1}{2}a_y t^2$$

Apply Newton's 2nd law to relate the acceleration of the particle to the net force acting on it:
$$a_y = \frac{F_{net,y}}{m} = \frac{qE_y}{m}$$

Substitute in the y-coordinate equation to obtain:
$$y = (v\sin\theta)t - \frac{qE_y}{2m}t^2$$

Eliminate the parameter t between the two equations to obtain:
$$y = (\tan\theta)x - \frac{qE_y}{2mv^2\cos^2\theta}x^2$$

Set $y = 0$ and solve for E_y:
$$E_y = \frac{mv^2 \sin 2\theta}{qx}$$

Substitute the non-particle specific data to obtain:
$$E_y = \frac{m(3\times 10^6 \text{ m/s})^2 \sin 70°}{q(0.015\text{ m})}$$
$$= (5.64\times 10^{14} \text{ m/s}^2)\frac{m}{q}$$

(a) Substitute for the mass and charge of an electron and evaluate E_y:
$$E_y = (5.64\times 10^{14} \text{ m/s}^2)\frac{9.11\times 10^{-31}\text{ kg}}{1.6\times 10^{-19}\text{ C}}$$
$$= \boxed{3.21\text{ kN/C}}$$

(b) Substitute for the mass and charge of a proton and evaluate E_y:
$$E_y = (5.64\times 10^{14} \text{ m/s}^2)\frac{1.67\times 10^{-27}\text{ kg}}{1.6\times 10^{-19}\text{ C}}$$
$$= \boxed{5.89\text{ MN/C}}$$

*58 • A dipole of moment 0.5 e·nm is placed in a uniform electric field with a magnitude of 4.0×10^4 N/C. What is the magnitude of the torque on the dipole when (a) the dipole is parallel to the electric field, (b) the dipole is perpendicular to the electric field, and (c) the dipole makes an angle of 30° with the electric field? (d) Find the potential energy of the dipole in the electric field for each case.

Picture the Problem The torque on an electric dipole in an electric field is given by $\vec{\tau} = \vec{p}\times\vec{E}$ and the potential energy of the dipole by $U = -\vec{p}\cdot\vec{E}$.

Using its definition, express the torque
$$\vec{\tau} = \vec{p}\times\vec{E}$$

18 Chapter 21

on a dipole moment in a uniform electric field:

and
$$\tau = pE\sin\theta$$
where θ is the angle between the electric dipole moment and the electric field.

(a) Evaluate τ for $\theta = 0°$:

$$\tau = pE\sin 0° = \boxed{0}$$

(b) Evaluate τ for $\theta = 90°$:

$$\tau = (0.5\,e\cdot\text{nm})(4.0\times 10^4\text{ N/C})\sin 90°$$
$$= \boxed{3.20\times 10^{-24}\text{ N}\cdot\text{m}}$$

(c) Evaluate τ for $\theta = 30°$:

$$\tau = (0.5\,e\cdot\text{nm})(4.0\times 10^4\text{ N/C})\sin 30°$$
$$= \boxed{1.60\times 10^{-24}\text{ N}\cdot\text{m}}$$

(d) Using its definition, express the potential energy of a dipole in an electric field:

$$U = -\vec{p}\cdot\vec{E} = -pE\cos\theta$$

Evaluate U for $\theta = 0°$:

$$U = -(0.5\,e\cdot\text{nm})(4.0\times 10^4\text{ N/C})\cos 0°$$
$$= \boxed{-3.20\times 10^{-24}\text{ J}}$$

Evaluate U for $\theta = 90°$:

$$U = -(0.5\,e\cdot\text{nm})(4.0\times 10^4\text{ N/C})\cos 90°$$
$$= \boxed{0}$$

Evaluate U for $\theta = 30°$:

$$U = -(0.5\,e\cdot\text{nm})(4.0\times 10^4\text{ N/C})\cos 30°$$
$$= \boxed{-2.77\times 10^{-24}\text{ J}}$$

*59 •• For a dipole oriented along the x axis, the electric field falls off as $1/x^3$ in the x direction and $1/y^3$ in the y direction. Use dimensional analysis to prove that, in any direction, the field far from the dipole falls off as $1/r^3$.

Picture the Problem We can combine the dimension of an electric field with the dimension of an electric dipole moment to prove that, in any direction, the dimension of the far field is proportional to $1/[L]^3$ and, hence, the electric field far from the dipole falls off as $1/r^3$.

Express the dimension of an electric field:

$$[E] = \frac{[kQ]}{[L]^2}$$

Express the dimension an electric dipole moment:

$$[p] = [Q][L]$$

Write the dimension of charge in terms of the dimension of an electric dipole moment:

$$[Q] = \frac{[p]}{[L]}$$

Substitute to obtain:

$$[E] = \frac{[k][p]}{[L]^2[L]} = \boxed{\frac{[k][p]}{[L]^3}}$$

This shows that the field E due to a dipole p falls off as $1/r^3$.

General Problems

*63 • (a) What mass would a proton have if its gravitational attraction to another proton exactly balanced out the electrostatic repulsion between them?
(b) What is the true ratio of these two forces?

Picture the Problem We can equate the gravitational force and the electric force acting on a proton to find the mass of the proton under the given condition.

(a) Express the condition that must be satisfied if the net force on the proton is zero:

$$F_g = F_e$$

Use Newton's law of gravity and Coulomb's law to substitute for F_g and F_e:

$$\frac{Gm^2}{r^2} = \frac{ke^2}{r^2}$$

Solve for m to obtain:

$$m = e\sqrt{\frac{k}{G}}$$

Substitute numerical values and evaluate m:

$$m = (1.6 \times 10^{-19}\,\text{C})\sqrt{\frac{8.99 \times 10^9\,\text{N} \cdot \text{m}^2/\text{C}^2}{6.67 \times 10^{-11}\,\text{N} \cdot \text{m}^2/\text{kg}^2}} = \boxed{1.86 \times 10^{-9}\,\text{kg}}$$

(b) Express the ratio of F_e and Fg:

$$\frac{\dfrac{ke^2}{r^2}}{\dfrac{Gm_p^2}{r^2}} = \frac{ke^2}{Gm_p^2}$$

20 Chapter 21

Substitute numerical values to obtain:

$$\frac{ke^2}{Gm_p^2} = \frac{(8.99\times 10^9 \text{ N}\cdot\text{m}^2/\text{C}^2)(1.6\times 10^{-19}\text{ C})^2}{(6.67\times 10^{-11} \text{ N}\cdot\text{m}^2/\text{kg}^2)(1.67\times 10^{-27}\text{ kg})^2} = \boxed{1.24\times 10^{36}}$$

*66 •• In copper, about one electron per atom is free to move about. A copper penny has a mass of 3 g. (*a*) What percentage of the free charge would have to be removed to give the penny a charge of 15 µC? (*b*) What would be the force of repulsion between two pennies carrying this charge if they were 25 cm apart? Assume that the pennies are point charges.

Picture the Problem We can find the percentage of the free charge that would have to be removed by finding the ratio of the number of free electrons n_e to be removed to give the penny a charge of 15 µC to the number of free electrons in the penny. Because we're assuming the pennies to be point charges, we can use Coulomb's law to find the force of repulsion between them.

(*a*) Express the fraction *f* of the free charge to be removed as the quotient of the number of electrons to be removed and the number of free electrons:

$$f = \frac{n_e}{N}$$

Relate N to Avogadro's number, the mass of the copper penny, and the molecular mass of copper:

$$\frac{N}{N_A} = \frac{m}{M} \Rightarrow N = N_A \frac{m}{M}$$

Relate n_e to the free charge Q to be removed from the penny:

$$Q = n_e[-e] \Rightarrow n_e = \frac{Q}{-e}$$

$$f = \frac{\frac{Q}{-e}}{N_A \frac{m}{M}} = -\frac{QM}{meN_A}$$

Substitute numerical values and evaluate *f*:

$$f = -\frac{(-15\,\mu\text{C})(63.5\text{ g/mol})}{(3\text{ g})(1.6\times 10^{-19}\text{ C})(6.02\times 10^{23}\text{ mol}^{-1})} = 3.29\times 10^{-9} = \boxed{3.29\times 10^{-7}\%}$$

(b) Use Coulomb's law to express the force of repulsion between the two pennies:

$$F = \frac{kq^2}{r^2} = \frac{k(n_e e)^2}{r^2}$$

Substitute numerical values and evaluate F:

$$F = \frac{(8.99\times 10^9 \text{ N}\cdot\text{m}^2/\text{C}^2)(9.38\times 10^{13})^2(1.6\times 10^{-19}\text{ C})^2}{(0.25\text{ m})^2} = \boxed{32.4\text{ N}}$$

*69 •• A positive charge Q is to be divided into two positive charges q_1 and q_2. Show that, for a given separation D, the force exerted by one charge on the other is greatest if $q_1 = q_2 = \tfrac{1}{2}Q$.

Picture the Problem We can use Coulomb's law to express the force exerted on one charge by the other and then set the derivative of this expression equal to zero to find the distribution of the charge that maximizes this force.

Using Coulomb's law, express the force that either charge exerts on the other:

$$F = \frac{kq_1 q_2}{D^2}$$

Express q_2 in terms of Q and q_1:

$$q_2 = Q - q_1$$

Substitute to obtain:

$$F = \frac{kq_1(Q-q_1)}{D^2}$$

Differentiate F with respect to q_1 and set this derivative equal to zero for extreme values:

$$\frac{dF}{dq_1} = \frac{k}{D^2}\frac{d}{dq_1}[q_1(Q-q_1)]$$

$$= \frac{k}{D^2}[q_1(-1)+Q-q_1]$$

$$= 0 \text{ for extrema}$$

Solve for q_1 to obtain:

$$q_1 = \tfrac{1}{2}Q \text{ and } q_2 = Q - q_1 = \tfrac{1}{2}Q$$

To determine whether a maximum or a minimum exists at $q_1 = \tfrac{1}{2}Q$, differentiate F a second time and evaluate this derivative at $q_1 = \tfrac{1}{2}Q$:

$$\frac{d^2 F}{dq_1^2} = \frac{k}{D^2}\frac{d}{dq_1}[Q-2q_1]$$

$$= \frac{k}{D^2}(-2)$$

$$< 0 \text{ independently of } q_1.$$

$$\therefore q_1 = q_2 = \tfrac{1}{2}Q \text{ maximizes } F.$$

*70 •• A charge Q is located at $x = 0$, and a charge $4Q$ is at $x = 12.0$ cm. The force on a charge of $-2\ \mu C$ is zero if that charge is placed at $x = 4.0$ cm, and is 126.4 N in the positive x direction if placed at $x = 8.0$ cm. Determine the charge Q.

Picture the Problem We can apply Coulomb's law and the superposition of forces to relate the net force acting on the charge $q = -2\ \mu C$ to x. Because Q divides out of our equation when $F(x) = 0$, we'll substitute the data given for $x = 8.0$ cm.

Using Coulomb's law, express the net force on q as a function of x:

$$F(x) = -\frac{kqQ}{x^2} + \frac{kq(4Q)}{(12\,\text{cm} - x)^2}$$

Simplify to obtain:

$$\frac{F(x)}{kq} = \left[-\frac{1}{x^2} + \frac{4}{(12\,\text{cm} - x)^2} \right] Q$$

Solve for Q:

$$Q = \frac{F(x)}{kq\left[-\dfrac{1}{x^2} + \dfrac{4}{(12\,\text{cm} - x)^2} \right]}$$

Evaluate Q for $x = 8$ cm:

$$Q = \frac{126.4\,\text{N}}{(8.99 \times 10^9\,\text{N} \cdot \text{m}^2/\text{C}^2)(2\,\mu C)\left[-\dfrac{1}{(8\,\text{cm})^2} + \dfrac{4}{(4\,\text{cm})^2} \right]} = \boxed{3.00\ \mu C}$$

*78 •• Two small spheres of mass m are suspended from a common point by threads of length L. When each sphere carries a charge q, each thread makes an angle θ with the vertical as shown in Figure 21-42. (a) Show that the charge q is given by

$$q = 2L\sin\theta\sqrt{\frac{mg\tan\theta}{k}}$$

where k is the Coulomb constant. (b) Find q if $m = 10$ g, $L = 50$ cm, and $\theta = 10°$.

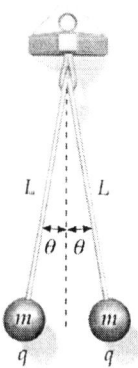

Figure 21-42 Problem 78

Picture the Problem Each sphere is in static equilibrium under the influence of the tension $\vec{T}$, the gravitational force $\vec{F}_g$, and the electric force $\vec{F}_E$. We can use Coulomb's law to relate the electric force to the charge on each sphere and their separation and the conditions for static equilibrium to relate these forces to the charge on each sphere.

(a) Apply the conditions for static equilibrium to the charged sphere:

$$\sum F_x = F_E - T\sin\theta = \frac{kq^2}{r^2} - T\sin\theta = 0$$

and

$$\sum F_y = T\cos\theta - mg = 0$$

Eliminate T between these equations to obtain:

$$\tan\theta = \frac{kq^2}{mgr^2}$$

Solve for q:

$$q = r\sqrt{\frac{mg\tan\theta}{k}}$$

Referring to the figure, relate the separation of the spheres r to the length of the pendulum L:

$$r = 2L\sin\theta$$

Substitute to obtain:

$$q = \boxed{2L\sin\theta\sqrt{\frac{mg\tan\theta}{k}}}$$

Chapter 21

(b) Evaluate q for $m = 10$ g, $L = 50$ cm, and $\theta = 10°$:

$$q = 2(0.5\,\text{m})\sin 10°\sqrt{\frac{(0.01\,\text{kg})(9.81\,\text{m/s}^2)\tan 10°}{8.99\times 10^9\,\text{N}\cdot\text{m}^2/\text{C}^2}} = \boxed{0.241\,\mu\text{C}}$$

*83 ••   An electron (charge $-e$, mass m) and a positron (charge $+e$, mass m) revolve around their common center of mass under the influence of their attractive coulomb force. Find the speed of each particle v in terms of e, m, k, and their separation r.

Picture the Problem The forces the electron and the proton exert on each other constitute an action-and-reaction pair. Because the magnitudes of their charges are equal and their masses are the same, we find the speed of each particle by finding the speed of either one. We'll apply Coulomb's force law for point charges and Newton's 2nd law to relate v to e, m, k, and r.

Apply Newton's 2nd law to the positron:
$$\frac{ke^2}{r^2} = m\frac{v^2}{\frac{1}{2}r} \Rightarrow \frac{ke^2}{r} = 2mv^2$$

Solve for v to obtain:
$$v = \boxed{\sqrt{\frac{ke^2}{2mr}}}$$

*90 •••   Two neutral polar molecules attract each other. Suppose that each molecule has a dipole moment $\vec{p}$, and that these dipoles are aligned along the x axis and separated by a distance d. Derive an expression for the force of attraction in terms of p and d.

Picture the Problem We can relate the force of attraction that each molecule exerts on the other to the potential energy function of either molecule using $F = -dU/dx$. We can relate U to the electric field at either molecule due to the presence of the other through $U = -pE$. Finally, the electric field at either molecule is given by $E = 2kp/x^3$.

Express the force of attraction between the dipoles in terms of the spatial derivative of the potential energy function of p_1:
$$F = -\frac{dU_1}{dx} \qquad (1)$$

Express the potential energy of the dipole p_1:
$$U_1 = -p_1 E_1$$
where E_1 is the field at p_1 due to p_2.

Express the electric field at p_1 due to p_2:
$$E_1 = \frac{2kp_2}{x^3}$$

where x is the separation of the dipoles.

Substitute to obtain:
$$U_1 = -\frac{2kp_1p_2}{x^3}$$

Substitute in equation (1) and differentiate with respect to x:
$$F = -\frac{d}{dx}\left[-\frac{2kp_1p_2}{x^3}\right] = \frac{6kp_1p_2}{x^4}$$

Evaluate F for $p_1 = p_2 = p$ and $x = d$ to obtain:
$$\boxed{F = \frac{6kp^2}{d^4}}$$

*93 ••• In Problem 92, there was a description of the Millikan experiment used to determine the charge on the electron. In the experiment, a switchable power supply is used so that the electrical field can point both up and down, but with the same magnitude, so that one can measure the terminal speed of the microsphere as it is pushed up (against the force of gravity) and down. Let v_u represent the terminal speed when the particle is moving up, and v_d the terminal speed when moving down. (a) If we let $v = v_u + v_d$, show that $v = qE/3\pi\eta r$, where q is the microsphere's net charge. What advantage does measuring both v_u and v_d give over measuring only one? (b) Because charge is quantized, v can only change by steps of magnitude Δv. Using the data from Problem 92, calculate Δv.

Picture the Problem The free body diagram shows the forces acting on the microsphere of mass m and having an excess charge of $q = Ne$ when the electric field is downward. Under terminal-speed conditions the sphere is in equilibrium under the influence of the electric force $\vec{F}_e$, its weight $m\vec{g}$, and the drag force $\vec{F}_d$. We can apply Newton's 2nd law, under terminal-speed conditions, to relate the number of excess charges N on the sphere to its mass and, using Stokes' law, to its terminal speed.

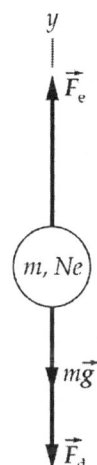

(a) Apply $\sum F_y = ma_y$ to the microsphere when the electric field is downward:
$$F_e - mg - F_d = ma_y$$
or, because $a_y = 0$,
$$F_e - mg - F_{d,\text{terminal}} = 0$$

Substitute for F_e and $F_{d,\text{terminal}}$ to obtain:
$$qE - mg - 6\pi\eta r v_u = 0$$
or, because $q = Ne$,
$$NeE - mg - 6\pi\eta r v_u = 0$$

26 Chapter 21

Solve for v_u to obtain:

$$v_u = \frac{NeE - mg}{6\pi\eta r} \quad (1)$$

With the field pointing upward, the electric force is downward and the application of $\sum F_y = ma_y$ to the microsphere yields:

$$F_{d,\text{terminal}} - F_e - mg = 0$$

or

$$6\pi\eta r v_d - NeE - mg = 0$$

Solve for v_d to obtain:

$$v_d = \frac{NeE + mg}{6\pi\eta r} \quad (2)$$

Add equations (1) and (2) to obtain:

$$v = v_u + v_d = \frac{NeE - mg}{6\pi\eta r}$$
$$+ \frac{NeE + mg}{6\pi\eta r}$$
$$= \frac{NeE}{3\pi\eta r} = \boxed{\frac{qE}{3\pi\eta r}}$$

> This has the advantage that you don't need to know the mass of the microsphere.

(b) Letting Δv represent the change in the terminal speed of the microsphere due to a gain (or loss) of one electron we have:

$$\Delta v = v_{N+1} - v_N$$

Noting that Δv will be the same whether the microsphere is moving upward or downward, express its terminal speed when it is moving upward with N electronic charges on it:

$$v_N = \frac{NeE - mg}{6\pi\eta r}$$

Express its terminal speed upward when it has $N + 1$ electronic charges:

$$v_{N+1} = \frac{(N+1)eE - mg}{6\pi\eta r}$$

Substitute and simplify to obtain:

$$\Delta v_{N+1} = \frac{(N+1)eE - mg}{6\pi\eta r} - \frac{NeE - mg}{6\pi\eta r}$$
$$= \frac{eE}{6\pi\eta r}$$

Substitute numerical values and evaluate Δv:

$$\Delta v = \frac{(1.6 \times 10^{-19} \text{ C})(6 \times 10^4 \text{ V/m})}{6\pi(1.8 \times 10^{-5} \text{ Pa} \cdot \text{m})(5.5 \times 10^{-7} \text{ m})}$$

$$= \boxed{5.15 \times 10^{-5} \text{ m/s}}$$

Chapter 22
The Electric Field 2: Continuous Charge Distributions

Conceptual Problems

*1 •• True or false:

(a) Gauss's law holds only for symmetric charge distributions.
(b) The result that $E = 0$ inside a conductor can be derived from Gauss's law.

(a) False. Gauss's law states that the net flux through any surface is given by $\phi_{net} = \oint E_n dA = 4\pi k Q_{inside}$. While it is true that Gauss's law is easiest to apply to symmetric charge distributions, it holds for *any* surface.

(b) True

*5 • True or false:

(a) If there is no charge in a region of space, the electric field on a surface surrounding the region must be zero everywhere.
(b) The electric field inside a uniformly charged spherical shell is zero.
(c) In electrostatic equilibrium, the electric field inside a conductor is zero.
(d) If the net charge on a conductor is zero, the charge density must be zero at every point on the surface of the conductor.

(a) False. Consider a spherical shell, in which there is no charge, in the vicinity of an infinite sheet of charge. The electric field due to the infinite sheet would be non-zero everywhere on the spherical surface.

(b) True (assuming there are no charges inside the shell).

(c) True.

(d) False. Consider a spherical conducting shell. Such a surface will have equal charges on its inner and outer surfaces but, because their areas differ, so will their charge densities.

*9 •• Suppose that the total charge on the conducting shell of Figure 22-36 is zero. It follows that the electric field for $r < R_1$ and $r > R_2$ points

(a) away from the center of the shell in both regions.

(b) toward the center of the shell in both regions.
(c) toward the center of the shell for $r < R_1$ and is zero for $r > R_2$.
(d) away from the center of the shell for $r < R_1$ and is zero for $r > R_2$.

Determine the Concept We can apply Gauss's law to determine the electric field for $r < R_1$ and $r > R_2$. We also know that the direction of an electric field at any point is determined by the direction of the electric force acting on a positively charged object located at that point.

From the application of Gauss's law we know that the electric field in both of these regions is not zero and is given by:
$$E_n = \frac{kQ}{r^2}$$

A positively charged object placed in either of these regions would experience an attractive force from the charge $-Q$ located at the center of the shell. $\boxed{(b) \text{ is correct.}}$

*10 •• If the conducting shell in Figure 22-36 is grounded, which of the following statements is then correct?

(a) The charge on the inner surface of the shell is $+Q$ and that on the outer surface is $-Q$.
(b) The charge on the inner surface of the shell is $+Q$ and that on the outer surface is zero.
(c) The charge on both surfaces of the shell is $+Q$.
(d) The charge on both surfaces of the shell is zero.

Determine the Concept We can decide what will happen when the conducting shell is grounded by thinking about the distribution of charge on the shell before it is grounded and the effect on this distribution of grounding the shell.

The negative point charge at the center of the conducting shell induces a positive charge on the inner surface of the shell and a negative charge on the outer surface. Grounding the shell attracts positive charge from ground; resulting in the outer surface becoming electrically neutral. $\boxed{(b) \text{ is correct.}}$

Estimation and Approximation

*14 •• Given that the maximum field sustainable in air without electrical discharge is approximately 3×10^6 N/C, estimate the total charge of a thundercloud. Make any assumptions that seem reasonable.

Picture the Problem We'll assume that the total charge is spread out uniformly (charge density = σ) in a thin layer at the bottom and top of the cloud and that the area of each surface of the cloud is 1 km². We can then use the definition of surface charge density and the expression for the electric field at the surface of a charged plane surface to estimate the total charge of the cloud.

Express the total charge Q of a thundercloud in terms of the surface area A of the cloud and the charge density σ:

$$Q = \sigma A$$

Express the electric field just outside the cloud:

$$E = \frac{\sigma}{\epsilon_0}$$

Solve for σ:

$$\sigma = \epsilon_0 E$$

Substitute for σ to obtain:

$$Q = \epsilon_0 E A$$

Substitute numerical values and evaluate Q:

$$Q = (8.85 \times 10^{-12} \, \text{C}^2/\text{N} \cdot \text{m}^2)(3 \times 10^6 \, \text{V/m})(1 \, \text{km}^2) = \boxed{26.6 \, \text{C}}$$

Remarks: This charge is in reasonably good agreement with the total charge transferred in a lightning strike of approximately 30 C.

Calculating $\vec{E}$ From Coulomb's Law

*17 • A uniform line charge of linear charge density $\lambda = 3.5$ nC/m extends from $x = 0$ to $x = 5$ m. (a) What is the total charge? Find the electric field on the x axis at (b) $x = 6$ m, (c) $x = 9$ m, and (d) $x = 250$ m. (e) Find the field at $x = 250$ m, using the approximation that the charge is a point charge at the origin, and compare your result with that for the exact calculation in Part (d).

Picture the Problem We can use the definition of λ to find the total charge of the line of charge and the expression for the electric field on the axis of a finite line of charge to evaluate E_x at the given locations along the x axis. In part (d) we can apply Coulomb's law for the electric field due to a point charge to approximate the electric field at $x = 250$ m.

(a) Use the definition of linear charge density to express Q in terms of λ:

$$Q = \lambda L = (3.5 \, \text{nC/m})(5 \, \text{m}) = \boxed{17.5 \, \text{nC}}$$

Express the electric field on the axis of a finite line charge:

$$E_x(x_0) = \frac{kQ}{x_0(x_0 - L)}$$

Chapter 22

(b) Substitute numerical values and evaluate E_x at $x = 6$ m:

$$E_x(6\,\text{m}) = \frac{(8.99\times10^9\,\text{N}\cdot\text{m}^2/\text{C}^2)(17.5\,\text{nC})}{(6\,\text{m})(6\,\text{m}-5\,\text{m})}$$
$$= \boxed{26.2\,\text{N/C}}$$

(c) Substitute numerical values and evaluate E_x at $x = 9$ m:

$$E_x(9\,\text{m}) = \frac{(8.99\times10^9\,\text{N}\cdot\text{m}^2/\text{C}^2)(17.5\,\text{nC})}{(9\,\text{m})(9\,\text{m}-5\,\text{m})}$$
$$= \boxed{4.37\,\text{N/C}}$$

(d) Substitute numerical values and evaluate E_x at $x = 250$ m:

$$E_x(250\,\text{m}) = \frac{(8.99\times10^9\,\text{N}\cdot\text{m}^2/\text{C}^2)(17.5\,\text{nC})}{(250\,\text{m})(250\,\text{m}-5\,\text{m})} = \boxed{2.57\,\text{mN/C}}$$

(e) Use Coulomb's law for the electric field due to a point charge to obtain:

$$E_x(x) = \frac{kQ}{x^2}$$

Substitute numerical values and evaluate $E_x(250\text{ m})$:

$$E_x(250\,\text{m}) = \frac{(8.99\times10^9\,\text{N}\cdot\text{m}^2/\text{C}^2)(17.5\,\text{nC})}{(250\,\text{m})^2} = \boxed{2.52\,\text{mN/C}}$$

Note that this result agrees to within 2% with the exact value obtained in (d).

*25 •• (a) Using a spreadsheet program or graphing calculator, make a graph of the electric field on the axis of a disk of radius $r = 30$ cm carrying a surface charge density $\sigma = 0.5$ nC/m^2. (b) Compare the field to the approximation $E = 2\pi k\sigma$. At what distance does the approximation differ from the exact solution by 10 percent?

Picture the Problem

(a) The electric field on the x axis of a disk of radius r carrying a surface charge density σ is given by:

$$E_x = 2\pi k\sigma\left(1 - \frac{x}{\sqrt{x^2+r^2}}\right)$$

(b) The electric field due to an infinite sheet of charge density σ is independent of the distance from the plane and is given by:

$$E_{\text{plate}} = 2\pi k\sigma$$

A spreadsheet solution is shown below. The formulas used to calculate the quantities in the columns are as follows:

Cell	Content/Formula	Algebraic Form
B3	9.00E+09	k
B4	5.00E−10	σ
B5	0.3	r
A8	0	x_0
A9	0.01	$x_0 + 0.01$
B8	2*PI()*$B$3*$B$4*(1−A8/(A8^2+$B$5^2)^2)^0.5)	$2\pi k\sigma\left(1 - \dfrac{x}{\sqrt{x^2+r^2}}\right)$
C8	2*PI()*$B$3*$B$4	$2\pi k\sigma$

	A	B	C
1			
2			
3	k=	9.00E+09	Nm^2/C^2
4	sigma=	5.00E-10	C/m^2
5	r=	0.3	m
6			
7	x	E(x)	E plate
8	0.00	28.27	28.3
9	0.01	27.33	28.3
10	0.02	26.39	28.3
11	0.03	25.46	28.3
12	0.04	24.54	28.3
13	0.05	23.63	28.3
14	0.06	22.73	28.3
15	0.07	21.85	28.3
73	0.65	2.60	28.3
74	0.66	2.53	28.3
75	0.67	2.47	28.3
76	0.68	2.41	28.3
77	0.69	2.34	28.3
78	0.70	2.29	28.3

The following graph shows E as a function of x. The electric field from an infinite sheet with the same charge density is shown for comparison – the magnitude of the electric fields differ by more than 10 percent for $x = 0.03$ m.

34 Chapter 22

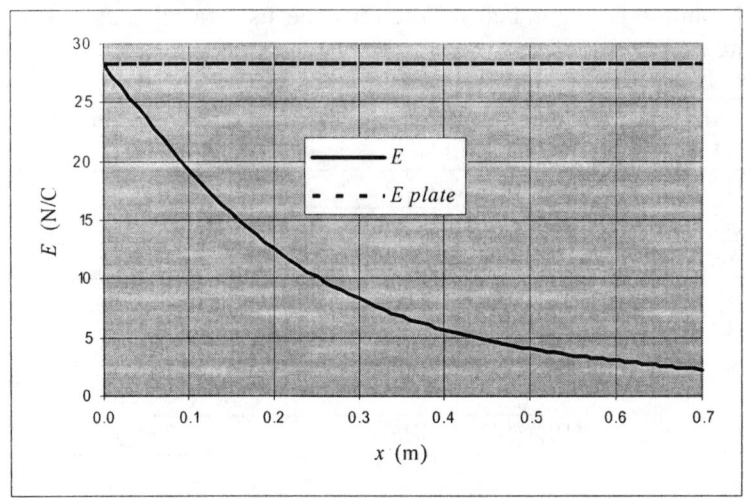

*30 ••• A hemispherical thin shell of radius R carries a uniform surface charge σ. Find the electric field at the center of the hemispherical shell ($r = 0$).

Picture the Problem Consider the ring with its axis along the z direction shown in the diagram. Its radius is $z = r\cos\theta$ and its width is $rd\theta$. We can use the equation for the field on the axis of a ring charge and then integrate to express the field at the center of the hemispherical shell.

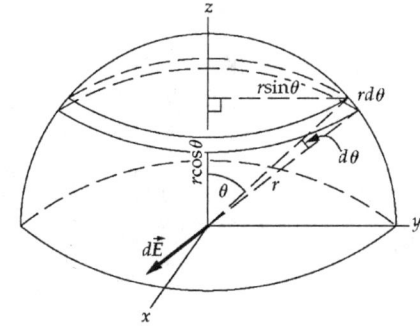

Express the field on the axis of the ring charge:

$$dE = \frac{kzdq}{\left(r^2 \sin^2\theta + r^2 \cos^2\theta\right)^{3/2}}$$

$$= \frac{kzdq}{r^3}$$

where $z = r\cos\theta$

Express the charge dq on the ring:

$$dq = \sigma dA = \sigma(2\pi r \sin\theta)rd\theta$$
$$= 2\pi\sigma r^2 \sin\theta d\theta$$

Substitute to obtain:

$$dE = \frac{k(r\cos\theta)2\pi\sigma r^2 \sin\theta d\theta}{r^3}$$
$$= 2\pi k\sigma \sin\theta \cos\theta d\theta$$

Integrate dE from $\theta = 0$ to $\pi/2$ to obtain:

$$E = 2\pi k\sigma \int_0^{\pi/2} \sin\theta \cos\theta \, d\theta$$

$$= 2\pi k\sigma \left[\tfrac{1}{2}\sin^2\theta\right]_0^{\pi/2} = \boxed{\pi k\sigma}$$

Gauss's Law

*32 • A single point charge $q = +2\ \mu C$ is at the origin. A spherical surface of radius 3.0 m has its center on the x axis at $x = 5$ m. (*a*) Sketch electric field lines for the point charge. Do any lines enter the spherical surface? (*b*) What is the net number of lines that cross the spherical surface, counting those that enter as negative? (*c*) What is the net flux of the electric field due to the point charge through the spherical surface?

Determine the Concept While the number of field lines that we choose to draw radially outward from q is arbitrary, we must show them originating at q and, in the absence of other charges, radially symmetric. The number of lines that we draw is, by agreement, in proportion to the magnitude of q.

(*a*) The sketch of the field lines and of the sphere is shown in the diagram to the right.

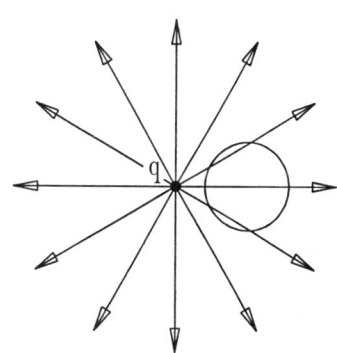

Given the number of field lines drawn from q, 3 lines enter the sphere.
Had we chosen to draw 24 field lines, 6 would have entered the spherical surface.

(*b*) The net number of lines crossing the surface is zero.

(*c*) The net flux is zero.

*36 • Since Newton's law of gravity and Coulomb's law have the same inverse-square dependence on distance, an expression analogous in form to Gauss's law can be found for gravity. The gravitational field $\vec{g}$ is the force per unit mass on a test mass m_0. Then, for a point mass m at the origin, the gravitational field g at some position r is

$$\vec{g} = -\frac{Gm}{r^2}\hat{r}$$

Compute the flux of the gravitational field through a spherical surface of radius r centered at the origin, and show that the gravitational analog of Gauss's law is
$\phi_{\text{net}} = -4\pi G m_{\text{inside}}$

Picture the Problem We'll define the flux of the gravitational field in a manner that is analogous to the definition of the flux of the electric field and then substitute for the gravitational field and evaluate the integral over the closed spherical surface.

Define the gravitational flux as:
$$\phi_g = \oint_S \vec{g} \cdot \hat{n}\, dA$$

Substitute for $\vec{g}$ and evaluate the integral to obtain:
$$\phi_g = \oint_S \left(-\frac{Gm}{r^2}\hat{r}\right) \cdot \hat{n}\, dA = -\frac{Gm}{r^2} \oint_S dA$$
$$= \left(-\frac{Gm}{r^2}\right)(4\pi r^2) = \boxed{-4\pi Gm}$$

Spherical Symmetry

*42 ••   Consider two concentric conducting spheres (Figure 22-38). The outer sphere is hollow and initially has a charge $-7Q$ deposited on it. The inner sphere is solid and has a charge $+2Q$ on it. (a) How is the charge distributed on the outer sphere? That is, how much charge is on the outer surface and how much charge is on the inner surface? (b) Suppose a wire is connected between the inner and outer spheres. After electrostatic equilibrium is established, how much total charge is on the outside sphere? How much charge is on the outer surface of the outside sphere, and how much charge is on the inner surface? Does the electric field at the surface of the inside sphere change when the wire is connected? If so, how? (c) Suppose we return to the original conditions in Part (a), with $+2Q$ on the inner sphere and $-7Q$ on the outer. We now connect the outer sphere to ground with a wire and then disconnect it. How much total charge will be on the outer sphere? How much charge will be on the inner surface of the outer sphere and how much will be on the outer surface?

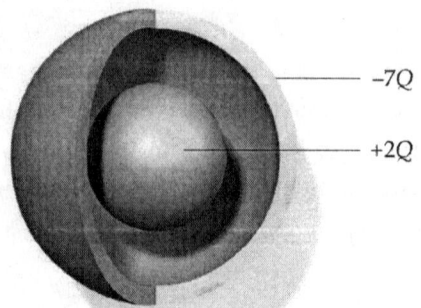

Figure 22-38 Problem 42

The Electric Field 2: Continuous Charge Distributions   37

Determine the Concept The charges on a conducting sphere, in response to the repulsive Coulomb forces each experiences, will separate until electrostatic equilibrium conditions exit. The use of a wire to connect the two spheres or to ground the outer sphere will cause additional redistribution of charge.

(*a*) Because the outer sphere is conducting, the field in the thin shell must vanish. Therefore, $-2Q$, uniformly distributed, resides on the inner surface, and $-5Q$, uniformly distributed, resides on the outer surface.

(*b*) Now there is no charge on the inner surface and $-5Q$ on the outer surface of the spherical shell. The electric field just outside the surface of the inner sphere changes from a finite value to zero.

(*c*) In this case, the $-5Q$ is drained off, leaving no charge on the outer surface and $-2Q$ on the inner surface. The total charge on the outer sphere is then $-2Q$.

*46 ••   Repeat Problem 44 for a sphere with volume charge density $\rho = C/r^2$ for $r < R$; $\rho = 0$ for $r > R$.

Picture the Problem We can find the total charge on the sphere by expressing the charge dq in a spherical shell and integrating this expression between $r = 0$ and $r = R$. By symmetry, the electric fields must be radial. To find E_r inside the charged sphere we choose a spherical Gaussian surface of radius $r < R$. To find E_r outside the charged sphere we choose a spherical Gaussian surface of radius $r > R$. On each of these surfaces, E_r is constant. Gauss's law then relates E_r to the total charge inside the surface.

(*a*) Express the charge dq in a shell of thickness dr and volume $4\pi r^2\, dr$:

$$dq = 4\pi r^2 \rho dr = 4\pi r^2 \frac{C}{r^2} dr$$
$$= 4\pi C dr$$

Integrate this expression from $r = 0$ to R to find the total charge on the sphere:

$$Q = 4\pi C \int_0^R dr = [4\pi Cr]_0^R$$
$$= \boxed{4\pi CR}$$

(*b*) Apply Gauss's law to a spherical surface of radius $r > R$ that is concentric with the nonconducting sphere to obtain:

$$\oint_S E_r dA = \frac{1}{\epsilon_0} Q_{\text{inside}}$$

or

$$4\pi r^2 E_r = \frac{Q_{\text{inside}}}{\epsilon_0}$$

Solve for E_r:

$$E_r(r > R) = \frac{Q_{inside}}{4\pi\epsilon_0}\frac{1}{r^2} = \frac{kQ_{inside}}{r^2}$$

$$= \frac{k4\pi CR}{r^2} = \boxed{\frac{CR}{\epsilon_0 r^2}}$$

Apply Gauss's law to a spherical surface of radius $r < R$ that is concentric with the nonconducting sphere to obtain:

$$\oint_S E_r dA = \frac{1}{\epsilon_0}Q_{inside}$$

or

$$4\pi r^2 E_r = \frac{Q_{inside}}{\epsilon_0}$$

Solve for E_r:

$$E_r(r < R) = \frac{Q_{inside}}{4\pi r^2 \epsilon_0} = \frac{4\pi Cr}{4\pi r^2 \epsilon_0}$$

$$= \boxed{\frac{C}{\epsilon_0 r}}$$

The graph of E_r versus r/R, with E_r in units of $C/\epsilon_0 R$, was plotted using a spreadsheet program.

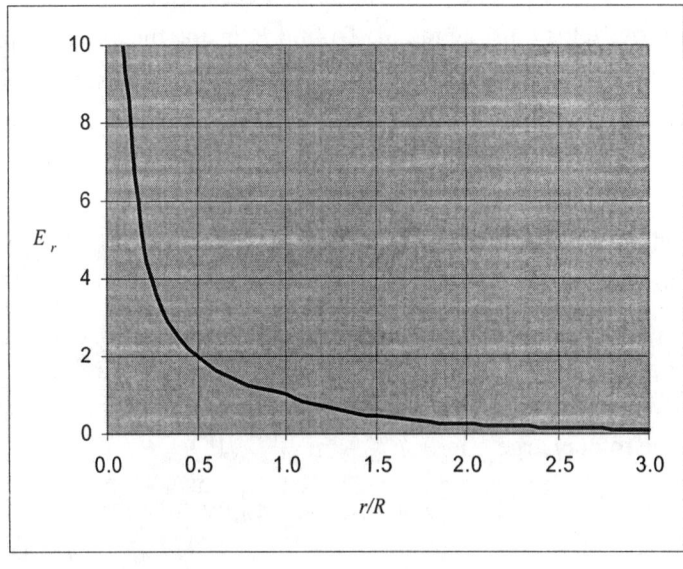

Cylindrical Symmetry

*52 •• Consider two infinitely long, concentric cylindrical shells. The inner shell has a radius R_1 and carries a uniform surface charge density of σ_1, and the outer shell has a radius R_2 and carries a uniform surface charge density of σ_2.

(a) Use Gauss's law to find the electric field in the regions $r < R_1$, $R_1 < r < R_2$, and $r > R_2$.

(b) What is the ratio of the surface charge densities σ_2/σ_1 and their relative signs if the electric field is zero at $r > R_2$? What would the electric field between the shells be in this case? (c) Sketch the electric field lines for the situation in Part (b) if σ_1 is positive.

Picture the Problem From symmetry; the field tangent to the surfaces of the shells must vanish. We can construct a Gaussian surface in the shape of a cylinder of radius r and length L and apply Gauss's law to find the electric field as a function of the distance from the centerline of the infinitely long, uniformly charged cylindrical shells.

(a) Apply Gauss's law to the cylindrical surface of radius r and length L that is concentric with the infinitely long, uniformly charged cylindrical shell:

$$\oint_S E_n dA = \frac{1}{\epsilon_0} Q_{inside}$$

or

$$2\pi r L E_n = \frac{Q_{inside}}{\epsilon_0}$$

where we've neglected the end areas because no flux crosses them.

Solve for E_n:

$$E_n = \frac{2kQ_{inside}}{Lr} \quad (1)$$

For $r < R_1$, $Q_{inside} = 0$ and:

$$E_n(r < R_1) = \boxed{0}$$

Express Q_{inside} for $R_1 < r < R_2$:

$$Q_{inside} = \sigma_1 A_1 = 2\pi \sigma_1 R_1 L$$

Substitute in equation (1) to obtain:

$$E_n(R_1 < r < R_2) = \frac{2k(2\pi\sigma_1 R_1 L)}{Lr}$$

$$= \boxed{\frac{\sigma_1 R_1}{\epsilon_0 r}}$$

Express Q_{inside} for $r > R_2$:

$$Q_{inside} = \sigma_1 A_1 + \sigma_2 A_2$$
$$= 2\pi \sigma_1 R_1 L + 2\pi \sigma_2 R_2 L$$

Substitute in equation (1) to obtain:

$$E_n(r > R_2) = \frac{2k(2\pi\sigma_1 R_1 L + 2\pi\sigma_2 R_2 L)}{Lr}$$

$$= \boxed{\frac{\sigma_1 R_1 + \sigma_2 R_2}{\epsilon_0 r}}$$

(b) Set $E = 0$ for $r > R_2$ to obtain:

$$\frac{\sigma_1 R_1 + \sigma_2 R_2}{\epsilon_0 r} = 0$$

or

40   Chapter 22

$$\sigma_1 R_1 + \sigma_2 R_2 = 0$$

Solve for the ratio of σ_1 to σ_2:

$$\boxed{\frac{\sigma_1}{\sigma_2} = -\frac{R_2}{R_1}}$$

Because the electric field is determined by the charge inside the Gaussian surface, the field under these conditions would be as given above:

$$E_n(R_1 < r < R_2) = \boxed{\frac{\sigma_1 R_1}{\epsilon_0 r}}$$

(c) Assuming that σ_1 is positive, the field lines would be directed as shown to the right.

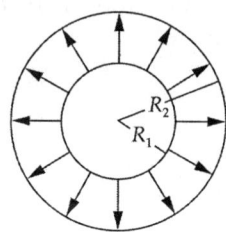

Charge and Field at Conductor Surfaces

*58 •   A penny is in an external electric field of magnitude 1.6 kN/C directed perpendicular to its faces. (a) Find the charge density on each face of the penny, assuming the faces are planes. (b) If the radius of the penny is 1 cm, find the total charge on one face.

Picture the Problem Because the penny is in an external electric field, it will have charges of opposite signs induced on its faces. The induced charge σ is related to the electric field by $E = \sigma/\epsilon_0$. Once we know σ, we can use the definition of surface charge density to find the total charge on one face of the penny.

(a) Relate the electric field to the charge density on each face of the penny:

$$E = \frac{\sigma}{\epsilon_0}$$

Solve for and evaluate σ:

$$\sigma = \epsilon_0 E$$
$$= (8.85 \times 10^{-12} \, C^2/N \cdot m^2)(1.6 \, kN/C)$$
$$= \boxed{14.2 \, nC/m^2}$$

(b) Use the definition of surface charge density to obtain:

$$\sigma = \frac{Q}{A} = \frac{Q}{\pi r^2}$$

Solve for and evaluate Q:

$$Q = \sigma \pi r^2 = \pi(14.2\,\text{nC/m}^2)(0.01\,\text{m})^2$$
$$= \boxed{4.45\,\text{pC}}$$

*63 •• A positive point charge of magnitude 2.5 μC is at the center of an uncharged spherical conducting shell of inner radius 60 cm and outer radius 90 cm. (*a*) Find the charge densities on the inner and outer surfaces of the shell and the total charge on each surface. (*b*) Find the electric field everywhere. (*c*) Repeat Part (*a*) and Part (*b*) with a net charge of +3.5 μC placed on the shell.

Picture the Problem Let the inner and outer radii of the uncharged spherical conducting shell be a and b and q represent the positive point charge at the center of the shell. The positive point charge at the center will induce a negative charge on the inner surface of the shell and, because the shell is uncharged, an equal positive charge will be induced on its outer surface. To solve part (*b*), we can construct a Gaussian surface in the shape of a sphere of radius r with the same center as the shell and apply Gauss's law to find the electric field as a function of the distance from this point. In part (*c*) we can use a similar strategy with the additional charge placed on the shell.

(*a*) Express the charge density on the inner surface:

$$\sigma_{\text{inner}} = \frac{q_{\text{inner}}}{A}$$

Express the relationship between the positive point charge q and the charge induced on the inner surface q_{inner}:

$$q + q_{\text{inner}} = 0$$

Substitute for q_{inner} to obtain:

$$\sigma_{\text{inner}} = \frac{-q}{4\pi a^2}$$

Substitute numerical values and evaluate σ_{inner}:

$$\sigma_{\text{inner}} = \frac{-2.5\,\mu\text{C}}{4\pi(0.6\,\text{m})^2} = \boxed{-0.553\,\mu\text{C/m}^2}$$

Express the charge density on the outer surface:

$$\sigma_{\text{outer}} = \frac{q_{\text{outer}}}{A}$$

Because the spherical shell is uncharged:

$$q_{\text{outer}} + q_{\text{inner}} = 0$$

Substitute for q_{outer} to obtain:

$$\sigma_{\text{outer}} = \frac{-q_{\text{inner}}}{4\pi b^2}$$

Substitute numerical values and evaluate σ_{outer}:

$$\sigma_{outer} = \frac{2.5\,\mu C}{4\pi(0.9\,m)^2} = \boxed{0.246\,\mu C/m^2}$$

(b) Apply Gauss's law to a spherical surface of radius r that is concentric with the point charge:

$$\oint_S E_n\,dA = \frac{1}{\epsilon_0} Q_{inside}$$

or

$$4\pi r^2 E_n = \frac{Q_{inside}}{\epsilon_0}$$

Solve for E_n:

$$E_n = \frac{Q_{inside}}{4\pi r^2 \epsilon_0} \quad (1)$$

For $r < a = 0.6$ m, $Q_{inside} = q$. Substitute in equation (1) and evaluate $E_n(r < 0.6$ m$)$ to obtain:

$$E_n(r < a) = \frac{q}{4\pi r^2 \epsilon_0} = \frac{kq}{r^2} = \frac{(8.99\times 10^9\,N\cdot m^2/C^2)(2.5\,\mu C)}{r^2}$$

$$= \boxed{(2.25\times 10^4\,N\cdot m^2/C)\frac{1}{r^2}}$$

Because the spherical shell is a conductor, a charge $-q$ will be induced on its inner surface. Hence, for 0.6 m $< r <$ 0.9 m:

$$Q_{inside} = 0$$
and
$$E_n(0.6\,m < r < 0.9\,m) = \boxed{0}$$

For $r > 0.9$ m, the net charge inside the Gaussian surface is q and:

$$E_n(r > 0.9\,m) = \frac{kq}{r^2} = \boxed{(2.25\times 10^4\,N\cdot m^2/C)\frac{1}{r^2}}$$

(c) Because $E = 0$ in the conductor:

$$q_{inner} = -2.5\,\mu C$$
and
$$\sigma_{inner} = \boxed{-0.553\,\mu C/m^2}$$
as before.

Express the relationship between the charges on the inner and outer surfaces of the spherical shell:

$$q_{outer} + q_{inner} = 3.5\,\mu C$$
and
$$q_{outer} = 3.5\,\mu C - q_{inner} = 6.0\,\mu C$$

σ_{outer} is now given by:

$$\sigma_{outer} = \frac{6\,\mu C}{4\pi(0.9\,\text{m})^2} = \boxed{0.589\,\mu C/m^2}$$

For $r < a = 0.6$ m, $Q_{inside} = q$ and $E_n(r < 0.6\,\text{m})$ is as it was in (a):

$$E_n(r < a) = \boxed{(2.25\times 10^4\,\text{N}\cdot\text{m}^2/\text{C})\frac{1}{r^2}}$$

Because the spherical shell is a conductor, a charge $-q$ will be induced on its inner surface. Hence, for 0.6 m $< r <$ 0.9 m:

$Q_{inside} = 0$

and

$E_n(0.6\,\text{m} < r < 0.9\,\text{m}) = \boxed{0}$

For $r > 0.9$ m, the net charge inside the Gaussian surface is 6 μC and:

$$E_n(r > 0.9\,\text{m}) = \frac{kq}{r^2} = (8.99\times 10^9\,\text{N}\cdot\text{m}^2/\text{C}^2)(6\,\mu C)\frac{1}{r^2} = \boxed{(5.39\times 10^4\,\text{N}\cdot\text{m}^2/\text{C})\frac{1}{r^2}}$$

General Problems

*69 •• A thin nonconducting uniformly charged spherical shell of radius r (Figure 22-41a) has a total charge of Q. A small circular plug is removed from the surface. (a) What is the magnitude and direction of the electric field at the center of the hole? (b) The plug is put back in the hole (Figure 22-41b). Using the result of Part (a), calculate the force acting on the plug. (c) From this, calculate the "electrostatic pressure" (force/unit area) tending to expand the sphere.

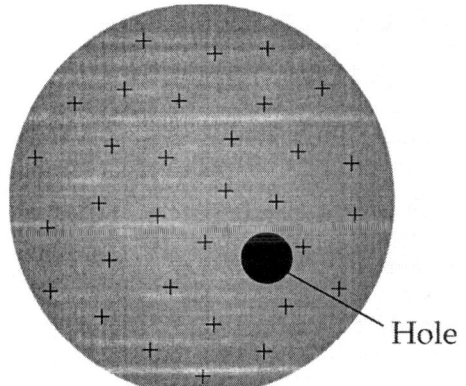
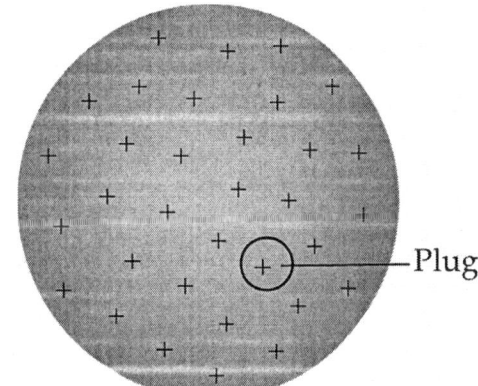

Figure 22-41a Problem 69   **Figure 22-41b** Problem 69

Picture the Problem If the patch is small enough, the field at the center of the patch comes from two contributions. We can view the field in the hole as the sum of the field from a uniform spherical shell of charge Q plus the field due to a small patch with surface charge density equal but opposite to that of the patch cut out.

(a) Express the magnitude of the electric field at the center of the hole:

$E = E_{spherical\,shell} + E_{hole}$

44   Chapter 22

Apply Gauss's law to a spherical gaussian surface just outside the given sphere:

$$E_{\text{spherical shell}}(4\pi r^2) = \frac{Q_{\text{enclosed}}}{\epsilon_0} = \frac{Q}{\epsilon_0}$$

Solve for $E_{\text{spherical shell}}$ to obtain:

$$E_{\text{spherical shell}} = \frac{Q}{4\pi \epsilon_0 r^2}$$

The electric field due to the small hole (small enough so that we can treat it as a plane surface) is:

$$E_{\text{hole}} = \frac{-\sigma}{2\epsilon_0}$$

Substitute and simplify to obtain:

$$E = \frac{Q}{4\pi \epsilon_0 r^2} + \frac{-\sigma}{2\epsilon_0}$$

$$= \frac{Q}{4\pi \epsilon_0 r^2} - \frac{Q}{2\epsilon_0 (4\pi r^2)}$$

$$= \boxed{\frac{Q}{8\pi \epsilon_0 r^2}}$$

(b) Express the force on the patch:

$$F = qE$$
where q is the charge on the patch.

Assuming that the patch has radius a, express the proportion between its charge and that of the spherical shell:

$$\frac{q}{\pi a^2} = \frac{Q}{4\pi r^2}$$

or

$$q = \frac{a^2}{4r^2} Q$$

Substitute for q and E in the expression for F to obtain:

$$F = \left(\frac{a^2}{4r^2} Q\right)\left(\frac{Q}{8\pi \epsilon_0 r^2}\right) = \boxed{\frac{Q^2 a^2}{32\pi \epsilon_0 r^4}}$$

(c) The pressure is the force divided by the area of the patch:

$$P = \frac{\frac{Q^2 a^2}{32\pi \epsilon_0 r^4}}{\pi a^2} = \boxed{\frac{Q^2}{32\pi^2 \epsilon_0 r^4}}$$

*73 ••   An infinitely long cylindrical shell is coaxial with the y axis and has a radius of 15 cm. It carries a uniform surface charge density $\sigma = 6\ \mu\text{C/m}^2$. A spherical shell of radius 25 cm is centered on the x axis at $x = 50$ cm and carries a uniform surface charge density $\sigma = -12\ \mu\text{C/m}^2$. Calculate the magnitude and direction of the electric field at (a) the origin; (b) $x = 20$ cm, $y = 10$ cm; and (c) $x = 50$ cm, $y = 20$ cm. (See Problem 48.)

Picture the Problem We can find the electric fields at the three points of interest by adding the electric fields due to the infinitely long cylindrical shell and the spherical shell. In Problem 42 it was established that, for an infinitely long cylindrical shell of radius R, $E_r(r < R) = 0$, and $E_r(r > R) = \sigma R/\epsilon_0 r$. We know that, for a spherical shell of radius R, $E_r(r < R) = 0$, and $E_r(r > R) = \sigma R^2/\epsilon_0 r^2$.

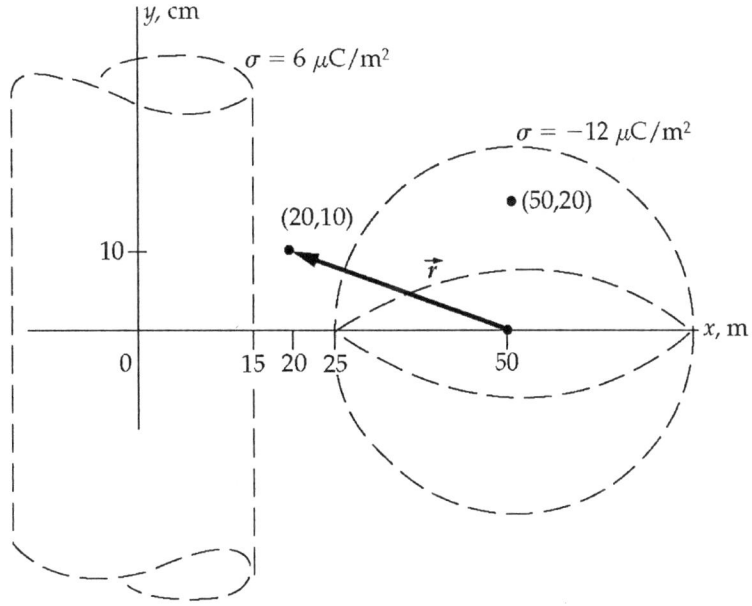

Express the resultant electric field as the sum of the fields due to the cylinder and sphere:

$$\vec{E} = \vec{E}_{cyl} + \vec{E}_{sph} \quad (1)$$

(a) Express and evaluate the electric field due to the cylindrical shell at the origin:

$$\vec{E}_{cyl}(0,0) = 0$$

because the origin is inside the cylindrical shell.

Express and evaluate the electric field due to the spherical shell at the origin:

$$\vec{E}_{sph}(0,0) = \frac{\sigma R^2}{\epsilon_0 r^2}(-\hat{i}) = \frac{(-12\,\mu C/m^2)(0.25\,m)^2}{(8.85 \times 10^{-12}\,C^2/N \cdot m^2)(0.5\,m)^2}(-\hat{i}) = (339\,kN/C)\hat{i}$$

Substitute in equation (1) to obtain:

$$\vec{E}(0,0) = 0 + (339\,kN/C)\hat{i}$$
$$= \boxed{(339\,kN/C)\hat{i}}$$

or

$$E(0,0) = \boxed{339\,kN/C}$$

and

$$\theta = \boxed{0°}$$

46   Chapter 22

(b) Express and evaluate the electric field due to the cylindrical shell at (0.2 m, 0.1 m):

$$\vec{E}_{cyl}(0.2\,\text{m}, 0.1\,\text{m}) = \frac{\sigma R}{\epsilon_0 r}\hat{i} = \frac{(6\,\mu\text{C/m}^2)(0.15\,\text{m})}{(8.85\times 10^{-12}\,\text{C}^2/\text{N}\cdot\text{m}^2)(0.2\,\text{m})}\hat{i} = (508\,\text{kN/C})\hat{i}$$

Express the electric field due to the charge on the spherical shell as a function of the distance from its center:

$$\vec{E}_{sph}(r) = \frac{\sigma R^2}{\epsilon_0 r^2}\hat{r}$$

where $\hat{r}$ is a unit vector pointing from (50 cm, 0) to (20 cm, 10 cm).

Referring to the diagram shown above, find r and $\hat{r}$:

$r = 0.316\,\text{m}$
and
$\hat{r} = -0.949\hat{i} + 0.316\hat{j}$

Substitute to obtain:

$$\vec{E}_{sph}(0.2\,\text{m}, 0.1\,\text{m}) = \frac{(-12\,\mu\text{C/m}^2)(0.25\,\text{m})^2}{(8.85\times 10^{-12}\,\text{C}^2/\text{N}\cdot\text{m}^2)(0.316\,\text{m})^2}(-0.949\hat{i} + 0.316\hat{j})$$
$$= (-849\,\text{kN/C})(-0.949\hat{i} + 0.316\hat{j})$$
$$= (806\,\text{kN/C})\hat{i} + (-268\,\text{kN/C})\hat{j}$$

Substitute in equation (1) to obtain:

$$\vec{E}(0.2\,\text{m}, 0.1\,\text{m}) = (508\,\text{kN/C})\hat{i} + (806\,\text{kN/C})\hat{i} + (-268\,\text{kN/C})\hat{j}$$
$$= \boxed{(1310\,\text{kN/C})\hat{i} + (-268\,\text{kN/C})\hat{j}}$$

or

$$E(0.2\,\text{m}, 0.1\,\text{m}) = \sqrt{(1310\,\text{kN/C})^2 + (-268\,\text{kN/C})^2} = \boxed{1340\,\text{kN/C}}$$

and

$$\theta = \tan^{-1}\left(\frac{-268\,\text{kN/C}}{1310\,\text{kN/C}}\right) = \boxed{348°}$$

(c) Express and evaluate the electric field due to the cylindrical shell at (0.5 m, 0.2 m):

$$\vec{E}_{cyl}(0.5\,\text{m}, 0.2\,\text{m}) = \frac{(6\,\mu\text{C/m}^2)(0.15\,\text{m})}{(8.85\times 10^{-12}\,\text{C}^2/\text{N}\cdot\text{m}^2)(0.5\,\text{m})}\hat{i} = (203\,\text{kN/C})\hat{i}$$

Express and evaluate the electric field due to the spherical shell at (0.5 m, 0.5 m):

$\vec{E}_{sph}(0.5\,m, 0.2\,m) = 0$

because (0.5 m, 0.2 m) is inside the spherical shell.

Substitute in equation (1) to obtain:

$\vec{E}(0.5\,m, 0.2\,m) = (203\,kN/C)\hat{i} + 0$
$= \boxed{(203\,kN/C)\hat{i}}$

or

$E(0.5\,m, 0.2\,m) = \boxed{203\,kN/C}$

and

$\theta = \boxed{0°}$

*76 •• Using the results of Problem 75, if we placed a proton above the nucleus of a hydrogen atom, at what distance r would the electric force on the proton balance the gravitational force mg acting on it? From this result, explain why even though the electrostatic force is enormously stronger than the gravitational force, it is the gravitational force we notice more.

Picture the Problem We will assume that the radius at which they balance is large enough that only the third term in the expression matters. Apply a condition for equilibrium will yield an equation that we can solve for the distance r.

Apply $\sum F = 0$ to the proton:

$$\frac{2ke^2}{a^2}e^{-2r/a} - mg = 0$$

To solve for r, isolate the exponential factor and take the natural logarithm of both sides of the equation:

$$r = \frac{a}{2}\ln\left(\frac{2ke^2}{mga^2}\right)$$

Substitute numerical values and evaluate r:

$$r = \frac{0.0529\,nm}{2}\ln\left[\frac{2(8.99\times10^9\,N\cdot m^2/C^2)(1.60\times10^{-19}\,C)^2}{(1.67\times10^{-27}\,kg)(9.81\,m/s^2)(0.0529\,nm)^2}\right] = \boxed{1.16\,nm}$$

Thus, even though the unscreened electrostatic force is 40 orders of magnitude larger than the gravitational force, screening reduces it to smaller than the gravitational force within a few nanometers.

Remarks: Note that the argument of the logarithm contains the ratio between the gravitational potential energy of a mass held a distance a_0 above the surface of the

48 Chapter 22

earth and the electrostatic potential energy for two unscreened charges a distance a_0 apart.

*84 •• A ring of radius R that lies in the yz plane carries a positive charge Q uniformly distributed over its length. A particle of mass m that carries a negative charge of magnitude q is at the center of the ring. (*a*) Show that if $x \ll R$, the electric field along the axis of the ring is proportional to x. (*b*) Find the force on the particle of mass m as a function of x. (*c*) Show that if m is given a small displacement in the x direction, it will perform simple harmonic motion. Calculate the period of that motion.

Picture the Problem Starting with the equation for the electric field on the axis of ring charge, we can factor the denominator of the expression to show that, for $x \ll R$, E_x is proportional to x. We can use $F_x = qE_x$ to express the force acting on the particle and apply Newton's 2nd law to show that, for small displacements from equilibrium, the particle will execute simple harmonic motion. Finally, we can find the period of the motion from its angular frequency, which we can obtain from the differential equation of motion.

(*a*) Express the electric field on the axis of the ring of charge:

$$E_x = \frac{kQx}{(x^2 + R^2)^{3/2}}$$

Factor R^2 from the denominator of E_x to obtain:

$$E_x = \frac{kQx}{\left[R^2\left(1 + \frac{x^2}{R^2}\right)\right]^{3/2}}$$

$$= \frac{kQx}{R^3\left(1 + \frac{x^2}{R^2}\right)^{3/2}} \approx \boxed{\frac{kQ}{R^3}x}$$

provided $x \ll R$.

(*b*) Express the force acting on the particle as a function of its charge and the electric field:

$$F_x = qE_x = \boxed{\frac{kqQ}{R^3}x}$$

(*c*) Because the negatively charged particle experiences a linear restoring force, we know that its motion will be simple harmonic. Apply Newton's 2nd law to the negatively charged particle to obtain:

$$m\frac{d^2x}{dt^2} = -\frac{kqQ}{R^3}x$$

or

$$\boxed{\frac{d^2x}{dt^2} + \frac{kqQ}{mR^3}x = 0}$$

the differential equation of simple harmonic motion.

Relate the period T of the simple harmonic motion to its angular frequency ω:	$T = \dfrac{2\pi}{\omega}$
From the differential equation we have:	$\omega^2 = \dfrac{kqQ}{mR^3}$
Substitute to obtain:	$\boxed{T = 2\pi\sqrt{\dfrac{mR^3}{kqQ}}}$

*94 ••• A dipole $\vec{p}$ is located at a distance r from an infinitely long line charge with a uniform linear charge density λ. Assume that the dipole is aligned with the field due to the line charge. Determine the force that acts on the dipole.

Picture the Problem We can find the field due to the infinitely long line charge from $E = 2k\lambda/r$ and the force that acts on the dipole using $F = p\, dE/dr$.

Express the force acting on the dipole:	$F = p\dfrac{dE}{dr}$
The electric field at the location of the dipole is given by:	$E = \dfrac{2k\lambda}{r}$
Substitute to obtain:	$F = p\dfrac{d}{dr}\left[\dfrac{2k\lambda}{r}\right] = \boxed{-\dfrac{2k\lambda p}{r^2}}$

where the minus sign indicates that the dipole is attracted to the line charge.

Chapter 23
Electric Potential

Conceptual Problems

*1 • A positive charge is released from rest in an electric field. Will it move toward a region of greater or smaller electric potential?

Determine the Concept A positive charge will move in whatever direction reduces its potential energy. The positive charge will reduce its potential energy if it moves toward a region of lower electric potential.

*7 •• Figure 23-26 shows a metal sphere carrying a charge $-Q$ and a point charge $+Q$. Sketch the electric field lines and equipotential surfaces in the vicinity of this charge system.

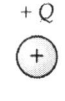

Figure 23-26 Problems 7 and 8

Picture the Problem The electric field lines, shown as solid lines, and the equipotential surfaces (intersecting the plane of the paper), shown as dashed lines, are sketched in the adjacent figure. The point charge $+Q$ is the point at the right, and the metal sphere with charge $-Q$ is at the left. Near the two charges the equipotential surfaces are spheres, and the field lines are normal to the metal sphere at the sphere's surface.

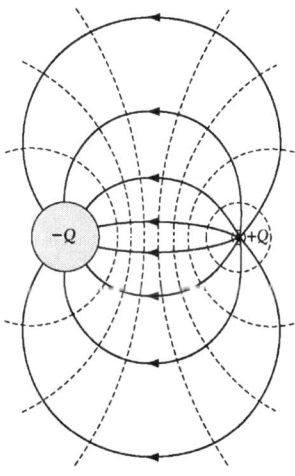

*11 • Two equal positive point charges $+Q$ are on the x axis. One is at $x = -a$ and the other is at $x = +a$. At the origin, (a) $E = 0$ and $V = 0$, (b) $E = 0$ and $V = 2kQ/a$, (c) $\vec{E} = (2kQ^2/a^2)\hat{i}$ and $V = 0$, (d) $\vec{E} = (2kQ^2/a^2)\hat{i}$ and $V = 2kQ/a$, or (e) none of the above is correct.

Picture the Problem We can use Coulomb's law and the superposition of fields to find E at the origin and the definition of the electric potential due to a point charge to find V at

52   Chapter 23

the origin.

Apply Coulomb's law and the superposition of fields to find the electric field E at the origin:

$$\vec{E} = \vec{E}_{+Q\,\text{at}-a} + \vec{E}_{+Q\,\text{at}\,a}$$
$$= \frac{kQ}{a^2}\hat{i} - \frac{kQ}{a^2}\hat{i} = 0$$

Express the potential V at the origin:

$$V = V_{+Q\,\text{at}-a} + V_{+Q\,\text{at}\,a}$$
$$= \frac{kQ}{a} + \frac{kQ}{a} = \frac{2kQ}{a}$$

and $\boxed{(b) \text{ is correct.}}$

*16 •   Two charged metal spheres are connected by a wire, and sphere A is larger than sphere B (Figure 23-27). The magnitude of the electric potential of sphere A is (a) greater than that at the surface of sphere B; (b) less than that at the surface of sphere B; (c) the same as that at the surface of sphere B; (d) greater than or less than that at the surface of sphere B, depending on the radii of the spheres, or (e) greater than or less than that at the surface of sphere B, depending on the charge on the spheres.

Figure 23-27 Problem 16

Determine the Concept When the two spheres are connected, their charges will redistribute until the two-sphere system is in electrostatic equilibrium. Consequently, the entire system must be an equipotential. $\boxed{(c) \text{ is corrent.}}$

Estimation and Approximation Problems

*18 •   Estimate the potential difference across the spark gap in a typical automobile spark plug. Because of the high compression of the gas in the piston, the electric field at which the gas sparks is roughly 2×10^7 V/m.

Picture the Problem The potential difference between the electrodes of the spark plug is the product of the electric field in the gap and the separation of the electrodes. We'll assume that the separation of the electrodes is 1 mm.

Express the potential difference between the electrodes of the spark

$V = Ed$

plug as a function of their separation d and electric field E between them:

Substitute numerical values and evaluate V:

$$V = (2\times 10^7 \text{ V/m})(10^{-3} \text{ m})$$
$$= \boxed{20.0 \text{ kV}}$$

Potential Difference

*26 •• An electron gun fires electrons at the screen of a television tube. The electrons start from rest and are accelerated through a potential difference of 30,000 V. What is the energy of the electrons when they hit the screen (*a*) in electron volts and (*b*) in joules? (*c*) What is the speed of impact of electrons with the screen of the picture tube?

Picture the Problem The work done on the electrons by the electric field changes their kinetic energy. Hence we can use the work-kinetic energy theorem to find the kinetic energy and the speed of impact of the electrons.

Use the work-kinetic energy theorem to relate the work done by the electric field to the change in the kinetic energy of the electrons:

$$W = \Delta K = K_f$$
or
$$K_f = e\Delta V \qquad (1)$$

(*a*) Substitute numerical values and evaluate K_f:

$$K_f = (1e)(30 \text{ kV}) = \boxed{3\times 10^4 \text{ eV}}$$

(*b*) Convert this energy to eV:

$$K_f = (3\times 10^4 \text{ eV})\left(\frac{1.6\times 10^{-19} \text{ J}}{\text{eV}}\right)$$
$$= \boxed{4.80\times 10^{-15} \text{ J}}$$

(*c*) From equation (1) we have:

$$\tfrac{1}{2} m v_f^2 = e\Delta V$$

Solve for v_f to obtain:

$$v_f = \sqrt{\frac{2e\Delta V}{m}}$$

Substitute numerical values and evaluate v_f:

$$v_f = \sqrt{\frac{2(1.6\times 10^{-19} \text{ C})(30 \text{ kV})}{9.11\times 10^{-31} \text{ kg}}}$$
$$= \boxed{1.03\times 10^8 \text{ m/s}}$$

Remarks: Note that this speed is about one-third that of light.

Potential Due to a System of Point Charges

*32 • Two point charges q and q' are separated by a distance a. At a point $a/3$ from q and along the line joining the two charges the potential is zero. Find the ratio q/q'.

Picture the Problem We can use the fact that the electric potential at the point of interest is the algebraic sum of the potentials at that point due to the charges q and q' to find the ratio q/q'.

Express the potential at the point of interest as the sum of the potentials due to the two charges:
$$\frac{kq}{a/3} + \frac{kq'}{2a/3} = 0$$

Simplify to obtain:
$$q + \frac{q'}{2} = 0$$

Solve for the ratio q/q':
$$\frac{q}{q'} = \boxed{-\frac{1}{2}}$$

*34 •• A point charge of $+3e$ is at the origin and a second point charge of $-2e$ is on the x axis at $x = a$. (a) Sketch the potential function $V(x)$ versus x for all x. (b) At what point or points is $V(x)$ zero? (c) How much work is needed to bring a third charge $+e$ to the point $x = \frac{1}{2}a$ on the x axis?

Picture the Problem For the two charges, $r = |x - a|$ and $|x|$ respectively and the electric potential at x is the algebraic sum of the potentials at that point due to the charges at $x = a$ and $x = 0$. We can use the graph and the function found in part (a) to identify the points at which $V(x) = 0$. We can find the work needed to bring a third charge $+e$ to the point $x = \frac{1}{2}a$ on the x axis from the change in the potential energy of this third charge.

Express the potential at x:
$$V(x) = \frac{k(3e)}{|x|} + \frac{k(-2e)}{|x-a|}$$

The following graph of $V(x)$ for $ke = 1$ and $a = 1$ was plotted using a spreadsheet program.

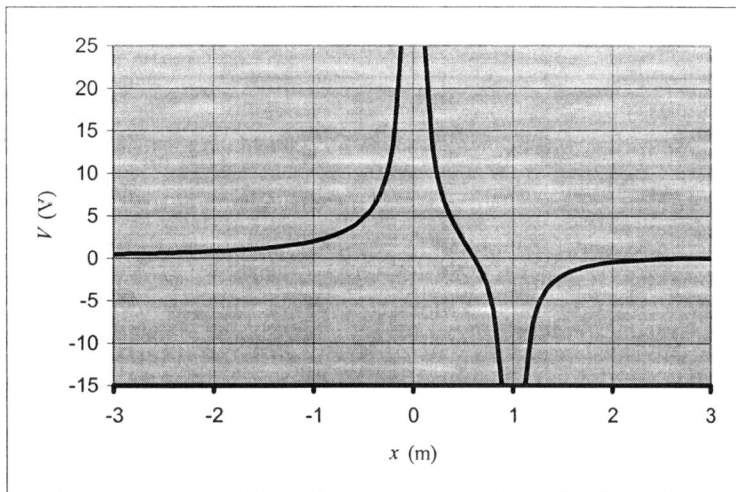

(*b*) From the graph we can see that $V(x) = 0$ when: $\quad x = \boxed{\pm\infty}$

Examining the function, we see that $V(x)$ is also zero provided: $\quad \dfrac{3}{|x|} - \dfrac{2}{|x-a|} = 0$

For $x > 0$, $V(x) = 0$ when: $\quad x = \boxed{3a}$

For $0 < x < a$, $V(x) = 0$ when: $\quad x = \boxed{0.6a}$

(*c*) Express the work that must be done in terms of the change in potential energy of the charge:
$$W = \Delta U = qV\!\left(\tfrac{1}{2}a\right)$$

Evaluate the potential at $x = \tfrac{1}{2}a$:
$$V\!\left(\tfrac{1}{2}a\right) = \dfrac{k(3e)}{\left|\tfrac{1}{2}a\right|} + \dfrac{k(-2e)}{\left|\tfrac{1}{2}a - a\right|}$$
$$= \dfrac{6ke}{a} - \dfrac{4ke}{a} = \dfrac{2ke}{a}$$

Substitute to obtain:
$$W = e\!\left(\dfrac{2ke}{a}\right) = \boxed{\dfrac{2ke^2}{a}}$$

Computing the Electric Field from the Potential

*36 • The potential due to a particular charge distribution is measured at several points along the *x* axis, as shown in Figure 23-28. For what value(s) in the range $0 < x < 10$ m is $E_x = 0$?

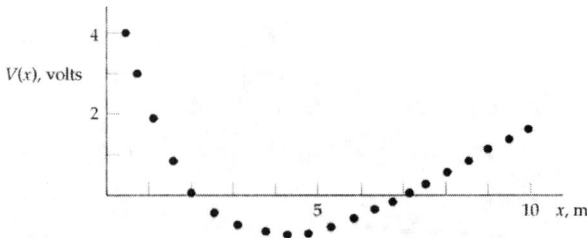

Figure 23-28 Problem 36

Picture the Problem Because $E_x = -dV/dx$, we can find the point(s) at which $E_x = 0$ by identifying the values for x for which $dV/dx = 0$.

Examination of the graph indicates that $dV/dx = 0$ at $x = 4.5$ m. Thus $E_x = 0$ at: $\qquad x = \boxed{4.5\,\text{m}}$

*42 ••   An electric field is given by $E_x = 2.0x^3$ kN/C. Find the potential difference between the points on the x axis at $x = 1$ m and $x = 2$ m.

Picture the Problem Because $V(x)$ and E_x are related through $E_x = -dV/dx$, we can find V from E by integration.

Separate variables to obtain: $\qquad dV = -E_x dx = -(2.0x^3\ \text{kN/C})dx$

Integrate V from V_1 to V_2 and x from 1 m to 2 m:

$$\int_{V_1}^{V_2} dV = -(2.0\,\text{kN/C})\int_{x_1}^{x_2} x^3\,dx$$

$$= -(2.0\,\text{kN/C})\left[\tfrac{1}{4}x^4\right]_{1\,\text{m}}^{2\,\text{m}}$$

Simplify to obtain: $\qquad V_2 - V_1 = \boxed{-7.50\,\text{kV}}$

Calculations of V for Continuous Charge Distributions

*47 ••   A rod of length L carries a charge Q uniformly distributed along its length. The rod lies along the y-axis with its center at the origin. (a) Find the potential as a function of position along the x-axis. (b) Show that the result obtained in Part (a) reduces to $V = kQ/x$ for $x \gg L$.

Electric Potential 57

Picture the Problem Let the charge per unit length be $\lambda = Q/L$ and dy be a line element with charge λdy. We can express the potential dV at any point on the x axis due to λdy and integrate of find $V(x, 0)$.

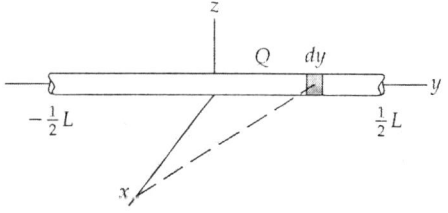

(*a*) Express the element of potential dV due to the line element dy:

$$dV = \frac{k\lambda}{r} dy$$

where $r = \sqrt{x^2 + y^2}$

Integrate dV from $y = -L/2$ to $y = L/2$:

$$V(x,0) = \frac{kQ}{L} \int_{-L/2}^{L/2} \frac{dy}{\sqrt{x^2 + y^2}}$$

$$= \boxed{\frac{kQ}{L} \ln\left(\frac{\sqrt{x^2 + L^2/4} + L/2}{\sqrt{x^2 + L^2/4} - L/2}\right)}$$

(*b*) Factor x from the numerator and denominator within the parentheses to obtain:

$$V(x,0) = \frac{kQ}{L} \ln\left(\frac{\sqrt{1 + \frac{L^2}{4x^2}} + \frac{L}{2x}}{\sqrt{1 + \frac{L^2}{4x^2}} - \frac{L}{2x}}\right)$$

Use $\ln \frac{a}{b} = \ln a - \ln b$ to obtain:

$$V(x,0) = \frac{kQ}{L} \left\{ \ln\left(\sqrt{1 + \frac{L^2}{4x^2}} + \frac{L}{2x}\right) - \ln\left(\sqrt{1 + \frac{L^2}{4x^2}} - \frac{L}{2x}\right) \right\}$$

Let $\varepsilon = \frac{L^2}{4x^2}$ and use $(1+\varepsilon)^{1/2} = 1 + \tfrac{1}{2}\varepsilon - \tfrac{1}{8}\varepsilon^2 + ...$ to expand $\sqrt{1 + \frac{L^2}{4x^2}}$:

$$\left(1 + \frac{L^2}{4x^2}\right)^{1/2} = 1 + \frac{1}{2}\frac{L^2}{4x^2} - \frac{1}{8}\left(\frac{L^2}{4x^2}\right)^2 + ... \approx 1 \text{ for } x \gg L.$$

Substitute to obtain:

$$V(x,0) = \frac{kQ}{L}\left\{\ln\left(1 + \frac{L}{2x}\right) - \ln\left(1 - \frac{L}{2x}\right)\right\}$$

58 Chapter 23

Let $\delta = \dfrac{L}{2x}$ and use $\ln(1+\delta) = \delta - \tfrac{1}{2}\delta^2 + \ldots$ to expand $\ln\left(1 \pm \dfrac{L}{2x}\right)$:

$$\ln\left(1 + \dfrac{L}{2x}\right) \approx \dfrac{L}{2x} - \dfrac{L^2}{4x^2} \quad \text{and} \quad \ln\left(1 - \dfrac{L}{2x}\right) \approx -\dfrac{L}{2x} - \dfrac{L^2}{4x^2} \quad \text{for } x \gg L.$$

Substitute and simplify to obtain:

$$V(x,0) = \dfrac{kQ}{L}\left\{\dfrac{L}{2x} - \dfrac{L^2}{4x^2} - \left(-\dfrac{L}{2x} - \dfrac{L^2}{4x^2}\right)\right\} = \boxed{\dfrac{kQ}{x}}$$

*51 •• A disk of radius R carries a charge density $+\sigma_0$ for $r < a$ and an equal but opposite charge density $-\sigma_0$ for $a < r < R$. The total charge carried by the disk is zero. (*a*) Find the potential a distance x along the axis of the disk. (*b*) Obtain an approximate expression for $V(x)$ when $x \gg R$.

Picture the Problem The potential at any location on the axis of the disk is the sum of the potentials due to the positive and negative charge distributions on the disk. Knowing that the total charge on the disk is zero and the charge densities are equal in magnitude will allow us to find the radius of the region that is positively charged. We can then use the expression derived in the text to find the potential due to this charge closest to the axis and integrate dV from $r = R/\sqrt{2}$ to $r = R$ to find the potential at x due to the negative charge distribution.

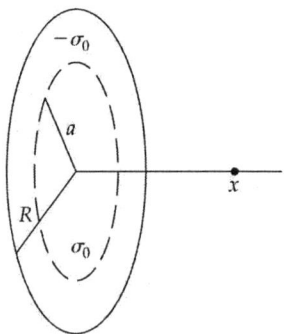

(*a*) Express the potential at a distance x along the axis of the disk as the sum of the potentials due to the positively and negatively charged regions of the disk:

$$V(x) = V_+(x) + V_-(x)$$

We know that the charge densities are equal in magnitude and that the total charge carried by the disk is zero. Express this condition in terms

$$Q_{r<a} = Q_{r>a}$$

or

$$\sigma_0 \pi a^2 = \sigma_0 \pi R^2 - \sigma_0 \pi a^2$$

of the charge in each of two regions of the disk:

Solve for a to obtain:
$$a = \frac{R}{\sqrt{2}}$$

Use this result and the general expression for the potential on the axis of a charged disk to express $V_+(x)$:
$$V_+(x) = 2\pi k\sigma_0 \left(\sqrt{x^2 + \frac{R^2}{2}} - x \right)$$

Express the potential on the axis of the disk due to a ring of charge a distance $r > a$ from the axis of the ring:
$$dV_-(x) = -2\pi k\sigma_0 \frac{r}{r'} dr$$
where $r' = \sqrt{x^2 + r^2}$.

Integrate this expression from $r = R/\sqrt{2}$ to $r = R$ to obtain:
$$V_-(x) = -2\pi k\sigma_0 \int_{R/\sqrt{2}}^{R} \frac{r}{\sqrt{x^2 + r^2}} dr$$
$$= -2\pi k\sigma_0 \left(\sqrt{x^2 + R^2} - \sqrt{x^2 + \frac{R^2}{2}} \right)$$

Substitute and simplify to obtain:

$$V(x) = 2\pi k\sigma_0 \left(\sqrt{x^2 + \frac{R^2}{2}} - x \right) - 2\pi k\sigma_0 \left(\sqrt{x^2 + R^2} - \sqrt{x^2 + \frac{R^2}{2}} \right)$$
$$= 2\pi k\sigma_0 \left(\sqrt{x^2 + \frac{R^2}{2}} - x - \sqrt{x^2 + R^2} + \sqrt{x^2 + \frac{R^2}{2}} \right)$$
$$= \boxed{2\pi k\sigma_0 \left(2\sqrt{x^2 + \frac{R^2}{2}} - \sqrt{x^2 + R^2} - x \right)}$$

(b) To determine V for $x \gg R$, factor x from the square roots and expand using the binomial expansion:
$$\sqrt{x^2 + \frac{R^2}{2}} = x\left(1 + \frac{R^2}{2x^2}\right)^{1/2}$$
$$\approx x\left(1 + \frac{R^2}{4x^2} - \frac{R^4}{32x^4}\right)$$

and

60   Chapter 23

$$\sqrt{x^2 + R^2} = x\left(1 + \frac{R^2}{x^2}\right)^{1/2}$$

$$\approx x\left(1 + \frac{R^2}{2x^2} - \frac{R^4}{8x^4}\right)$$

Substitute to obtain:

$$V(x) \approx 2\pi k\sigma_0\left(2x\left(1 + \frac{R^2}{4x^2} - \frac{R^4}{32x^4}\right) - x\left(1 + \frac{R^2}{2x^2} - \frac{R^4}{8x^4}\right) - x\right) = \boxed{\frac{\pi k\sigma_0 R^4}{8x^3}}$$

*58 •• Show that for $x \gg R$ the potential on the axis of a disk charge approaches kQ/x, where $Q = \sigma\pi R^2$ is the total charge on the disk. [*Hint:* Write $(x^2 + R^2)^{1/2} = x(1 + R^2/x^2)^{1/2}$ and use the binomial expansion.]

Picture the Problem The potential on the axis of a disk charge of radius R and charge density σ is given by $V = 2\pi k\sigma\left[(x^2 + R^2)^{1/2} - x\right]$.

Express the potential on the axis of the disk charge:

$$V = 2\pi k\sigma\left[(x^2 + R^2)^{1/2} - x\right]$$

Factor x from the radical and use the binomial expansion to obtain:

$$(x^2 + R^2)^{1/2} = x\left(1 + \frac{R^2}{x^2}\right)^{1/2} = x\left[1 + \frac{R^2}{2x^2} + \left(\frac{1}{2}\right)\left(-\frac{1}{2}\right)\left(\frac{1}{2}\right)\frac{R^4}{x^4} + \ldots\right]$$

$$\approx x\left[1 + \frac{R^2}{2x^2} - \frac{R^4}{8x^4}\right]$$

Substitute for the radical term to obtain:

$$V = 2\pi k\sigma\left\{x\left[1 + \frac{R^2}{2x^2} - \frac{R^4}{8x^4}\right] - x\right\}$$

$$= 2\pi k\sigma\left(\frac{R^2}{2x} - \frac{R^4}{8x^3}\right)$$

$$\approx 2\pi k\sigma\left(\frac{R^2}{2x}\right) = \boxed{\frac{kQ}{x}}$$

provided $x \gg R$.

*63 • Find the greatest surface charge density σ_{max} that can exist on a conductor before dielectric breakdown of the air occurs.

Electric Potential 61

Picture the Problem We can solve the equation giving the electric field at the surface of a conductor for the greatest surface charge density that can exist before dielectric breakdown of the air occurs.

Relate the electric field at the surface of a conductor to the surface charge density:

$$E = \frac{\sigma}{\epsilon_0}$$

Solve for σ under dielectric breakdown of the air conditions:

$$\sigma_{max} = \epsilon_0 E_{breakdown}$$

Substitute numerical values and evaluate σ_{max}:

$$\sigma_{max} = (8.85 \times 10^{-12} \, C^2/N \cdot m^2)(3 \, MV/m)$$
$$= \boxed{26.6 \, \mu C/m^2}$$

*66 ••• Calculate the potential relative to infinity at the center of a uniformly charged sphere of radius R and charge Q.

Picture the Problem We can find the potential relative to infinity at the center of the sphere by integrating the electric field for 0 to ∞. We can apply Gauss's law to find the electric field both inside and outside the spherical shell.

The potential relative to infinity the center of the spherical shell is:

$$V = \int_0^R E_{r<R} dr + \int_R^\infty E_{r>R} dr \qquad (1)$$

Apply Gauss's law to a spherical surface of radius $r < R$ to obtain:

$$\oint_S E_n dA = E_{r<R}(4\pi r^2) = \frac{Q_{inside}}{\epsilon_0}$$

Using the fact that the sphere is uniformly charged, express Q_{inside} in terms of Q:

$$\frac{Q_{inside}}{\frac{4}{3}\pi r^3} = \frac{Q}{\frac{4}{3}\pi R^3} \Rightarrow Q_{inside} = \frac{r^3}{R^3} Q$$

Substitute for Q_{inside} to obtain:

$$E_{r<R}(4\pi r^2) = \frac{r^3}{\epsilon_0 R^3} Q$$

Solve for $E_{r<R}$:

$$E_{r<R} = \frac{r}{4\pi \epsilon_0 R^3} Q = \frac{kQ}{R^3} r$$

Apply Gauss's law to a spherical surface of radius $r > R$ to obtain:

$$\oint_S E_n dA = E_{r>R}(4\pi r^2) = \frac{Q_{inside}}{\epsilon_0} = \frac{Q}{\epsilon_0}$$

Solve for $E_{r>R}$ to obtain:

$$E_{r>R} = \frac{Q}{4\pi \epsilon_0 r^2} = \frac{kQ}{r^2}$$

62 Chapter 23

Substitute for $E_{r<R}$ and $E_{r>R}$ in equation (1) and evaluate the resulting integral:

$$V = \frac{kQ}{R^3}\int_0^R r\,dr + kQ\int_R^\infty \frac{dr}{r^2}$$

$$= \frac{kQ}{R^3}\left[\frac{r^2}{2}\right]_0^R + kQ\left[-\frac{1}{r}\right]_R^\infty = \boxed{\frac{3kQ}{2R}}$$

General Problems

*71 •• Two infinitely long parallel wires carry a uniform charge per unit length λ and $-\lambda$ respectively. The wires are in the xz plane, parallel with the z axis. The positively charged wire intersects the x axis at $x = -a$, and the negatively charged wire intersects the x axis at $x = +a$. (a) Choose the origin as the reference point where the potential is zero, and express the potential at an arbitrary point (x,y) in the xy plane in terms of x, y, λ, and a. Use this expression to solve for the potential everywhere on the y axis. (b) Use a spreadsheet program to plot the equipotential curve in the xy plane that passes through the point $x = \frac{1}{4}a, y = 0$. Use $a = 5$ cm and $\lambda = 5$ nC/m.

Picture the Problem The geometry of the wires is shown to the right. The potential at the point whose coordinates are (x, y) is the sum of the potentials due to the charge distributions on the wires.

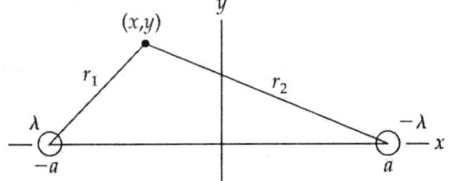

(a) Express the potential at the point whose coordinates are (x, y):

$$V(x,y) = V_{\text{wire at} -a} + V_{\text{wire at } a}$$

$$= 2k\lambda \ln\left(\frac{r_{\text{ref}}}{r_1}\right) + 2k(-\lambda)\ln\left(\frac{r_{\text{ref}}}{r_2}\right)$$

$$= 2k\lambda\left[\ln\left(\frac{r_{\text{ref}}}{r_1}\right) - \ln\left(\frac{r_{\text{ref}}}{r_2}\right)\right]$$

$$= \frac{\lambda}{2\pi\epsilon_0}\ln\left(\frac{r_2}{r_1}\right)$$

where $V(0) = 0$.

Because $r_1 = \sqrt{(x+a)^2 + y^2}$ and $r_2 = \sqrt{(x-a)^2 + y^2}$:

$$V(x,y) = \boxed{\frac{\lambda}{2\pi\epsilon_0}\ln\left(\frac{\sqrt{(x-a)^2+y^2}}{\sqrt{(x+a)^2+y^2}}\right)}$$

On the y-axis, $x = 0$ and:

$$V(0,y) = \frac{\lambda}{2\pi\epsilon_0}\ln\left(\frac{\sqrt{a^2+y^2}}{\sqrt{a^2+y^2}}\right)$$

$$= \frac{\lambda}{2\pi\epsilon_0}\ln(1) = \boxed{0}$$

(b) Evaluate the potential at $\left(\tfrac{1}{4}a,0\right) = (1.25\,\text{cm}, 0)$:

$$V\left(\tfrac{1}{4}a,0\right) = \frac{\lambda}{2\pi\epsilon_0} \ln\left(\frac{\sqrt{\left(\tfrac{1}{4}a - a\right)^2}}{\sqrt{\left(\tfrac{1}{4}a + a\right)^2}}\right)$$

$$= \frac{\lambda}{2\pi\epsilon_0} \ln\left(\frac{3}{5}\right)$$

Equate $V(x,y)$ and $V\left(\tfrac{1}{4}a,0\right)$:

$$\frac{3}{5} = \frac{\sqrt{(x-5)^2 + y^2}}{\sqrt{(x+5)^2 + y^2}}$$

Solve for y to obtain:

$$y = \pm\sqrt{21.25x - x^2 - 25}$$

A spreadsheet program to plot $y = \pm\sqrt{21.25x - x^2 - 25}$ is shown below. The formulas used to calculate the quantities in the columns are as follows:

Cell	Content/Formula	Algebraic Form
A2	1.25	$\tfrac{1}{4}a$
A3	A2 + 0.05	$x + \Delta x$
B2	SQRT(21.25*A2 − A2^2 − 25)	$y = \sqrt{21.25x - x^2 - 25}$
B4	−SQRT(21.25*A2 − A2^2 − 25)	$y = -\sqrt{21.25x - x^2 - 25}$

	A	B	C
1	x	y_pos	y_neg
2	1.25	0.00	0.00
3	1.30	0.97	−0.97
4	1.35	1.37	−1.37
5	1.40	1.67	−1.67
6	1.45	1.93	−1.93
7	1.50	2.15	−2.15
370	19.65	2.54	−2.54
371	19.70	2.35	−2.35
372	19.75	2.15	−2.15
373	19.80	1.93	−1.93
374	19.85	1.67	−1.67
375	19.90	1.37	−1.37
376	19.95	0.97	−0.97

The following graph shows the equipotential curve in the xy plane for

$$V\left(\tfrac{1}{4}a,0\right) = \frac{\lambda}{2\pi\epsilon_0} \ln\left(\frac{3}{5}\right).$$

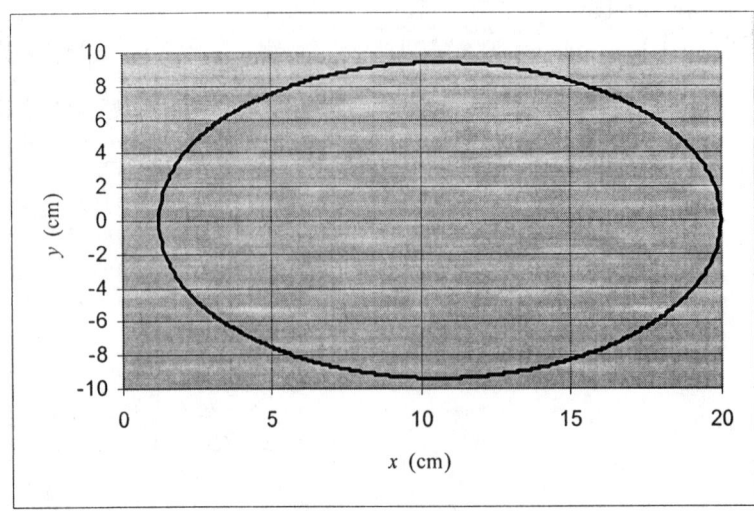

*76 •• A Van de Graaff generator has a potential difference of 1.25 MV between the belt and the outer shell. Charge is supplied at the rate of 200 μC/s. What minimum power is needed to drive the moving belt?

Picture the Problem We can use the definition of power and the expression for the work done in moving a charge through a potential difference to find the minimum power needed to drive the moving belt.

Relate the power need to drive the moving belt to the rate at which the generator is doing work:
$$P = \frac{dW}{dt}$$

Express the work done in moving a charge q through a potential difference ΔV:
$$W = q\Delta V$$

Substitute to obtain:
$$P = \frac{d}{dt}[q\Delta V] = \Delta V \frac{dq}{dt}$$

Substitute numerical values and evaluate P:
$$P = (1.25\,\text{MV})(200\,\mu\text{C/s}) = \boxed{250\,\text{W}}$$

*82 •• A metal sphere centered at the origin carries a surface charge of charge density $\sigma = 24.6$ nC/m². At $r = 2.0$ m, the potential is 500 V and the magnitude of the electric field is 250 V/m. Determine the radius of the metal sphere.

Picture the Problem We can use the definition of surface charge density to relate the radius R of the sphere to its charge Q and the potential function $V(r) = kQ/r$ to relate Q

to the potential at $r = 2$ m.
Use its definition, relate the surface charge density σ to the charge Q on the sphere and the radius R of the sphere:

$$\sigma = \frac{Q}{4\pi R^2}$$

Solve for R to obtain:

$$R = \sqrt{\frac{Q}{4\pi\sigma}}$$

Relate the potential at $r = 2.0$ m to the charge on the sphere:

$$V(r) = \frac{kQ}{r}$$

Solve for Q to obtain:

$$Q = \frac{rV(r)}{k}$$

Substitute to obtain:

$$R = \sqrt{\frac{rV(r)}{4\pi k\sigma}} = \sqrt{\frac{4\pi \epsilon_0\, rV(r)}{4\pi\sigma}}$$

$$= \sqrt{\frac{\epsilon_0\, rV(r)}{\sigma}}$$

Substitute numerical values and evaluate R:

$$R = \sqrt{\frac{(8.85 \times 10^{-12}\ \text{C}^2/\text{N}\cdot\text{m}^2)(2\,\text{m})(500\,\text{V})}{24.6\,\text{nC/m}^2}} = \boxed{0.600\,\text{m}}$$

*87 ••• A point charge q is a distance d away from a grounded conducting plane of infinite extent (Figure 23-31a). For this configuration the potential V is zero, both at all points infinitely far from the particle in all directions, and at all points on the conducting plane. Consider a set of coordinate axes with the particle located on the x axis at $x = d$. A second configuration (Figure 23-31b) has the conducting plane replaced by a particle of charge $-q$ located on the x axis at $x = -d$. (*a*) Show that for the second configuration the potential function is zero at all points infinitely far from the particle in all directions, and at all points on the yz plane–just as was the case for the first configuration. (*b*) A theorem, called the uniqueness theorem, shows that throughout the half-space $x > 0$ the potential function V–and thus the electric field $\vec{E}$ –for the two configurations are identical. Using this result, obtain the electric field $\vec{E}$ at every point in the yz plane in the second configuration. (The uniqueness theorem tells us that in the first configuration the electric field at each point in the yz plane is the same as it is in the second configuration.) Use this result to find the surface charge density σ at each point in the conducting plane (in the first configuration).

66   Chapter 23

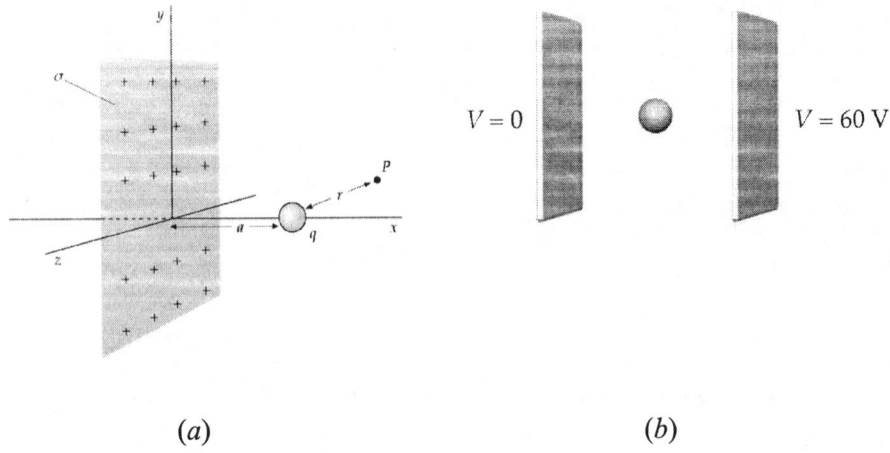

(a)   (b)

Figure 23-31 Problem 87

Picture the Problem We can consider the relationship between the potential and the electric field to show that this arrangement is equivalent to replacing the plane by a point charge of magnitude $-q$ located a distance d beneath the plane. In (b) we can first find the field at the plane surface and then use $\sigma = \epsilon_0 E$ to find the surface charge density. In (c) the work needed to move the charge to a point $2d$ away from the plane is the product of the potential difference between the points at distances $2d$ and $3d$ from $-q$ multiplied by the separation Δx of these points.

(a) | The potential anywhere on the plane is 0 in either arrangement and the electric field is perpendicular to the plane in both arrangements, so they must give the same potential everywhere in the xy plane. Also, because the net charge is zero, the potential at infinity is zero.

(b) The surface charge density is given by:

$$\sigma = \epsilon_0 E \quad (1)$$

At any point on the plane, the electric field points in the negative x direction and has magnitude:

$$E = \frac{kq}{d^2 + r^2} \cos\theta$$

where θ is the angle between the horizontal and a vector pointing from the positive charge to the point of interest on the xz plane and r is the distance along the plane from the origin (i.e., directly to the left of the charge).

Electric Potential 67

Because $\cos\theta = \dfrac{d}{\sqrt{d^2+r^2}}$:

$$E = \dfrac{kq}{d^2+r^2}\dfrac{d}{\sqrt{d^2+r^2}}$$

$$= \dfrac{kqd}{(d^2+r^2)^{3/2}}$$

$$= \dfrac{qd}{4\pi\epsilon_0 (d^2+r^2)^{3/2}}$$

Substitute for E in equation (1) to obtain:

$$\boxed{\sigma = \dfrac{qd}{4\pi(d^2+r^2)^{3/2}}}$$

*90 ••• Consider two concentric spherical metal shells of radii a and b, where $b > a$. The outer shell has a charge Q, but the inner shell is grounded. This means that the inner shell is at zero potential and that electric field lines leave the outer shell and go to infinity, but other electric field lines leave the outer shell and end on the inner shell. Find the charge on the inner shell.

Picture the Problem The diagram shows a cross-sectional view of a portion of the concentric spherical shells. Let the charge on the inner shell be q. The dashed line represents a spherical Gaussian surface over which we can integrate $\vec{E}\cdot\hat{n}dA$ in order to find E_r for $r \geq b$. We can find $V(b)$ from the integral of E_r between $r = \infty$ and $r = b$. We can obtain a second expression for $V(b)$ by considering the potential difference between a and b and solving the two equations simultaneously for the charge q on the inner shell.

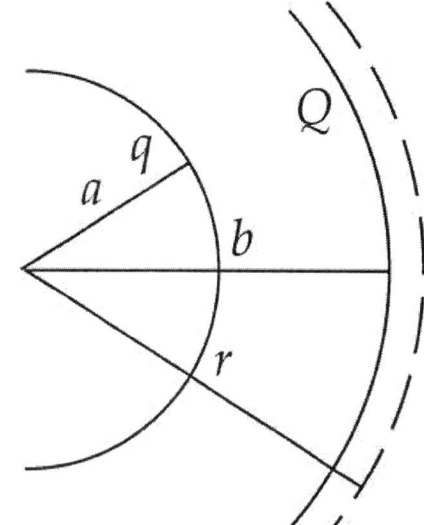

Apply Gauss's law to a spherical surface of radius $r \geq b$:

$$E_r(4\pi r^2) = \dfrac{Q+q}{\varepsilon_0}$$

Solve for E_r to obtain:

$$E_r = \dfrac{k(Q+q)}{r^2}$$

Use E_r to find $V(b)$:

$$V(b) = -k(Q+q)\int_\infty^b \dfrac{dr}{r^2}$$

$$= \dfrac{k(Q+q)}{b}$$

68   Chapter 23

We can also determine $V(b)$ by considering the potential difference between a, i.e., 0 and b:

$$V(b) = kq\left(\frac{1}{b} - \frac{1}{a}\right)$$

Equate these expressions for $V(b)$ to obtain:

$$\frac{k(Q+q)}{b} = ka\left(\frac{1}{b} - \frac{1}{a}\right)$$

Solve for q to obtain:

$$\boxed{q = -\frac{a}{b}Q}$$

*94 ••• Problem 93 can be modified to be used as a very simple model for nuclear fission. When a ^{235}U nucleus absorbs a neutron, it can fission into the fragments ^{140}Xe and ^{94}Sr, plus 2 neutrons ejected. The ^{235}U has 92 protons, while ^{140}Xe has 54 and ^{94}Sr has 38. Estimate the energy liberated by this fission process (in MeV), assuming that the mass density of the nucleus is constant and has a value $\rho \sim 4\times10^{17}$ kg/m^3.

Picture the Problem We can use the definition of density to express the radius R of a nucleus as a function of its atomic mass N. We can then use the result derived in Problem 91 to express the electrostatic energies of the ^{235}U nucleus and the nuclei of the fission fragments ^{140}Xe and ^{94}Sr.

The energy released by this fission process is:

$$\Delta E = U_{^{235}U} - \left(U_{^{140}Xe} + U_{^{94}Sr}\right) \quad (1)$$

Express the mass of a nucleus in terms of its density and volume:

$$Nm = \tfrac{4}{3}\rho\pi R^3$$

where N is the nuclear number.

Solve for R to obtain:

$$R = \sqrt[3]{\frac{3Nm}{4\pi\rho}}$$

Substitute numerical values and evaluate R as a function of N:

$$R = \sqrt[3]{\frac{3(1.660\times10^{-27}\text{ kg})}{4\pi(4\times10^{17}\text{ kg/m}^3)}}N^{1/3}$$
$$= (9.97\times10^{-16}\text{ m})N^{1/3}$$

The 'radius' of the ^{235}U nucleus is therefore:

$$R_U = (9.97\times10^{-16}\text{ m})(235)^{1/3}$$
$$= 6.15\times10^{-15}\text{ m}$$

From Problem 91 we have:

$$U = \frac{3Q^2}{20\pi\epsilon_0 R}$$

Substitute numerical values and evaluate the electrostatic energy of the ^{235}U nucleus:

$$U_{^{235}U} = \frac{3(92 \times 1.6 \times 10^{-19}\,\text{C})^2}{20\pi(8.85 \times 10^{-12}\,\text{C}^2/\text{N}\cdot\text{m}^2)(6.15 \times 10^{-15}\,\text{m})}$$

$$= 1.91 \times 10^{-10}\,\text{J} \times \frac{1\,\text{eV}}{1.6 \times 10^{-19}\,\text{J/eV}} = 1189\,\text{MeV}$$

Proceed as above to find the electrostatic energy of the fission fragments ^{140}Xe and ^{94}Sr:

$$U_{^{140}Xe} = \frac{3(54 \times 1.6 \times 10^{-19}\,\text{C})^2}{20\pi(8.85 \times 10^{-12}\,\text{C}^2/\text{N}\cdot\text{m}^2)(6.15 \times 10^{-15}\,\text{m})}$$

$$= 6.57 \times 10^{-11}\,\text{J} \times \frac{1\,\text{eV}}{1.6 \times 10^{-19}\,\text{J/eV}} = 410\,\text{MeV}$$

and

$$U_{^{94}Sr} = \frac{3(38 \times 1.6 \times 10^{-19}\,\text{C})^2}{20\pi(8.85 \times 10^{-12}\,\text{C}^2/\text{N}\cdot\text{m}^2)(6.15 \times 10^{-15}\,\text{m})}$$

$$= 3.25 \times 10^{-11}\,\text{J} \times \frac{1\,\text{eV}}{1.602 \times 10^{-19}\,\text{J/eV}} = 203\,\text{MeV}$$

Substitute for $U_{^{235}U}$, $U_{^{140}Xe}$, and $U_{^{94}Sr}$ in equation (1) and evaluate ΔE:

$$\Delta E = 1189\,\text{MeV} - (410\,\text{MeV} + 203\,\text{MeV})$$

$$= \boxed{576\,\text{MeV}}$$

Chapter 24
Electrostatic Energy and Capacitance

Conceptual Problems

*1 • If the voltage across a parallel-plate capacitor is doubled, its capacitance (a) doubles. (b) drops by half. (c) remains the same.

Determine the Concept The capacitance of a parallel-plate capacitor is a function of the surface area of its plates, the separation of these plates, and the electrical properties of the matter between them. The capacitance is, therefore, independent of the voltage across the capacitor. $\boxed{(c) \text{ is correct.}}$

*5 •• A parallel-plate air capacitor is connected to a constant-voltage battery. If the separation between the capacitor plates is doubled while the capacitor remains connected to the battery, the energy stored in the capacitor (a) quadruples. (b) doubles. (c) remains unchanged. (d) drops to half its previous value. (e) drops to one-fourth its previous value.

Picture the Problem The energy stored in a capacitor is given by $U = \tfrac{1}{2}QV$ and the capacitance of a parallel-plate capacitor by $C = \epsilon_0 A/d$. We can combine these relationships, using the definition of capacitance and the condition that the potential difference across the capacitor is constant, to express U as a function of d.

Express the energy stored in the capacitor:	$U = \tfrac{1}{2}QV$
Use the definition of capacitance to express the charge of the capacitor:	$Q = CV$
Substitute to obtain:	$U = \tfrac{1}{2}CV^2$
Express the capacitance of a parallel-plate capacitor in terms of the separation d of its plates:	$C = \dfrac{\epsilon_0 A}{d}$ where A is the area of one plate.
Substitute to obtain:	$U = \dfrac{\epsilon_0 A V^2}{2d}$

Because $U \propto \dfrac{1}{d}$, doubling the separation of the plates will reduce

72 Chapter 24

the energy stored in the capacitor to 1/2 its previous value:

(d) is correct.

*10 •• Two capacitors half-filled with a dielectric are shown in Figure 24-28. The area and separation of each capacitor is the same. Which has the higher capacitance, that shown in Figure (a) or in Figure (b)?

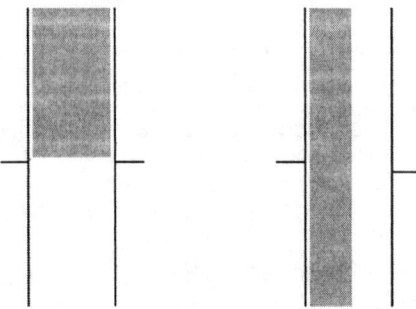

Figure 24-28 Problem 10

Picture the Problem We can treat the configuration in (a) as two capacitors in parallel and the configuration in (b) as two capacitors in series. Finding the equivalent capacitance of each configuration and examining their ratio will allow us to decide whether (a) or (b) has the greater capacitance. In both cases, we'll let C_1 be the capacitance of the dielectric-filled capacitor and C_2 be the capacitance of the air capacitor.

In configuration (a) we have:
$$C_a = C_1 + C_2$$

Express C_1 and C_2:
$$C_1 = \frac{\kappa \epsilon_0 A_1}{d_1} = \frac{\kappa \epsilon_0 \frac{1}{2} A}{d} = \frac{\kappa \epsilon_0 A}{2d}$$

and

$$C_2 = \frac{\epsilon_0 A_2}{d_2} = \frac{\epsilon_0 \frac{1}{2} A}{d} = \frac{\epsilon_0 A}{2d}$$

Substitute for C_1 and C_2 and simplify to obtain:
$$C_a = \frac{\kappa \epsilon_0 A}{2d} + \frac{\epsilon_0 A}{2d} = \frac{\epsilon_0 A}{2d}(\kappa + 1)$$

In configuration (b) we have:
$$\frac{1}{C_b} = \frac{1}{C_1} + \frac{1}{C_2} \Rightarrow C_b = \frac{C_1 C_2}{C_1 + C_2}$$

Express C_1 and C_2:
$$C_1 = \frac{\epsilon_0 A_1}{d_1} = \frac{\epsilon_0 A}{\frac{1}{2}d} = \frac{2\epsilon_0 A}{d}$$

and

Electrostatic Energy and Capacitance 73

$$C_2 = \frac{\kappa \epsilon_0 A_2}{d_2} = \frac{\kappa \epsilon_0 A}{\frac{1}{2}d} = \frac{2\kappa \epsilon_0 A}{d}$$

Substitute for C_1 and C_2 and simplify to obtain:

$$C_b = \frac{\left(\frac{2\epsilon_0 A}{d}\right)\left(\frac{2\kappa \epsilon_0 A}{d}\right)}{\frac{2\epsilon_0 A}{d} + \frac{2\kappa \epsilon_0 A}{d}}$$

$$= \frac{\left(\frac{2\epsilon_0 A}{d}\right)\left(\frac{2\kappa \epsilon_0 A}{d}\right)}{\frac{2\epsilon_0 A}{d}(\kappa+1)}$$

$$= \frac{2\epsilon_0 A}{d}\left(\frac{\kappa}{\kappa+1}\right)$$

Divide C_b by C_a:

$$\frac{C_b}{C_a} = \frac{\frac{2\epsilon_0 A}{d}\left(\frac{\kappa}{\kappa+1}\right)}{\frac{\epsilon_0 A}{2d}(\kappa+1)} = \frac{4\kappa}{(\kappa+1)^2}$$

Because $\frac{4\kappa}{(\kappa+1)^2} < 1$ for $\kappa > 1$:

$$\boxed{C_a > C_b}$$

Estimation and Approximation

*14 •• To create the high-energy densities needed to operate a pulsed nitrogen laser, the discharge from a high-capacitance capacitor is used. Typically, the energy requirement per pulse (i.e., per discharge) is 100 J. Estimate the capacitance required if the discharge is applied through a spark gap of 1 mm width. Assume that the dielectric breakdown of nitrogen occurs at $E \approx 3 \times 10^6$ V/m.

Picture the Problem The energy stored in a capacitor is given by $U = \frac{1}{2}CV^2$.

Relate the energy stored in a capacitor to its capacitance and the potential difference across it:

$$U = \tfrac{1}{2}CV^2$$

Solve for C:

$$C = \frac{2U}{V^2}$$

The potential difference across the spark gap is related to the width of the gap d and the electric field E in

$$V = Ed$$

74   Chapter 24

the gap:

Substitute for V in the expression for C to obtain:	$C = \dfrac{2U}{E^2 d^2}$
Substitute numerical values and evaluate C:	$C = \dfrac{2(100\,\text{J})}{(3\times 10^6\,\text{V/m})^2 (0.001\,\text{m})^2}$ $= \boxed{22.2\,\mu\text{F}}$

Electrostatic Potential Energy

*19 • What is the electrostatic potential energy of an isolated spherical conductor of 10 cm radius that is charged to 2 kV?

Picture the Problem The potential of an isolated spherical conductor is given by $V = kQ/r$, where Q is its charge and r its radius, and its electrostatic potential energy by $U = \tfrac{1}{2}QV$. We can combine these relationships to find the sphere's electrostatic potential energy.

Express the electrostatic potential energy of the isolated spherical conductor as a function of its charge Q and potential V:	$U = \tfrac{1}{2}QV$
Express the potential of the spherical conductor:	$V = \dfrac{kQ}{r}$
Solve for Q to obtain:	$Q = \dfrac{rV}{k}$
Substitute to obtain:	$U = \tfrac{1}{2}\left(\dfrac{rV}{k}\right)V = \dfrac{rV^2}{2k}$
Substitute numerical values and evaluate U:	$U = \dfrac{(0.1\,\text{m})(2\,\text{kV})^2}{2(8.99\times 10^9\,\text{N}\cdot\text{m}^2/\text{C}^2)}$ $= \boxed{22.2\,\mu\text{J}}$

Electrostatic Energy and Capacitance 75

Capacitance

*22 • An isolated spherical conductor of 10 cm radius is charged to 2 kV. (*a*) How much charge is on the conductor? (*b*) What is the capacitance of the sphere? (*c*) How does the capacitance change if the sphere is charged to 6 kV?

Picture the Problem The charge on the spherical conductor is related to its radius and potential according to $V = kQ/r$ and we can use the definition of capacitance to find the capacitance of the sphere.

(*a*) Relate the potential V of the spherical conductor to the charge on it and to its radius:

$$V = \frac{kQ}{r}$$

Solve for and evaluate Q:

$$Q = \frac{rV}{k}$$

$$= \frac{(0.1\,\text{m})(2\,\text{kV})}{8.99 \times 10^9\,\text{N}\cdot\text{m}^2/\text{C}^2} = \boxed{22.2\,\text{nC}}$$

(*b*) Use the definition of capacitance to relate the capacitance of the sphere to its charge and potential:

$$C = \frac{Q}{V} = \frac{22.2\,\text{nC}}{2\,\text{kV}} = \boxed{11.1\,\text{pF}}$$

(*c*) $\boxed{\text{It doesn't. The capacitance of a sphere is a function of its radius.}}$

The Storage of Electrical Energy

*28 • Find the energy per unit volume in an electric field that is equal to 3 MV/m, which is the dielectric strength of air.

Picture the Problem The energy per unit volume in an electric field varies with the square of the electric field according to $u = \epsilon_0 E^2/2$.

Express the energy per unit volume in an electric field:

$$u = \tfrac{1}{2}\epsilon_0 E^2$$

Substitute numerical values and evaluate u:

$$u = \tfrac{1}{2}(8.85 \times 10^{-12}\,\text{C}^2/\text{N}\cdot\text{m}^2)(3\,\text{MV/m})^2$$

$$= \boxed{39.8\,\text{J/m}^3}$$

76   Chapter 24

*31 •• A parallel-plate capacitor with plates of area 500 cm^2 is charged to a potential difference V and is then disconnected from the voltage source. When the plates are moved 0.4 cm farther apart, the voltage between the plates increases by 100 V. (*a*) What is the charge Q on the positive plate of the capacitor? (*b*) How much does the energy stored in the capacitor increase due to the movement of the plates?

Picture the Problem We can relate the charge Q on the positive plate of the capacitor to the charge density of the plate σ using its definition. The charge density, in turn, is related to the electric field between the plates according to $\sigma = \epsilon_0 E$ and the electric field can be found from $E = \Delta V/\Delta d$. We can use $\Delta U = \frac{1}{2}Q\Delta V$ in part (*b*) to find the increase in the energy stored due to the movement of the plates.

(*a*) Express the charge Q on the positive plate of the capacitor in terms of the plate's charge density σ and surface area A:

$$Q = \sigma A$$

Relate σ to the electric field E between the plates of the capacitor:

$$\sigma = \epsilon_0 E$$

Express E in terms of the change in V as the plates are separated a distance Δd:

$$E = \frac{\Delta V}{\Delta d}$$

Substitute for σ and E to obtain:

$$Q = \epsilon_0 EA = \epsilon_0 A \frac{\Delta V}{\Delta d}$$

Substitute numerical values and evaluate Q:

$$Q = (8.85 \times 10^{-12}\, \mathrm{C^2/N \cdot m^2})(500\,\mathrm{cm^2})\frac{100\,\mathrm{V}}{0.4\,\mathrm{cm}} = \boxed{11.1\,\mathrm{nC}}$$

(*b*) Express the change in the electrostatic energy in terms of the change in the potential difference:

$$\Delta U = \tfrac{1}{2}Q\Delta V$$

Substitute numerical values and evaluate ΔU:

$$\Delta U = \tfrac{1}{2}(11.1\,\mathrm{nC})(100\,\mathrm{V}) = \boxed{0.553\,\mu\mathrm{J}}$$

Electrostatic Energy and Capacitance 77

Combinations of Capacitors

*35 • Three capacitors are connected in a triangle as shown in Figure 24-29. Find the equivalent capacitance between points *a* and *c*.

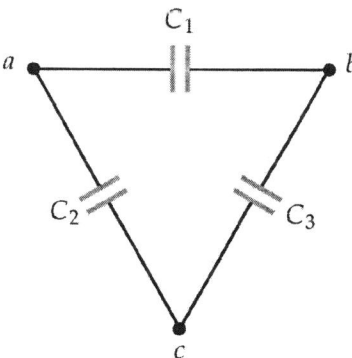

Figure 24-29 Problem 35

Picture the Problem Because we're interested in the equivalent capacitance across terminals *a* and *c*, we need to recognize that capacitors C_1 and C_3 are in series with each other and in parallel with capacitor C_2.

Find the equivalent capacitance of C_1 and C_3 in series:

$$\frac{1}{C_{1+3}} = \frac{1}{C_1} + \frac{1}{C_3}$$

Solve for C_{1+3}:

$$C_{1+3} = \frac{C_1 C_3}{C_1 + C_3}$$

Find the equivalent capacitance of C_{1+3} and C_2 in parallel:

$$C_{eq} = C_2 + C_{1+3} = \boxed{C_2 + \frac{C_1 C_3}{C_1 + C_3}}$$

*38 •• Three identical capacitors are connected so that their maximum equivalent capacitance is 15 μF. (*a*) Describe how the capacitors are combined. (*b*) There are three other ways to combine all three capacitors in a circuit. What are the equivalent capacitances for each arrangement?

Picture the Problem We can use the properties of capacitors connected in series and in parallel to find the equivalent capacitances for various connection combinations.

(*a*) $\boxed{\text{If their capacitance is to be a maximum, they must be connected in parallel.}}$

Find the capacitance of each capacitor:

$C_{eq} = 3C = 15\,\mu\text{F}$

and

78   Chapter 24

$$C = 5\,\mu F$$

(b) (1) Connect the three capacitors in series:

$$\frac{1}{C_{eq}} = \frac{3}{5\,\mu F} \text{ and } C_{eq} = \boxed{1.67\,\mu F}$$

(2) Connect two in parallel, with the third in series with that combination:

$$C_{eq,\,two\,in\,parallel} = 2(5\,\mu F) = 10\,\mu F$$

and

$$\frac{1}{C_{eq}} = \frac{1}{10\,\mu F} + \frac{1}{5\,\mu F}$$

Solve for C_{eq}:

$$C_{eq} = \frac{(10\,\mu F)(5\,\mu F)}{10\,\mu F + 5\,\mu F} = \boxed{3.33\,\mu F}$$

(3) Connect two in series, with the third in parallel with that combination:

$$\frac{1}{C_{eq,\,two\,in\,series}} = \frac{2}{5\,\mu F}$$

or

$$C_{eq,\,two\,in\,series} = 2.5\,\mu F$$

Find the capacitance equivalent to 2.5 μF and 5 μF in parallel:

$$C_{eq} = 2.5\,\mu F + 5\,\mu F = \boxed{7.50\,\mu F}$$

*44 ••   Find all the different possible equivalent capacitances that can be obtained using a 1-μF, a 2-μF, and a 4-μF capacitor in any combination that includes all three, or any two, of the capacitors.

Picture the Problem We can connect two capacitors in parallel, all three in parallel, two in series, three in series, two in parallel in series with the third, and two in series in parallel with the third.

Connect 2 in parallel to obtain:

$$C_{eq} = 1\,\mu F + 2\,\mu F = \boxed{3\,\mu F}$$

or

$$C_{eq} = 1\,\mu F + 4\,\mu F = \boxed{5\,\mu F}$$

or

$$C_{eq} = 2\,\mu F + 4\,\mu F = \boxed{6\,\mu F}$$

Connect all three in parallel to obtain:

$$C_{eq} = 1\,\mu F + 2\,\mu F + 4\,\mu F = \boxed{7\,\mu F}$$

Connect two in series:

$$C_{eq} = \frac{(1\,\mu F)(2\,\mu F)}{1\,\mu F + 2\,\mu F} = \boxed{\frac{2}{3}\,\mu F}$$

or

$$C_{eq} = \frac{(1\,\mu F)(4\,\mu F)}{1\,\mu F + 4\,\mu F} = \boxed{\frac{4}{5}\,\mu F}$$

or

$$C_{eq} = \frac{(2\,\mu F)(4\,\mu F)}{2\,\mu F + 4\,\mu F} = \boxed{\frac{4}{3}\,\mu F}$$

Connect all three in series:

$$C_{eq} = \frac{(1\,\mu F)(2\,\mu F)(4\,\mu F)}{(1\,\mu F)(2\,\mu F)+(2\,\mu F)(4\,\mu F)+(1\,\mu F)(4\,\mu F)} = \boxed{\frac{4}{7}\,\mu F}$$

Connect two in parallel, in series with the third:

$$C_{eq} = \frac{(4\,\mu F)(1\,\mu F + 2\,\mu F)}{1\,\mu F + 2\,\mu F + 4\,\mu F} = \boxed{\frac{12}{7}\,\mu F}$$

or

$$C_{eq} = \frac{(1\,\mu F)(4\,\mu F + 2\,\mu F)}{1\,\mu F + 2\,\mu F + 4\,\mu F} = \boxed{\frac{6}{7}\,\mu F}$$

or

$$C_{eq} = \frac{(2\,\mu F)(4\,\mu F + 1\,\mu F)}{1\,\mu F + 2\,\mu F + 4\,\mu F} = \boxed{\frac{10}{7}\,\mu F}$$

Connect two in series, in parallel with the third:

$$C_{eq} = \frac{(1\,\mu F)(2\,\mu F)}{1\,\mu F + 2\,\mu F} + 4\,\mu F = \boxed{\frac{14}{3}\,\mu F}$$

or

$$C_{eq} = \frac{(4\,\mu F)(2\,\mu F)}{4\,\mu F + 2\,\mu F} + 1\,\mu F = \boxed{\frac{7}{3}\,\mu F}$$

or

$$C_{eq} = \frac{(1\,\mu F)(4\,\mu F)}{1\,\mu F + 4\,\mu F} + 2\,\mu F = \boxed{\frac{14}{5}\,\mu F}$$

Parallel-Plate Capacitors

*49 •• Design a 0.1-μF parallel-plate capacitor with air between the plates that can be charged to a maximum potential difference of 1000 V. (*a*) What is the minimum possible separation between the plates? (*b*) What minimum area must the plates of the capacitor have?

Picture the Problem The potential difference across the capacitor plates V is related to their separation d and the electric field between them according to $V = Ed$. We can use this equation with $E_{max} = 3$ MV/m to find d_{min}. In part (b) we can use the expression for the capacitance of a parallel-plate capacitor to find the required area of the plates.

(a) Use the relationship between the potential difference across the plates and the electric field between them to find the minimum separation of the plates:

$$d_{min} = \frac{V}{E_{max}} = \frac{1000\,\text{V}}{3\,\text{MV/m}} = \boxed{0.333\,\text{mm}}$$

(b) Use the expression for the capacitance of a parallel-plate capacitor to relate the capacitance to the area of a plate:

$$C = \frac{\epsilon_0 A}{d}$$

Solve for A:

$$A = \frac{Cd}{\epsilon_0}$$

Substitute numerical values and evaluate A:

$$A = \frac{(0.1\,\mu\text{F})(0.333\,\text{mm})}{8.85 \times 10^{-12}\,\text{C}^2/\text{N}\cdot\text{m}^2} = \boxed{3.76\,\text{m}^2}$$

Cylindrical Capacitors

*53 •• A goniometer is a precise instrument for measuring angles. A capacitive goniometer is shown in Figure 24-34. Each plate of the variable capacitor consists of a flat metal semicircle with inner radius R_1 and outer radius R_2. The plates share a common rotation axis, and the width of the air gap separating the plates is d. Calculate the capacitance as a function of the angle θ and the parameters given.

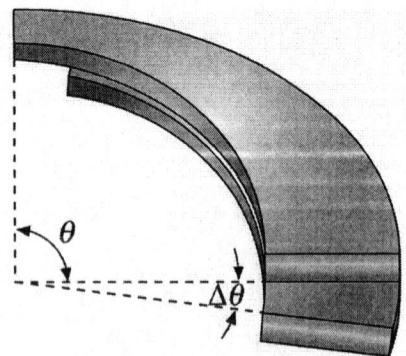

Figure 24-34a Problem 53

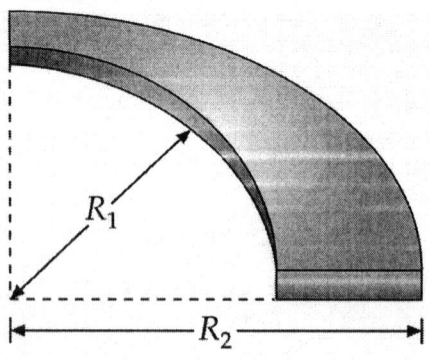

Figure 24-34b Problem 53

Electrostatic Energy and Capacitance 81

Picture the Problem We can use the expression for the capacitance of a parallel-plate capacitor of variable area and the geometry of the figure to express the capacitance of the goniometer.

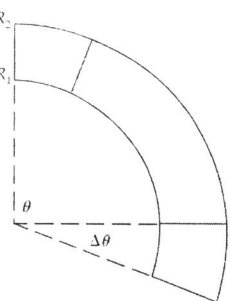

The capacitance of the parallel-plate capacitor is given by:
$$C = \frac{\epsilon_0 (A - \Delta A)}{d}$$

The area of the plates is:
$$A = \pi(R_2^2 - R_1^2)\frac{\theta}{2\pi} = (R_2^2 - R_1^2)\frac{\theta}{2}$$

If the top plate rotates through an angle $\Delta\theta$, then the area is reduced by:
$$\Delta A = \pi(R_2^2 - R_1^2)\frac{\Delta\theta}{2\pi} = (R_2^2 - R_1^2)\frac{\Delta\theta}{2}$$

Substitute for A and ΔA in the expression for C to obtain:
$$C = \frac{\epsilon_0}{d}\left[(R_2^2 - R_1^2)\frac{\theta}{2} - (R_2^2 - R_1^2)\frac{\Delta\theta}{2}\right]$$

$$= \boxed{\frac{\epsilon_0 (R_2^2 - R_1^2)}{2d}(\theta - \Delta\theta)}$$

Spherical Capacitors

*55 •• A spherical capacitor consists of two thin, concentric spherical shells of radii R_1 and R_2. (*a*) Show that the capacitance is given by $C = 4\pi \epsilon_0 R_1 R_2 / (R_2 - R_1)$. (*b*) Show that when the radii of the shells are nearly equal, the capacitance is given approximately by the expression for the capacitance of a parallel-plate capacitor, $C = \epsilon_0 A/d$, where A is the area of the sphere and $d = R_2 - R_1$.

Picture the Problem We can use the definition of capacitance and the expression for the potential difference between charged concentric spherical shells to show that $C = 4\pi \epsilon_0 R_1 R_2 / (R_2 - R_1)$.

(*a*) Using its definition, relate the capacitance of the concentric spherical shells to their charge Q and the potential difference V between their surfaces:
$$C = \frac{Q}{V}$$

82 Chapter 24

Express the potential difference between the conductors:

$$V = kQ\left(\frac{1}{R_1} - \frac{1}{R_2}\right) = kQ\frac{R_2 - R_1}{R_1 R_2}$$

Substitute to obtain:

$$C = \frac{Q}{kQ\frac{R_2 - R_1}{R_1 R_2}} = \frac{R_1 R_2}{k(R_2 - R_1)}$$

$$= \boxed{\frac{4\pi \epsilon_0 R_1 R_2}{R_2 - R_1}}$$

(b) Because $R_2 = R_1 + d$:

$$R_1 R_2 = R_1(R_1 + d)$$
$$= R_1^2 + R_1 d$$
$$\approx R_1^2 = R^2$$

because d is small.

Substitute to obtain:

$$C \approx \frac{4\pi \epsilon_0 R^2}{d} = \boxed{\frac{\epsilon_0 A}{d}}$$

Disconnected and Reconnected Capacitors

*60 ••   Two capacitors, $C_1 = 4\ \mu F$ and $C_2 = 12\ \mu F$, are connected in series across a 12-V battery. They are carefully disconnected so that they are not discharged and they are then reconnected to each other, with positive plate to positive plate and negative plate to negative plate. (*a*) Find the potential difference across each capacitor after they are connected. (*b*) Find the initial energy stored and the final energy stored in the capacitors.

Picture the Problem When the capacitors are reconnected, each will have the charge it acquired while they were connected in series across the 12-V battery and we can use the definition of capacitance and their equivalent capacitance to find the common potential difference across them. In part (*b*) we can use $U = \frac{1}{2}CV^2$ to find the initial and final energy stored in the capacitors.

(*a*) Using the definition of capacitance, express the potential difference across each capacitor when they are reconnected:

$$V = \frac{2Q}{C_{eq}} \quad (1)$$

where Q is the charge on each capacitor *before* they are disconnected.

Find the equivalent capacitance of the two capacitors after they are connected in parallel:

$$C_{eq} = C_1 + C_2$$
$$= 4\ \mu F + 12\ \mu F$$
$$= 16\ \mu F$$

Electrostatic Energy and Capacitance   83

Express the charge Q on each capacitor before they are disconnected:	$Q = C'_{eq}V$
Express the equivalent capacitance of the two capacitors connected in series:	$C'_{eq} = \dfrac{C_1 C_2}{C_1 + C_2} = \dfrac{(4\,\mu\text{F})(12\,\mu\text{F})}{4\,\mu\text{F} + 12\,\mu\text{F}} = 3\,\mu\text{F}$
Substitute to find Q:	$Q = (3\,\mu\text{F})(12\,\text{V}) = 36\,\mu\text{C}$
Substitute in equation (1) and evaluate V:	$V = \dfrac{2(36\,\mu\text{C})}{16\,\mu\text{F}} = \boxed{4.50\,\text{V}}$
(b) Express and evaluate the energy stored in the capacitors initially:	$U_i = \tfrac{1}{2} C'_{eq} V_i^2 = \tfrac{1}{2}(3\,\mu\text{F})(12\,\text{V})^2$ $= \boxed{216\,\mu\text{J}}$
Express and evaluate the energy stored in the capacitors when they have been reconnected:	$U_f = \tfrac{1}{2} C_{eq} V_f^2 = \tfrac{1}{2}(16\,\mu\text{F})(4.5\,\text{V})^2$ $= \boxed{162\,\mu\text{J}}$

*64 ••   A 20-pF capacitor is charged to 3 kV and then removed from the battery and connected to an uncharged 50-pF capacitor. (a) What is the new charge on each capacitor? (b) Find the initial energy stored in the 20-pF capacitor, and find the final energy stored in the two capacitors. Is electrostatic potential energy gained or lost when the two capacitors are connected?

Picture the Problem Let the numeral 1 refer to the 20-pF capacitor and the numeral 2 to the 50-pF capacitor. We can use conservation of charge and the fact that the connected capacitors will have the same potential difference across them to find the charge on each capacitor. We can decide whether electrostatic potential energy is gained or lost when the two capacitors are connected by calculating the change ΔU in the electrostatic energy during this process.

(a) Using the fact that no charge is lost in connecting the capacitors, relate the charge Q initially on the 20-pF capacitor to the charges on the two capacitors when they have been connected:	$Q = Q_1 + Q_2$	(1)

84 Chapter 24

Because the capacitors are in parallel, the potential difference across them is the same:

$$V_1 = V_2 \Rightarrow \frac{Q_1}{C_1} = \frac{Q_2}{C_2}$$

Solve for Q_1 to obtain:

$$Q_1 = \frac{C_1}{C_2} Q_2$$

Substitute in equation (1) and solve for Q_2 to obtain:

$$Q_2 = \frac{Q}{1 + C_1/C_2} \quad (2)$$

Use the definition of capacitance to find the charge Q initially on the 20-pF capacitor:

$$Q = C_1 V = (20\,\text{pF})(3\,\text{kV}) = 60\,\text{nC}$$

Substitute in equation (2) and evaluate Q_2:

$$Q_2 = \frac{60\,\text{nC}}{1 + 20\,\text{pF}/50\,\text{pF}} = \boxed{42.9\,\text{nC}}$$

Substitute in equation (1) to obtain:

$$Q_1 = Q - Q_2$$
$$= 60\,\text{nC} - 42.9\,\text{nC} = \boxed{17.1\,\text{nC}}$$

(b) Express the change in the electrostatic potential energy of the system when the two capacitors are connected:

$$\Delta U = U_f - U_i$$
$$= \frac{Q^2}{2C_{eq}} - \frac{Q^2}{2C_1}$$
$$= \frac{Q^2}{2}\left(\frac{1}{C_{eq}} - \frac{1}{C_1}\right)$$

Substitute numerical values and evaluate ΔU:

$$\Delta U = \frac{(60\,\text{nC})^2}{2}\left(\frac{1}{70\,\text{pF}} - \frac{1}{20\,\text{pF}}\right)$$
$$= -64.3\,\mu\text{J}$$

Because $\Delta U < 0$, electrostatic energy is lost when the two capacitors are connected.

*66 •• A capacitor of capacitance C has a charge Q. A student connects one terminal of the capacitor to a terminal of an identical uncharged capacitor. When the remaining two terminals are connected, charge flows until electrostatic equilibrium is reestablished and both capacitors have charge $Q/2$ on them. Compare the total energy initially stored in the one capacitor to the total energy stored in the two after the second has been charged. Where did the missing energy go? This energy was dissipated in the connecting wires via Joule heating, which is discussed in Chapter 25.

Electrostatic Energy and Capacitance 85

Picture the Problem We can use the expression for the energy stored in a capacitor to express the ratio of the energy stored in the system after the discharge of the first capacitor to the energy stored in the system prior to the discharge.

Express the energy U initially stored in the capacitor whose capacitance is C:

$$U = \frac{Q^2}{2C}$$

The energy U' stored in the two capacitors after the first capacitor has discharged is:

$$U' = \frac{\left(\frac{Q}{2}\right)^2}{2C} + \frac{\left(\frac{Q}{2}\right)^2}{2C} = \frac{Q^2}{4C}$$

Express the ratio of U' to U:

$$\frac{U'}{U} = \frac{\frac{Q^2}{4C}}{\frac{Q^2}{2C}} = \frac{1}{2} \Rightarrow U' = \boxed{\tfrac{1}{2}U}$$

Dielectrics

*72 •• Two capacitors, each consisting of two conducting plates of surface area A, with an air gap of width d. They are connected in parallel, as shown in Figure 24-37, and each has a charge Q. A slab of width d and area A with dielectric constant κ is inserted between the plates of one of the capacitors. Calculate the new charge Q' on that capacitor.

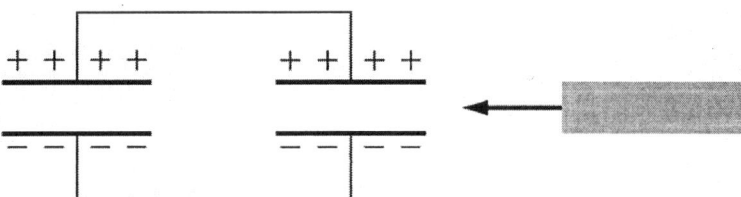

Figure 24-37 Problem 72

Picture the Problem Let the charge on the capacitor with the air gap be Q_1 and the charge on the capacitor with the dielectric gap be Q_2. If the capacitances of the capacitors were initially C, then the capacitance of the capacitor with the dielectric inserted is $C' = \kappa C$. We can use the conservation of charge and the equivalence of the potential difference across the capacitors to obtain two equations that we can solve simultaneously for Q_1 and Q_2.

Apply conservation of charge during the insertion of the dielectric to obtain:

$$Q_1 + Q_2 = 2Q \quad (1)$$

Because the capacitors have the same potential difference across

$$\frac{Q_1}{C} = \frac{Q_2}{\kappa C} \quad (2)$$

86 Chapter 24

them:

Solve equations (1) and (2) simultaneously to obtain:
$$Q_1 = \boxed{\dfrac{2Q}{1+\kappa}} \text{ and } Q_2 = \boxed{\dfrac{2Q\kappa}{1+\kappa}}$$

*75 •• What is the dielectric constant of a dielectric on which the induced bound charge density is (a) 80 percent of the free-charge density on the plates of a capacitor filled by the dielectric, (b) 20 percent of the free-charge density, and (c) 98 percent of the free-charge density?

Picture the Problem The bound charge density is related to the dielectric constant and the free charge density according to $\sigma_b = \left(1 - \dfrac{1}{\kappa}\right)\sigma_f$.

Solve the equation relating σ_b, σ_f, and κ for κ to obtain:
$$\kappa = \dfrac{1}{1-\sigma_b/\sigma_f}$$

(a) Evaluate this expression for $\sigma_b/\sigma_f = 0.8$:
$$\kappa = \dfrac{1}{1-0.8} = \boxed{5.00}$$

(b) Evaluate this expression for $\sigma_b/\sigma_f = 0.2$:
$$\kappa = \dfrac{1}{1-0.2} = \boxed{1.25}$$

(c) Evaluate this expression for $\sigma_b/\sigma_f = 0.98$:
$$\kappa = \dfrac{1}{1-0.98} = \boxed{50.0}$$

*77 •• Find the capacitance of the parallel-plate capacitor shown in Figure 24-38.

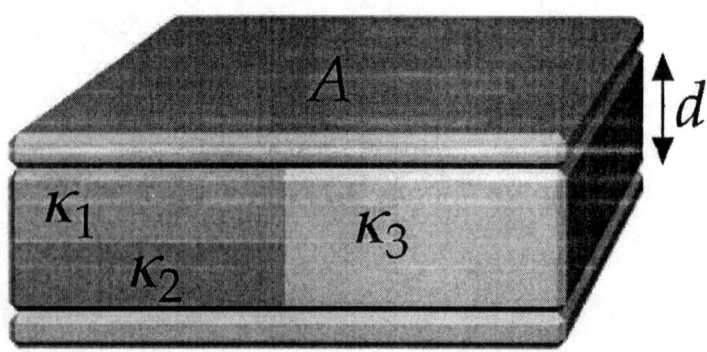

Figure 24-38 Problem 77

Picture the Problem We can model this parallel-plate capacitor as a combination of two capacitors C_1 and C_2 in series with capacitor C_3 in parallel.

Express the capacitance of two series-connected capacitors in parallel with a third:

$$C = C_3 + C_s \quad (1)$$

where

$$C_s = \frac{C_1 C_2}{C_1 + C_2} \quad (2)$$

Express each of the capacitances C_1, C_2, and C_3 in terms of the dielectric constants, plate areas, and plate separations:

$$C_1 = \frac{\kappa_1 \epsilon_0 \left(\frac{1}{2}A\right)}{\frac{1}{2}d} = \frac{\kappa_1 \epsilon_0 A}{d},$$

$$C_2 = \frac{\kappa_2 \epsilon_0 \left(\frac{1}{2}A\right)}{\frac{1}{2}d} = \frac{\kappa_2 \epsilon_0 A}{d},$$

and

$$C_3 = \frac{\kappa_3 \epsilon_0 \left(\frac{1}{2}A\right)}{d} = \frac{\kappa_3 \epsilon_0 A}{2d}$$

Substitute in equation (2) to obtain:

$$C_s = \frac{\left(\frac{\kappa_1 \epsilon_0 A}{d}\right)\left(\frac{\kappa_2 \epsilon_0 A}{d}\right)}{\frac{\kappa_1 \epsilon_0 A}{d} + \frac{\kappa_2 \epsilon_0 A}{d}}$$

$$= \frac{\kappa_1 \kappa_2}{\kappa_1 + \kappa_2}\left(\frac{\epsilon_0 A}{d}\right)$$

Substitute in equation (1) to obtain:

$$C = \frac{\kappa_3 \epsilon_0 A}{2d} + \frac{\kappa_1 \kappa_2}{\kappa_1 + \kappa_2}\left(\frac{\epsilon_0 A}{d}\right)$$

$$= \boxed{\left(\kappa_3 + \frac{2\kappa_1 \kappa_2}{\kappa_1 + \kappa_2}\right)\left(\frac{\epsilon_0 A}{2d}\right)}$$

General Problems

*81 • Three capacitors have capacitances of 2 μF, 4 μF, and 8 μF. Find the equivalent capacitance if (a) the capacitors are connected in parallel and (b) if the capacitors are connected in series.

Picture the Problem We can use the equations for the equivalent capacitance of three capacitors connected in parallel and in series to find these equivalent capacitances.

(a) Express the equivalent capacitance of three capacitors connected in parallel:

$$C_{eq} = C_1 + C_2 + C_3$$

88 Chapter 24

Substitute numerical values and evaluate C_{eq}:

$$C_{eq} = 2.0\,\mu F + 4.0\,\mu F + 8.0\,\mu F$$
$$= \boxed{14.0\,\mu F}$$

(b) Express the equivalent capacitance of the three capacitors connected in series:

$$C_{eq} = \frac{C_1 C_2 C_3}{C_1 C_2 + C_2 C_3 + C_1 C_3}$$

Substitute numerical values and evaluate C_{eq}:

$$C_{eq} = \frac{(2\,\mu F)(4\,\mu F)(8\,\mu F)}{(2\,\mu F)(4\,\mu F) + (4\,\mu F)(8\,\mu F) + (2\,\mu F)(8\,\mu F)} = \boxed{1.14\,\mu F}$$

*85 •• Figure 24-40 shows four capacitors connected in the arrangement known as a capacitance bridge. The capacitors are initially uncharged. What must the relation between the four capacitances be so that the potential between points c and d is zero when a voltage V is applied between points a and b?

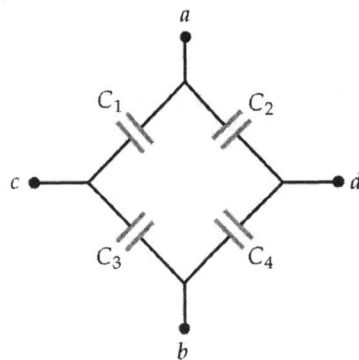

Figure 24-40 Problem 85

Picture the Problem Note that with V applied between a and b, C_1 and C_3 are in series, and so are C_2 and C_4. Because in a series combination the potential differences across the two capacitors are inversely proportional to the capacitances, we can establish proportions involving the capacitances and potential differences for the left- and right-hand side of the network and then use the condition that $V_c = V_d$ to eliminate the potential differences and establish the relationship between the capacitances.

Letting Q represent the charge on capacitors 1 and 2, relate the potential differences across the capacitors to their common charge and capacitances:

$$V_1 = \frac{Q}{C_1}$$

and

$$V_3 = \frac{Q}{C_3}$$

Electrostatic Energy and Capacitance 89

| Divide the first of these equations by the second to obtain: | $\dfrac{V_1}{V_3} = \dfrac{C_3}{C_1}$ | (1) |

| Proceed similarly to obtain: | $\dfrac{V_2}{V_4} = \dfrac{C_4}{C_2}$ | (2) |

| Divide equation (1) by equation (2) to obtain: | $\dfrac{V_1 V_4}{V_3 V_2} = \dfrac{C_3 C_2}{C_1 C_4}$ | (3) |

If $V_c = V_d$ then we must have: $\quad V_1 = V_2$ and $V_3 = V_4$

Substitute in equation (3) and rearrange to obtain: $\quad \boxed{C_2 C_3 = C_1 C_4}$

*90 •• A parallel-plate capacitor is constructed from a layer of silicon dioxide of thickness 5×10^{-6} m between two conducting films. The dielectric constant of silicon dioxide is 3.8 and its dielectric strength is 8×10^6 V/m. (*a*) What voltage can be applied across this capacitor without dielectric breakdown? (*b*) What should the surface area of the layer of silicon dioxide be for a 10-pF capacitor? (*c*) Estimate the number of these capacitors that can fit into a square 1 cm by 1 cm.

Picture the Problem The maximum voltage is related to the dielectric strength of the medium according to $V_{max} = E_{max} d$ and we can use the expression for the capacitance of a parallel-plate capacitor to determine the required area of the plates.

(*a*) Relate the maximum voltage that can be applied across this capacitor to the dielectric strength of silicon dioxide:
$$V_{max} = E_{max} d$$

Substitute numerical values and evaluate V_{max}:
$$V_{max} = (8\times 10^6 \text{ V/m})(5\times 10^{-6} \text{ m})$$
$$= \boxed{40.0 \text{ V}}$$

(*b*) Relate the capacitance of a parallel-plate capacitor to area A of its plates:
$$C = \dfrac{\kappa \epsilon_0 A}{d}$$

Solve for A to obtain:
$$A = \dfrac{Cd}{\kappa \epsilon_0}$$

Substitute numerical values and evaluate A:

$$A = \frac{(10\,\text{pF})(5\times 10^{-6}\,\text{m})}{3.8(8.85\times 10^{-12}\,\text{C}^2/\text{N}\cdot\text{m}^2)}$$

$$= 1.49\times 10^{-6}\,\text{m}^2$$

$$= \boxed{1.49\,\text{mm}^2}$$

(c) Express the number of capacitors n in terms of the area of a square 1 cm by 1cm and the area required for each capacitor:

$$n = \frac{(1\,\text{cm})^2}{A} = \frac{100\,\text{mm}^2}{1.49\,\text{mm}^2} \approx \boxed{67}$$

*94 •• A parallel-plate capacitor is filled with two dielectrics of equal size, as shown in Figure 24-42. (a) Show that this system can be considered to be two capacitors of area $\tfrac{1}{2}A$ connected in parallel. (b) Show that the capacitance is increased by the factor $(\kappa_1 + \kappa_2)/2$.

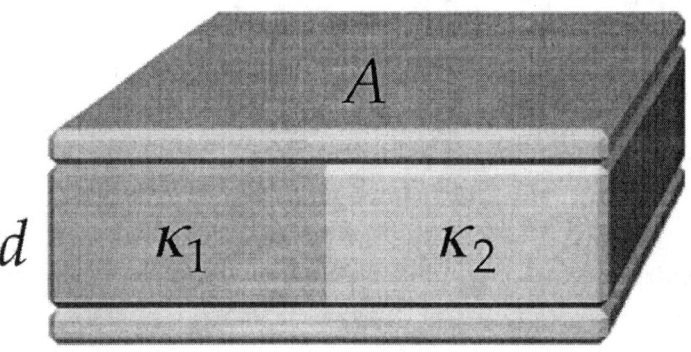

Figure 24-42 Problem 94

Picture the Problem We can express the ratio of C_{eq} to C_0 to show that the capacitance with the dielectrics in place is $(\kappa_1 + \kappa_2)/2$ times greater than that of the capacitor in the absence of the dielectrics.

(a) | Because the capacitor plates are conductors, the potentials are the same across the entire upper and lower plates. Hence, the system is equivalent to two capacitors, each of area $A/2$, in parallel.

(b) Relate the capacitance C_0, in the absence of the dielectrics, to the plate area and separation:

$$C_0 = \frac{\epsilon_0 A}{d}$$

Electrostatic Energy and Capacitance 91

Express the equivalent capacitance of capacitors C_1 and C_2, each with plate area $A/2$, connected in parallel:

$$C_{eq} = C_1 + C_2$$

$$= \frac{\kappa_1 \epsilon_0 \left(\tfrac{1}{2}A\right)}{d} + \frac{\kappa_1 \epsilon_0 \left(\tfrac{1}{2}A\right)}{d}$$

$$= \frac{\kappa_1 \epsilon_0 A}{2d}(\kappa_1 + \kappa_2)$$

Express the ratio of C_{eq} to C_0 and simplify to obtain:

$$\frac{C_{eq}}{C_0} = \frac{\dfrac{\kappa_1 \epsilon_0 A}{2d}(\kappa_1 + \kappa_2)}{\dfrac{\epsilon_0 A}{d}} = \boxed{\tfrac{1}{2}(\kappa_1 + \kappa_2)}$$

*97 ••• An electrically isolated capacitor with charge Q is partly filled with a dielectric substance as shown in Figure 24-43. The capacitor consists of two rectangular plates of edge lengths a and b separated by distance d. The distance which the dielectric is inserted is x. (*a*) What is the energy stored in the capacitor? (*Hint: the capacitor can be thought of as two capacitors connected in parallel.*) (*b*) Because the energy of the capacitor decreases as x increases, the electric field must be doing positive work on the dielectric, meaning that there must be an electric force pulling it in. Calculate the force by examining how the stored energy varies with x. (*c*) Express the force in terms of the capacitance and voltage. (*d*) What is the origin of this force?

Picture the Problem We can model this capacitor as the equivalent of two capacitors connected in parallel, one with an air gap and other filled with a dielectric of constant κ. Let the numeral 1 denote the capacitor with the dielectric material whose constant is κ and the numeral 2 the air-filled capacitor.

(*a*) Using the hint, express the energy stored in the capacitor as a function of the equivalent capacitance C_{eq}:

$$U = \frac{1}{2}\frac{Q^2}{C_{eq}}$$

The capacitances of the two capacitors are:

$$C_1 = \frac{\kappa \epsilon_0 a x}{d} \text{ and } C_2 = \frac{\epsilon_0 a(a-x)}{d}$$

Because the capacitors are in parallel, C_{eq} is the sum of C_1 and C_2:

$$C_{eq} = C_1 + C_2 = \frac{\kappa \epsilon_0 a x}{d} + \frac{\epsilon_0 a(a-x)}{d}$$

$$= \frac{\epsilon_0 a}{d}(\kappa x + a - x)$$

$$= \frac{\epsilon_0 a}{d}[(\kappa - 1)x + a]$$

92   Chapter 24

Substitute for C_{eq} in the expression for U and simplify to obtain:

$$U = \boxed{\frac{Q^2 d}{2\epsilon_0 a[(\kappa-1)x+a]}}$$

(b) The force exerted by the electric field is given by:

$$F = -\frac{dU}{dx}$$

$$= -\frac{d}{dx}\left[\frac{1}{2\epsilon_0}\frac{Q^2 d}{a[(\kappa-1)x+a]}\right]$$

$$= -\frac{Q^2 d}{2\epsilon_0 a}\frac{d}{dx}\{[(\kappa-1)x+a]^{-1}\}$$

$$= \boxed{\frac{(\kappa-1)Q^2 d}{2a\epsilon_0[(\kappa-1)x+a]^2}}$$

(c) Rewrite our result in (b) to obtain:

$$F = \frac{(\kappa-1)Q^2\left(\frac{a\epsilon_0}{d}\right)}{2\left(\frac{a\epsilon_0}{d}\right)^2[(\kappa-1)x+a]^2}$$

$$= \frac{(\kappa-1)Q^2\left(\frac{a\epsilon_0}{d}\right)}{2C_{eq}^2}$$

$$= \boxed{\frac{(\kappa-1)a\epsilon_0 V^2}{2d}}$$

Note that this expression is independent of x.

(d) | This force originates from the fringing fields around the edges of the capacitor. The effect of the force is to pull the dielectric into the space between the capacitor plates. |

*102 ••• You are asked to construct a parallel-plate, air-gap capacitor that will store 100 kJ of energy. (a) What minimum volume is required between the plates of the capacitor? (b) Suppose you have developed a dielectric that can withstand 3×10^8 V/m and has a dielectric constant of $\kappa = 5$. What volume of this dielectric, between the plates of the capacitor, is required for it to be able to store 100 kJ of energy?

Picture the Problem Recall that the dielectric strength of air is 3 MV/m. We can express the maximum energy to be stored in terms of the capacitance of the air-gap capacitor and the maximum potential difference between its plates. This maximum potential can, in turn, be expressed in terms of the maximum electric field (dielectric strength) possible in

Electrostatic Energy and Capacitance 93

the air gap. We can solve the resulting equation for the volume of the space between the plates. In part (*b*) we can modify the equation we derive in part (*a*) to accommodate a dielectric with a constant other than 1.

(*a*) Express the energy stored in the capacitor in terms of its capacitance and the potential difference across it:

$$U_{max} = \tfrac{1}{2} C V_{max}^2$$

Express the capacitance of the air-gap parallel-plate capacitor:

$$C = \frac{\epsilon_0 A}{d}$$

Relate the maximum potential difference across the plates to the maximum electric field between them:

$$V_{max} = E_{max} d$$

Substitute to obtain:

$$U_{max} = \tfrac{1}{2} \left(\frac{\epsilon_0 A}{d} \right) (E_{max} d)^2 = \tfrac{1}{2} \epsilon_0 (Ad) E_{max}^2$$

$$= \tfrac{1}{2} \epsilon_0 \upsilon E^2$$

where $\upsilon = Ad$ is the volume between the plates.

Solve for υ:

$$\upsilon = \frac{2 U_{max}}{\epsilon_0 E_{max}^2} \quad (1)$$

Substitute numerical values and evaluate υ:

$$\upsilon = \frac{2(100 \text{kJ})}{(8.85 \times 10^{-12} \text{C}^2/\text{N} \cdot \text{m}^2)(3 \text{MV/m})^2}$$

$$= \boxed{2.51 \times 10^3 \text{ m}^3}$$

(*b*) With the dielectric in place equation (1) becomes:

$$\upsilon = \frac{2 U_{max}}{\kappa \epsilon_0 E_{max}^2} \quad (2)$$

Evaluate equation (2) with $\kappa = 5$ and $E_{max} = 3 \times 10^8$ V/m:

$$\upsilon = \frac{2(100 \text{kJ})}{5(8.85 \times 10^{-12} \text{C}^2/\text{N} \cdot \text{m}^2)(3 \times 10^8 \text{ V/m})^2}$$

$$= \boxed{5.02 \times 10^{-2} \text{ m}^3}$$

*106 ••• The two capacitors shown in Figure 24-45 have capacitances $C_1 = 0.4 \ \mu\text{F}$ and $C_2 = 1.2 \ \mu\text{F}$. The voltages across the two capacitors are V_1 and V_2, respectively, and the total stored energy in the two capacitors is 1.14 mJ. If terminals *b* and *c* are connected together, the voltage is $V_a - V_d = 80$ V; if terminal *a* is connected to terminal b, and terminal c is connected to terminal d, the voltage $V_a - V_d = 20$ V. Find the initial voltages V_1 and V_2.

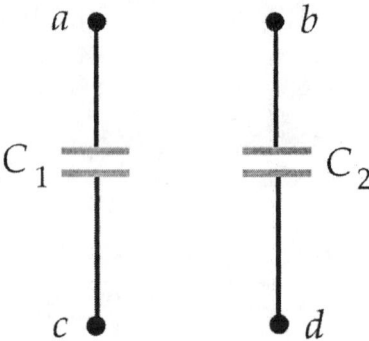

Figure 24-45 Problem 106

Picture the Problem We can express the two conditions on the voltage in terms of the charges Q_1 and Q_2 and the capacitances C_1 and C_2 and solve the equations simultaneously to find Q_1 and Q_2. We can then use the definition of capacitance to find the initial voltages V_1 and V_2.

Express the condition for the series connection:

$V_1 + V_2 = 80$ V

or

$$\frac{Q_1}{C_1} + \frac{Q_2}{C_2} = 80 \text{ V}$$

Substitute numerical values to obtain:

$$\frac{Q_1}{0.4\,\mu F} + \frac{Q_2}{1.2\,\mu F} = 80 \text{ V}$$

or

$3Q_1 + Q_2 = 96\,\mu C$ \hfill (1)

Use the definition of capacitance to express the condition for the parallel connection:

$$\frac{Q_1 + Q_2}{C_{eq}} = 20 \text{ V}$$

Because the capacitors are now connected in parallel:

$C_{eq} = C_1 + C_2 = 0.4\,\mu F + 1.2\,\mu F = 1.6\,\mu F$

Substitute to obtain:

$$\frac{Q_1 + Q_2}{1.6\,\mu F} = 20 \text{ V}$$

or

$Q_1 + Q_2 = 32\,\mu C$ \hfill (2)

Solve equations (1) and (2)

$Q_1 = 32\,\mu C$ and $Q_2 = 0$

Electrostatic Energy and Capacitance   95

simultaneously to obtain:

Use the definition of capacitance to obtain:

$$V_1 = \frac{Q_1}{C_1} = \frac{32\,\mu C}{0.4\,\mu F} = \boxed{80.0\,V}$$

and

$$V_2 = \frac{Q_2}{C_2} = \frac{0}{0.4\,\mu F} = \boxed{0}$$

*110 ••• A capacitor has rectangular plates of length a and width b. The top plate is inclined at a small angle, as shown in Figure 24-48. The plate separation varies from $d = y_0$ at the left to $d = 2y_0$ at the right, where y_0 is much less than a or b. Calculate the capacitance using strips of width dx and length b to approximate differential capacitors of area $b\,dx$ and separation $d = y_0 + (y_0/a)x$ that are connected in parallel.

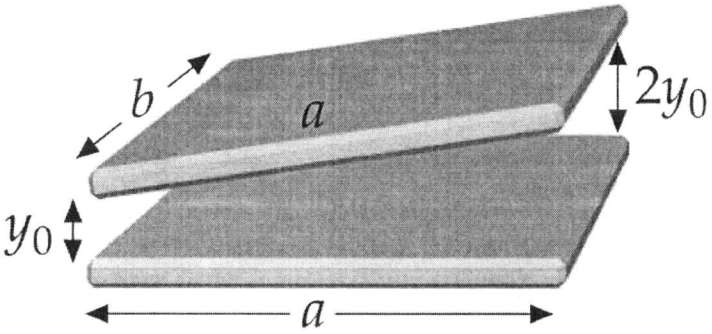

Figure 24-48 Problem 110

Picture the Problem Choose a coordinate system in which the positive x direction is the right and the origin is at the left edge of the capacitor. We can express an element of capacitance dC and then integrate this expression to find C for this capacitor.

Express an element of capacitance dC of length b, width dx and separation $d = y_0 + (y_0/a)x$:

$$dC = \frac{\epsilon_0 b}{d}dx = \frac{\epsilon_0 b}{y_0(1 + x/a)}dx$$

These elements are all in parallel, so the total capacitance is obtained by integration:

$$C = \frac{\epsilon_0 b}{y_0}\int_0^{y_0}\frac{1}{1+x/a}dx = \boxed{\frac{\epsilon_0 ab}{y_0}\ln 2}$$

Chapter 25
Electric Current and Direct-Current Circuits

Conceptual Problems

*1 • In our study of electrostatics, we concluded that there is no electric field within a conductor in electrostatic equilibrium. How is it that we can now discuss electric fields inside a conductor?

Determine the Concept When current flows, the charges are not in equilibrium. In that case, the electric field provides the force needed for the charge flow.

*5 •• A metal bar is to be used as a resistor. Its dimensions are 2 by 4 by 10 units. To get the smallest resistance from this bar, one should attach leads to the opposite sides that have the dimensions of

(*a*) 2 by 4 units.
(*b*) 2 by 10 units.
(*c*) 4 by 10 units.
(*d*) All connections will give the same resistance.
(*e*) None of the above is correct.

Picture the Problem The resistance of the metal bar varies directly with its length and inversely with its cross-sectional area. Hence, to minimize the resistance of the bar, we should connect to the surface for which the ratio of the length to the contact area is least.

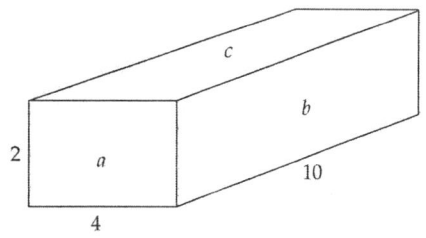

Denoting the surfaces as *a*, *b*, and *c*, complete the table to the right:

Surface	L	A	L/A
a	10	8	0.8
b	4	20	0.2
c	2	40	0.05

Because connecting to surface *c* minimizes *R*:

(*c*) is correct.

*10 • Two resistors with resistances R_1 and R_2 are connected in parallel. If $R_1 \gg R_2$, the equivalent resistance of the combination is approximately (*a*) R_1. (*b*) R_2. (*c*) 0. (*d*) infinite

Picture the Problem We can find the equivalent resistance of this two-resistor combination and then apply the condition that $R_1 \gg R_2$.

98   Chapter 25

Express the equivalent resistance of R_1 and R_2 in parallel:

$$\frac{1}{R_{eq}} = \frac{1}{R_1} + \frac{1}{R_2}$$

Solve for R_{eq} to obtain:

$$R_{eq} = \frac{R_1 R_2}{R_1 + R_2}$$

Factor R_1 from the denominator and simplify to obtain:

$$R_{eq} = \frac{R_1 R_2}{R_1\left(1+\frac{R_2}{R_1}\right)} = \frac{R_2}{1+\frac{R_2}{R_1}}$$

If $R_1 \gg R_2$, then:

$$R_{eq} = R_{eff} \approx R_2 \text{ and } \boxed{(b) \text{ is correct}}$$

*13 • Two resistors are connected in series across a potential difference. Resistor A has twice the resistance of resistor B. If the current carried by resistor A is I, then what is the current carried by resistor B? (*a*) I (*b*) $2I$ (*c*) $I/2$ (*d*) $4I$ (*e*) $I/4$

Determine the Concept In a series circuit, because there are no alternative pathways, all resistors carry the same current. The potential difference across each resistor, keeping with Ohm's law, is given by the product of the current and the resistance and, hence, is not the same across each resistor unless the resistors are identical. $\boxed{(a) \text{ is correct.}}$

*17 • An ideal ammeter should have _____ internal resistance.
(*a*) infinite
(*b*) zero

Determine the Concept An ideal ammeter would have zero resistance. An ammeter consists of a very small resistance in parallel with a galvanometer movement. The small resistance accomplishes two purposes: 1) It protects the galvanometer movement by shunting most of the current in the circuit around the galvanometer movement, and 2) It minimizes the loading of the circuit by the ammeter by minimizing the resistance of the ammeter. $\boxed{(b) \text{ is correct.}}$

*21 •• A battery is connected to a series combination of a switch, a resistor, and an initially uncharged capacitor. The switch is closed at $t = 0$. Which of the following statements is true?

(*a*) As the charge on the capacitor increases, the current increases.
(*b*) As the charge on the capacitor increases, the voltage drop across the resistor increases.
(*c*) As the charge on the capacitor increases, the current remains constant.

(d) As the charge on the capacitor increases, the voltage drop across the capacitor decreases.

(e) As the charge on the capacitor increases, the voltage drop across the resistor decreases.

Determine the Concept Applying Kirchhoff's loop rule to the circuit, we obtain $\mathcal{E} - V_R - V_C = 0$, where V_R is the voltage drop across the resistor. Applying Ohm's law to the resistor, we obtain $V_R = IR$. Because I decreases as the capacitor is charged, V_R decreases with time. $\boxed{(e) \text{ is correct.}}$

*24 • All voltage sources have some internal resistance, usually on the order of 100 Ω or less. From this fact, explain the following statement that appears in some electronics textbooks: "A voltage source likes to see a high resistance."

Determine the Concept The potential difference across an external resistor of resistance R is given by $\dfrac{R}{r+R} V$, where r is the internal resistance and V the voltage supplied by the source. The higher R is, the higher the voltage drop across R. Put differently, the higher the resistance a voltage source sees, the less its own resistance will change the circuit.

Estimation and Approximation

*30 •• Compact fluorescent lightbulbs cost $6 each and have an expected lifetime of 8000 h. These bulbs consume 20 W of power, but produce the illumination equivalent to 75-W incandescent bulbs. Incandescent bulbs cost approximately $1.50 each and have an expected lifetime of 1200 h. If the average household has, on the average, six 75-W incandescent lightbulbs on constantly, and if energy costs 11.5 cents per kilowatt-hour, how much money would a consumer save each year by installing the energy-efficient fluorescent lightbulbs?

Picture the Problem We can find the annual savings by taking into account the costs of the two types of bulbs, the rate at which they consume energy and the cost of that energy, and their expected lifetimes.

Express the yearly savings: $\Delta\$ = \text{Cost}_{\text{incandescent}} - \text{Cost}_{\text{fluorescent}}$ (1)

Express the annual cost with the incandescent bulbs: $\text{Cost}_{\text{incandescent}} = \text{Cost}_{\text{bulbs}} + \text{Cost}_{\text{energy}}$

Express and evaluate the annual cost of the incandescent bulbs:

$$\text{Cost}_{\text{bulbs}} = \text{number of bulbs in use} \times \text{annual consumption of bulbs} \times \text{cost per bulb}$$

$$= (6)\left(\frac{365.24\,\text{d} \times \frac{24\,\text{h}}{\text{d}}}{1200\,\text{h}}\right)(\$1.50) = \$65.74$$

Find the cost of operating the incandescent bulbs for one year:

$$\text{Cost}_{\text{energy}} = \text{energy consumed} \times \text{cost per unit of energy}$$
$$= 6(75\,\text{W})(365.25\,\text{d})(24\,\text{h/d})(\$0.115/\text{kW} \cdot \text{h})$$
$$= \$453.64$$

Express the annual cost with the fluorescent bulbs:

$$\text{Cost}_{\text{fluorescent}} = \text{Cost}_{\text{bulbs}} + \text{Cost}_{\text{energy}}$$

Express and evaluate the annual cost of the fluorescent bulbs:

$$\text{Cost}_{\text{bulbs}} = \text{number of bulbs in use} \times \text{annual consumption of bulbs} \times \text{cost per bulb}$$

$$= (6)\left(\frac{365.24\,\text{d} \times \frac{24\,\text{h}}{\text{d}}}{8000\,\text{h}}\right)(\$6) = \$39.45$$

Find the cost of operating the fluorescent bulbs for one year:

$$\text{Cost}_{\text{energy}} = \text{energy consumed} \times \text{cost per unit of energy}$$
$$= 6(20\,\text{W})\left(365.24\,\text{d} \times \frac{24\,\text{h}}{\text{d}}\right)(\$0.115/\text{kW} \cdot \text{h})$$
$$= \$120.97$$

Substitute in equation (1) and evaluate the cost savings $\Delta\$$:

$$\Delta\$ = \text{Cost}_{\text{incandescent}} - \text{Cost}_{\text{fluorescent}} = (\$65.74 + \$453.64) - (\$39.45 + \$120.97)$$
$$= \boxed{\$358.96}$$

*32 •• A laser diode used in making a laser pointer is a highly nonlinear circuit element. For a voltage drop across it less than approximately 2.3 V, it behaves as if it has effectively infinite internal resistance, but for voltages across it higher than this it has a very low internal resistance-effectively zero. (*a*) A laser pointer is made by putting two 1.55 V watch batteries in series across the laser diode. If the batteries each have an internal resistance between 100 Ω and 150 Ω, estimate the current in the laser diode.

(b) About half of the power delivered to the laser diode goes into radiant energy. Using this fact, estimate the power of the laser diode, and compare this to typical quoted values of about 3 mW. (c) If the batteries each have a capacity of 20-mA hours (i.e., they can deliver a constant current of 20 mA for approximately one hour before discharging), estimate how long one can continuously operate the laser pointer before replacing the batteries.

Picture the Problem Let r be the internal resistance of each battery and use Ohm's law to express the current in laser diode as a function of the potential difference across r. We can find the power of the laser diode from the product of the potential difference across the internal resistance of the batteries and the current delivered by them I and the time-to-discharge from the combined capacities of the two batteries and I.

(a) Use Ohm's law to express the current in the laser diode:

$$I = \frac{V_{\text{internal resistance}}}{2r}$$

The potential difference across the internal resistance is:

$$V_{\text{internal resistance}} = \mathcal{E} - 2.3\,\text{V}$$

Substitute to obtain:

$$I = \frac{\mathcal{E} - 2.3\,\text{V}}{2r}$$

Assuming that $r = 125\,\Omega$:

$$I = \frac{2(1.55) - 2.3\,\text{V}}{2(125\,\Omega)} = \boxed{3.20\,\text{mA}}$$

(b) The power delivered by the batteries is given by:

$$P = IV = (3.2\,\text{mA})(2.3\,\text{V}) = 7.36\,\text{mW}$$

The power of the laser is half this value:

$$P_{\text{laser}} = \tfrac{1}{2}P = \tfrac{1}{2}(7.36\,\text{mW}) = \boxed{3.68\,\text{mW}}$$

Express the ratio of P_{laser} to P_{quoted}:

$$\frac{P_{\text{laser}}}{P_{\text{quoted}}} = \frac{3.68\,\text{mW}}{3\,\text{mW}} = 1.23$$

or

$$P_{\text{laser}} = \boxed{123\% P_{\text{quoted}}}$$

(c) Express the time-to-discharge:

$$\Delta t = \frac{\text{Capacity}}{I}$$

Because each battery has a capacity of 20 mA·h, the series combination has a capacity of 40 mA·h and:

$$\Delta t = \frac{40\,\text{mA·h}}{3.20\,\text{mA}} = \boxed{12.5\,\text{h}}$$

Current and the Motion of Charges

*37 •• A 10-gauge copper wire and a 14-gauge copper wire are welded together end to end. The wires carry a current of 15 A. If there is one free electron per copper atom in each wire, find the drift velocity of the electrons in each wire.

Picture the Problem The current will be the same in the two wires and we can relate the drift velocity of the electrons in each wire to their current densities and the cross-sectional areas of the wires. We can find the number density of charge carriers n using $n = \rho N_A / M$, where ρ is the mass density, N_A Avogadro's number, and M the molar mass. We can find the cross-sectional area of 10- and 14-gauge wires in Table 25-2.

Relate the current density to the drift velocity of the electrons in the 10-gauge wire:

$$\frac{I_{10\,\text{gauge}}}{A_{10\,\text{gauge}}} = nev_d$$

Solve for v_d:

$$v_{d,10} = \frac{I_{10\,\text{gauge}}}{neA_{10\,\text{gauge}}}$$

The number density of charge carriers n is related to the mass density ρ, Avogadro's number N_A, and the molar mass M:

$$n = \frac{\rho N_A}{M}$$

For copper, $\rho = 8.93$ g/cm³ and $M = 63.5$ g/mol. Substitute and evaluate n:

$$n = \frac{(8.93\,\text{g/cm}^3)(6.02 \times 10^{23}\,\text{atoms/mol})}{63.5\,\text{g/mol}}$$

$$= 8.47 \times 10^{28}\,\text{atoms/m}^3$$

Use Table 25-2 to find the cross-sectional area of 10-gauge wire:

$$A_{10} = 5.261\,\text{mm}^2$$

Substitute numerical values and evaluate $v_{d,10}$:

$$v_{d,10} = \frac{15\,\text{A}}{(8.47 \times 10^{28}\,\text{m}^{-3})(1.60 \times 10^{-19}\,\text{C})(5.261\,\text{mm}^2)} = \boxed{0.210\,\text{mm/s}}$$

Express the continuity of the current in the two wires:

$$I_{10\,\text{gauge}} = I_{14\,\text{gauge}}$$

or

$$nev_{d,10}A_{10\,\text{gauge}} = nev_{d,14}A_{14\,\text{gauge}}$$

Electric Current and Direct-Current Circuits 103

Solve for $v_{d,14}$ to obtain:

$$v_{d,14} = v_{d,10} \frac{A_{10\,gauge}}{A_{14\,gauge}}$$

Use Table 25-2 to find the cross-sectional area of 14-gauge wire:

$$A_{14} = 2.081\,mm^2$$

Substitute numerical values and evaluate $v_{d,14}$:

$$v_{d,14} = (0.210\,mm/s)\frac{5.261\,mm^2}{2.081\,mm^2}$$

$$= \boxed{0.531\,mm/s}$$

*39 •• In a proton supercollider, the protons in a 5-mA beam move with nearly the speed of light. (*a*) How many protons are there per meter of the beam? (*b*) If the cross-sectional area of the beam is $10^{-6}\,m^2$, what is the number density of protons?

Picture the Problem We can relate the number of protons per meter N to the number n of free charge-carrying particles per unit volume in a beam of cross-sectional area A and then use the relation between current and drift velocity to relate n to I.

(*a*) Express the number of protons per meter N in terms of the number n of free charge-carrying particles per unit volume in a beam of cross-sectional area A:

$$N = nA \qquad (1)$$

Use the relation between current and drift velocity to relate I and n:

$$I = enAv$$

Solve for n to obtain:

$$n = \frac{I}{eAv}$$

Substitute to obtain:

$$N = \frac{IA}{eAv} = \frac{I}{ev}$$

Substitute numerical values and evaluate N:

$$N = \frac{5\,mA}{(1.60\times10^{-19}\,C)(3\times10^8\,m/s)}$$

$$= \boxed{1.04\times10^8\,m^{-1}}$$

(b) From equation (1) we have:
$$n = \frac{N}{A} = \frac{1.04 \times 10^8 \text{ m}^{-1}}{10^{-6} \text{ m}^2}$$
$$= \boxed{1.04 \times 10^{14} \text{ m}^{-3}}$$

Resistance and Ohm's Law

*44 • The third (current-carrying) rail of a subway track is made of steel and has a cross-sectional area of about 55 cm². The resistivity of steel is 10^{-7} Ω·m. What is the resistance of 10 km of this track?

Picture the Problem We can use $R = \rho L/A$ to find the resistance of the track.

(a) Relate the resistance of the track to its resistivity ρ, cross-sectional area A, and length L:
$$R = \rho \frac{L}{A}$$

Substitute numerical values and evaluate R:
$$R = (10^{-7} \, \Omega \cdot \text{m}) \frac{10 \text{ km}}{55 \text{ cm}^2} = \boxed{0.182 \, \Omega}$$

*49 •• A copper wire and an iron wire with the same length and diameter carry the same current I. (a) Find the ratio of the potential drops across these wires. (b) In which wire is the electric field greater?

Picture the Problem We can use Ohm's law to express the ratio of the potential differences across the two wires and $R = \rho L/A$ to relate the resistances of the wires to their lengths, resistivities, and cross-sectional areas. Once we've found the ratio of the potential differences across the wires, we can use $E = V/L$ to decide which wire has the greater electric field.

(a) Apply Ohm's law to express the potential drop across each wire:
$$V_{\text{Cu}} = IR_{\text{Cu}}$$
and
$$V_{\text{Fe}} = IR_{\text{Fe}}$$

Divide the first of these equations by the second to express the ratio of the potential drops across the wires:
$$\frac{V_{\text{Cu}}}{V_{\text{Fe}}} = \frac{IR_{\text{Cu}}}{IR_{\text{Fe}}} = \frac{R_{\text{Cu}}}{R_{\text{Fe}}} \quad (1)$$

Relate the resistances of the wires to their resistivity, cross-sectional area, and length:
$$R_{\text{Cu}} = \rho_{\text{Cu}} \frac{L_{\text{Cu}}}{A_{\text{Cu}}}$$
and

$$R_{Fe} = \rho_{Fe}\frac{L_{Fe}}{A_{Fe}}$$

Divide the first of these equations by the second to express the ratio of the resistances of the wires:

$$\frac{R_{Cu}}{R_{Fe}} = \frac{\rho_{Cu}\dfrac{L_{Cu}}{A_{Cu}}}{\rho_{Fe}\dfrac{L_{Fe}}{A_{Fe}}} = \frac{\rho_{Cu}}{\rho_{Fe}}$$

because $L_{Cu} = L_{Fe}$ and $A_{Cu} = A_{Fe}$.

Substitute in equation (1) to obtain:

$$\frac{V_{Cu}}{V_{Fe}} = \frac{\rho_{Cu}}{\rho_{Fe}}$$

Substitute numerical values (see Table 2625-1 for the resistivities of copper and iron) and evaluate the ratio of the potential differences:

$$\frac{V_{Cu}}{V_{Fe}} = \frac{1.7\times 10^{-8}\,\Omega\cdot m}{10\times 10^{-8}\,\Omega\cdot m} = \boxed{0.170}$$

(b) Express the electric field in each conductor in terms of its length and the potential difference across it:

$$E_{Cu} = \frac{V_{Cu}}{L_{Cu}}$$

and

$$E_{Fe} = \frac{V_{Fe}}{L_{Fe}}$$

Divide the first of these equations by the second to obtain:

$$\frac{E_{Cu}}{E_{Fe}} = \frac{\dfrac{V_{Cu}}{L_{Cu}}}{\dfrac{V_{Cu}}{L_{Cu}}} = \frac{V_{Cu}}{V_{Fe}} = 0.170$$

or

$$E_{Fe} = \frac{E_{Cu}}{0.17} = 5.88 E_{Cu}$$

Because $E_{Fe} = 5.88 E_{Cu}$: $\boxed{E \text{ is greater in the iron wire.}}$

*54 •• A diode is a circuit element with a very nonlinear IV curve. In a diode, $I = I_0\left(e^{V/25\text{ mV}} - 1\right)$, where $I_0 \sim 2\times 10^{-9}$ A. Using a spreadsheet program, make a graph of $I(V)$ for a typical diode, for both forward biasing ($V > 0$) and back-biasing ($V < 0$). Show that if you plot $\ln I$ vs V for forward biasing (using $V > 0.3$ V or so), you get nearly a straight line. What is the slope of the line?

106 Chapter 25

Picture the Problem A spreadsheet program to plot I as a function of V is shown below. The formulas used to calculate the quantities in the columns are as follows:

Cell	Content/Formula	Algebraic Form
B1	2	I_0
A5	−200	V (mV)
A6	A5 + 25	$V + \Delta V$
B5	$B$1*(EXP(A5/25) − 1)	$I_0\left(e^{V/25\,\text{mV}} - 1\right)$

	A	B	C
1	I_0=	2	nA
2			
3	V	I	
4	(mV)	(nA)	
5	−200.0	−2.00	
6	−175.0	−2.00	
7	−150.0	−2.00	
15	50.0	12.78	
16	75.0	38.17	
17	100.0	107.20	

The following graph was plotted using the data in spreadsheet table shown above.

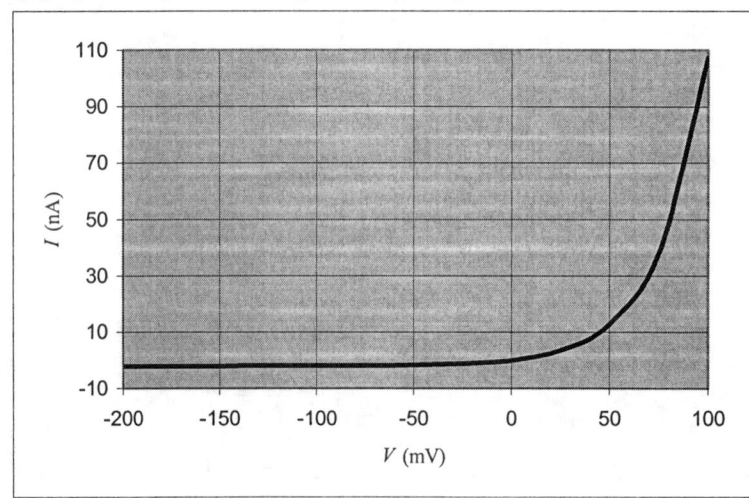

A spreadsheet program to plot $\ln(I)$ as a function of V for $V > 0.3$ V follows. The formulas used to calculate the quantities in the columns are as follows:

Cell	Content/Formula	Algebraic Form
B1	2	2 nA
A5	300	V
A6	A5 + 10	$V + \Delta V$
B5	LN($B$1*(EXP(A5/25) − 1))	$\ln\left[I_0\left(e^{V/25\,\text{mV}} - 1\right)\right]$

A	B	C
I_0=	2	nA
V (mV)	ln(I)	
300	12.69	
310	13.09	
320	13.49	
330	13.89	
340	14.29	
350	14.69	
970	39.49	
980	39.89	
990	40.29	
1000	40.69	

A graph of ln(I) as a function of V follows. Microsoft Excel's Trendline feature was used to obtain the equation of the line.

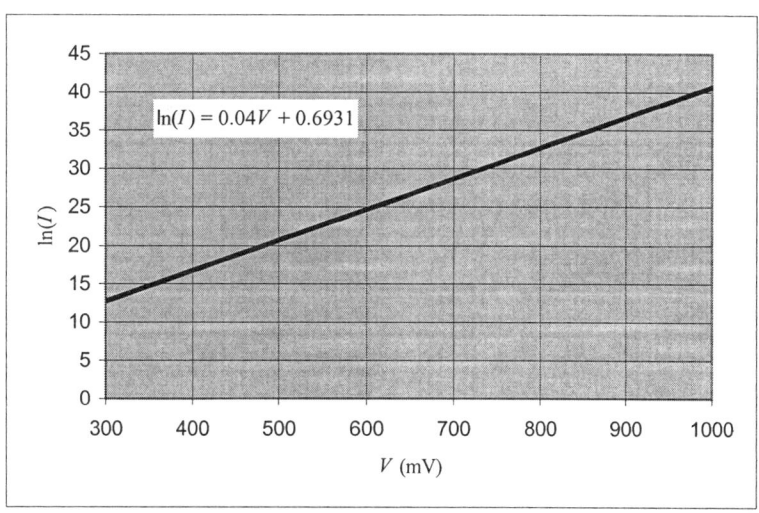

For $V \gg 25$ mV:

$$e^{V/25\,\text{mV}} - 1 \approx e^{V/25\,\text{mV}}$$
and
$$I \approx I_0 e^{V/25\,\text{mV}}$$

Take the natural logarithm of both sides of the equation to obtain:

$$\ln(I) = \ln\left(I_0 e^{V/25\,\text{mV}}\right)$$
$$= \ln(I_0) + \frac{1}{25\,\text{mV}} V$$

which is of the form $y = mx + b$, where

$$m = \frac{1}{25\,\text{mV}} = \boxed{0.04\,(\text{mV})^{-1}}$$

in agreement with our graphical result.

108   Chapter 25

*58 •••   The space between two concentric spherical-shell conductors is filled with a material that has a resistivity of 10^9 Ω·m. If the inner shell has a radius of 1.5 cm and the outer shell has a radius of 5 cm, what is the resistance between the conductors? (*Hint:* Find the resistance of a spherical-shell element of the material of area $4\pi r^2$ and length dr, and integrate to find the total resistance of the set of shells in series.)

Picture the Problem The diagram shows a cross-sectional view of the concentric spheres of radii a and b as well as a spherical-shell element of radius r. Using the *Hint* we can express the resistance dR of the spherical-shell element and then integrate over the volume filled with the material whose resistivity ρ is given to find the resistance between the conductors. Note that the elements of resistance are in series.

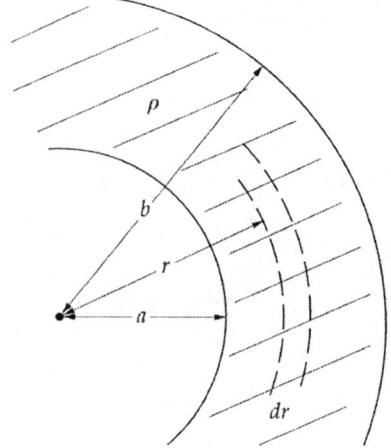

Express the element of resistance dR:

$$dR = \rho \frac{dr}{A} = \rho \frac{dr}{4\pi r^2}$$

Integrate dR from $r = a$ to $r = b$ to obtain:

$$R = \frac{\rho}{4\pi} \int_a^b \frac{dr}{r^2} = \frac{\rho}{4\pi}\left(\frac{1}{a} - \frac{1}{b}\right)$$

Substitute numerical values and evaluate R:

$$R = \frac{10^9\,\Omega\cdot\text{m}}{4\pi}\left(\frac{1}{1.5\,\text{cm}} - \frac{1}{5\,\text{cm}}\right)$$

$$= \boxed{3.71\times 10^9\,\Omega}$$

Temperature Dependence of Resistance

*60 •   A tungsten rod is 50 cm long and has a square cross-sectional area with sides of 1.0 mm. (*a*) What is its resistance at 20°C? (*b*) What is its resistance at 40°C?

Picture the Problem We can use $R = \rho L/A$ to find the resistance of the rod at 20°C. Ignoring the effects of thermal expansion, we can we apply the equation defining the temperature coefficient of resistivity, α, to relate the resistance at 40°C to the resistance at 20°C.

(*a*) Express the resistance of the rod at 20°C as a function of its

$$R_{20} = \rho_{20}\frac{L}{A}$$

Electric Current and Direct-Current Circuits 109

resistivity, length, and cross-sectional area:

Substitute numerical values and evaluate R_{20}:

$$R_{20} = (5.5 \times 10^{-8}\,\Omega \cdot \text{m})\frac{0.5\,\text{m}}{(1\,\text{mm})^2}$$

$$= \boxed{27.5\,\text{m}\Omega}$$

(b) Express the resistance of the rod at 40°C as a function of its resistance at 20°C and the temperature coefficient of resistivity α:

$$R_{40} = \rho_{40}\frac{L}{A}$$

$$= \rho_{20}[1+\alpha(t_C - 20\text{C}°)]\frac{L}{A}$$

$$= \rho_{20}\frac{L}{A} + \rho_{20}\frac{L}{A}\alpha(t_C - 20\text{C}°)$$

$$= R_{20}[1+\alpha(t_C - 20\text{C}°)]$$

Substitute numerical values (see Table 25-1 for the temperature coefficient of resistivity of tungsten) and evaluate R_{40}:

$$R_{40} = (27.5\,\text{m}\Omega)[1+(4.5\times10^{-3}\,\text{K}^{-1})(20\,\text{C}°)] = \boxed{30.0\,\text{m}\Omega}$$

*65 ••• A wire of cross-sectional area A, length L_1, resistivity ρ_1, and temperature coefficient α_1 is connected end to end to a second wire of the same cross-sectional area, length L_2, resistivity ρ_2, and temperature coefficient α_2, so that the wires carry the same current. (a) Show that if $\rho_1 L_1 \alpha_1 + \rho_2 L_2 \alpha_2 = 0$, the total resistance R is independent of temperature for small temperature changes.
(b) If one wire is made of carbon and the other is copper, find the ratio of their lengths for which R is approximately independent of temperature.

Picture the Problem Expressing the total resistance of the two current-carrying (and hence warming) wires connected in series in terms of their resistivities, temperature coefficients of resistivity, lengths and temperature change will lead us to an expression in which, if $\rho_1 L_1 \alpha_1 + \rho_2 L_2 \alpha_2 = 0$, the total resistance is temperature independent. In part (b) we can apply the condition that $\rho_1 L_1 \alpha_1 + \rho_2 L_2 \alpha_2 = 0$ to find the ratio of the lengths of the carbon and copper wires.

(a) Express the total resistance of these two wires connected in series:

$$R = R_1 + R_2$$
$$= \rho_1 \frac{L_1}{A}(1+\alpha_1 \Delta T) + \rho_2 \frac{L_2}{A}(1+\alpha_2 \Delta T) + \frac{1}{A}[\rho_1 L_1(1+\alpha_1 \Delta T) + \rho_2 L_2(1+\alpha_2 \Delta T)]$$

110 Chapter 25

Expand and simplify this expression to obtain:

$$R = \frac{1}{A}\left[\rho_1 L_1 + \rho_2 L_2 + (\rho_1 L_1 \alpha_1 + \rho_1 L_1 \alpha_2)\Delta T\right]$$

If $\rho_1 L_1 \alpha_1 + \rho_2 L_2 \alpha_2 = 0$, then:

$$R = \boxed{\frac{1}{A}\left[\rho_1 L_1 + \rho_2 L_2\right]} \text{ independently of}$$

the temperature.

(b) Apply the condition for temperature independence obtained in (a) to the carbon and copper wires:

$$\rho_C L_C \alpha_C + \rho_{Cu} L_{Cu} \alpha_{Cu} = 0$$

Solve for the ratio of L_{Cu} to L_C:

$$\frac{L_{Cu}}{L_C} = -\frac{\rho_C \alpha_C}{\rho_{Cu} \alpha_{Cu}}$$

Substitute numerical values (see Table 25-1 for the temperature coefficient of resistivity of carbon and copper) and evaluate the ratio of L_{Cu} to L_C:

$$\frac{L_{Cu}}{L_C} = -\frac{(3500 \times 10^{-8}\,\Omega \cdot m)(-0.5 \times 10^{-3}\,K^{-1})}{(1.7 \times 10^{-8}\,\Omega \cdot m)(3.9 \times 10^{-3}\,K^{-1})} = \boxed{264}$$

Energy in Electric Circuits

*68 • Find the power dissipated in a resistor connected across a constant potential difference of 120 V if its resistance is (a) 5 Ω and (b) 10 Ω.
Picture the Problem We can use $P = V^2/R$ to find the power dissipated by the two resistors.

Express the power dissipated in a resistor as a function of its resistance and the potential difference across it:

$$P = \frac{V^2}{R}$$

(a) Evaluate P for $V = 120$ V and $R = 5\,\Omega$:

$$P = \frac{(120\,V)^2}{5\,\Omega} = \boxed{2.88\,kW}$$

(b) Evaluate P for $V = 120$ V and $R = 10\,\Omega$:

$$P = \frac{(120\,V)^2}{10\,\Omega} = \boxed{1.44\,kW}$$

Electric Current and Direct-Current Circuits 111

*73 • (a) How much power is delivered by the emf of the battery in Problem 72 when it delivers a current of 20 A? (b) How much of this power is delivered to the starter? (c) By how much does the chemical energy of the battery decrease when it delivers a current of 20 A to the starter for 3 min? (d) How much heat is developed in the battery when it delivers a current of 20 A for 3 min?

Picture the Problem We can find the power delivered by the battery from the product of its emf and the current it delivers. The power delivered to the battery can be found from the product of the potential difference across the terminals of the starter (or across the battery when current is being drawn from it) and the current being delivered to it. In part (c) we can use the definition of power to relate the decrease in the chemical energy of the battery to the power it is delivering and the time during which current is drawn from it. In part (d) we can use conservation of energy to relate the energy delivered by the battery to the heat developed the battery and the energy delivered to the starter

(a) Express the power delivered by the battery as a function of its emf and the current it delivers:

$$P = \mathcal{E}I = (12\,\text{V})(20\,\text{A}) = \boxed{240\,\text{W}}$$

(b) Relate the power delivered to the starter to the potential difference across its terminals:

$$P_{\text{starter}} = V_{\text{starter}} I$$
$$= (11.4\,\text{V})(20\,\text{A}) = \boxed{228\,\text{W}}$$

(c) Use the definition of power to express the decrease in the chemical energy of the battery as it delivers current to the starter:

$$\Delta E = P\Delta t$$
$$= (240\,\text{W})(3\,\text{min}) = \boxed{43.2\,\text{kJ}}$$

(d) Use conservation of energy to relate the energy delivered by the battery to the heat developed in the battery and the energy delivered to the starter:

$$E_{\text{delivered by battery}} = E_{\text{transformed into heat}} + E_{\text{delivered to starter}}$$
$$= Q + E_{\text{delivered to starter}}$$

Express the energy delivered by the battery and the energy delivered to the starter in terms of the rate at which this energy is delivered:

$$P\Delta t = Q + P_s \Delta t$$

Solve for Q to obtain:

$$Q = (P - P_s)\Delta t$$

112 Chapter 25

Substitute numerical values and evaluate Q:

$$Q = (240\,\text{W} - 228\,\text{W})(3\,\text{min})$$
$$= \boxed{2.16\,\text{kJ}}$$

*77 •• A lightweight electric car is powered by ten 12-V batteries. At a speed of 80 km/h, the average frictional force is 1200 N. (a) What must be the power of the electric motor if the car is to travel at a speed of 80 km/h? (b) If each battery can deliver a total charge of 160 A·h before recharging, what is the total charge in coulombs that can be delivered by the 10 batteries before charging? (c) What is the total electrical energy delivered by the 10 batteries before recharging? (d) How far can the car travel at 80 km/h before the batteries must be recharged? (e) What is the cost per kilometer if the cost of recharging the batteries is 9 cents per kilowatt-hour?

Picture the Problem We can use $P = fv$ to find the power the electric motor must develop to move the car at 80 km/h against a frictional force of 1200 N. We can find the total charge that can be delivered by the 10 batteries using $\Delta Q = NI\Delta t$. The total electrical energy delivered by the 10 batteries before recharging can be found using the definition of emf. We can find the distance the car can travel from the definition of work and the cost per kilometer of driving the car this distance by dividing the cost of the required energy by the distance the car has traveled.

(a) Express the power the electric motor must develop in terms of the speed of the car and the friction force:

$$P = fv = (1200\,\text{N})(80\,\text{km/h})$$
$$= \boxed{26.7\,\text{kW}}$$

(b) Use the definition of current to express the total charge that can be delivered before charging:

$$\Delta Q = NI\Delta t = 10(160\,\text{A}\cdot\text{h})\left(\frac{3600\,\text{s}}{\text{h}}\right)$$
$$= \boxed{5.76\,\text{MC}}$$

where N is the number of batteries.

(c) Use the definition of emf to express the total electrical energy available in the batteries:

$$W = Q\mathcal{E} = (5.76\,\text{MC})(12\,\text{V})$$
$$= \boxed{69.1\,\text{MJ}}$$

(d) Relate the amount of work the batteries can do to the work required to overcome friction:

$$W = fd$$

Solve for and evaluate d:

$$d = \frac{W}{f} = \frac{69.1\,\text{MJ}}{1200\,\text{N}} = \boxed{57.6\,\text{km}}$$

(e) Express the cost per kilometer as the ratio of the ratio of the cost of the energy to the distance traveled before recharging:

$$\text{Cost/km} = \frac{(\$0.09/\text{kW} \cdot \text{h})\mathcal{E}It}{d} = \frac{(\$0.09/\text{kW} \cdot \text{h})(120\,\text{V})(160\,\text{A} \cdot \text{h})}{57.6\,\text{km}} = \boxed{\$0.03/\text{km}}$$

Combinations of Resistors

*79 • (a) Find the equivalent resistance between point a and point b in Figure 25-49. (b) If the potential drop between point a and point b is 12 V, find the current in each resistor.

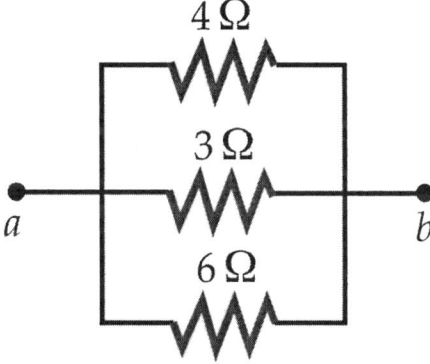

Figure 25-49 Problem 79

Picture the Problem We can either solve this problem by using the expression for the equivalent resistance of three resistors connected in parallel and then using Ohm's law to find the current in each resistor, or we can apply Ohm's law first to find the current through each resistor and then use Ohm's law a second time to find the equivalent resistance of the parallel combination. We'll follow the first procedure.

(a) Express the equivalent resistance of the three resistors in parallel and solve for R_{eq}:

$$\frac{1}{R_{eq}} = \frac{1}{4\,\Omega} + \frac{1}{3\,\Omega} + \frac{1}{6\,\Omega}$$

and

$$R_{eq} = \boxed{1.33\,\Omega}$$

(b) Apply Ohm's law to each of the resistors to find the current flowing through each:

$$I_4 = \frac{V}{4\,\Omega} = \frac{12\,\text{V}}{4\,\Omega} = \boxed{3.00\,\text{A}}$$

$$I_3 = \frac{V}{3\,\Omega} = \frac{12\,\text{V}}{3\,\Omega} = \boxed{4.00\,\text{A}}$$

and

$$I_6 = \frac{V}{6\,\Omega} = \frac{12\,\text{V}}{6\,\Omega} = \boxed{2.00\,\text{A}}$$

Remarks: You would find it instructive to use Kirchhoff's junction rule (conservation of charge) to confirm our values for the currents through the three resistors.

*83 •• A 5-V power supply has an internal resistance of 50 Ω. What is the smallest resistor that we can put in series with the power supply so that the voltage drop across the resistor is larger than 4.5 V?

Picture the Problem Let r represent the resistance of the internal resistance of the power supply, ε the emf of the power supply, R the resistance of the external resistor to be placed in series with the power supply, and I the current drawn from the power supply. We can use Ohm's law to express the potential difference across R and apply Kirchhoff's loop rule to express the current through R in terms of ε, r, and R.

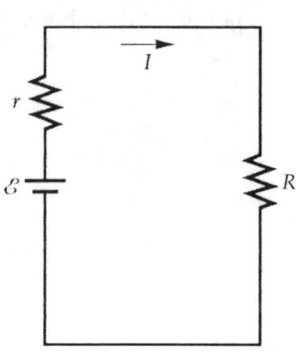

Express the potential difference across the resistor whose resistance is R:

$$V_R = IR \qquad (1)$$

Apply Kirchhoff's loop rule to the circuit to obtain:

$$\varepsilon - Ir - IR = 0$$

Solve for I to obtain:

$$I = \frac{\varepsilon}{r + R}$$

Substitute in equation (1) to obtain:

$$V_R = \left(\frac{\varepsilon}{r + R}\right) R$$

Solve for R to obtain:

$$R = \frac{V_R r}{\varepsilon - V_R}$$

Substitute numerical values and evaluate R:

$$R = \frac{(4.5\,\text{V})(50\,\Omega)}{5\,\text{V} - 4.5\,\text{V}} = \boxed{450\,\Omega}$$

*89 •• A length of wire has a resistance of 120 Ω. The wire is cut into N identical pieces that are then connected in parallel. The resistance of the parallel arrangement is 1.875 Ω. Find N.

Picture the Problem We can use the equation for N identical resistors connected in parallel to relate N to the resistance R of each piece of wire and the equivalent resistance

Express the resistance of the N pieces connected in parallel:

$$\frac{1}{R_{eq}} = \frac{N}{R}$$

where R is the resistance of one of the N pieces.

Relate the resistance of one of the N pieces to the resistance of the wire:

$$R = \frac{R_{wire}}{N}$$

Substitute to obtain:

$$\frac{1}{R_{eq}} = \frac{N^2}{R_{wire}}$$

Solve for N:

$$N = \sqrt{\frac{R_{wire}}{R_{eq}}}$$

Substitute numerical values and evaluate N:

$$N = \sqrt{\frac{120\,\Omega}{1.875\,\Omega}} = \boxed{8}$$

Kirchhoff's Rules

*93 • In Figure 25-56, the emf is 6 V and $R = 0.5$ Ω. The rate of joule heating in R is 8 W. (*a*) What is the current in the circuit? (*b*) What is the potential difference across R? (*c*) What is r?

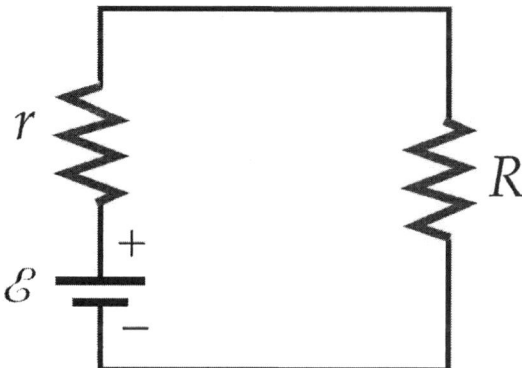

Figure 25-56 Problem 93

116   Chapter 25

Picture the Problem We can relate the current provided by the source to the rate of Joule heating using $P = I^2R$ and use Ohm's law and Kirchhoff's rules to find the potential difference across R and the value of r.

(a) Relate the current I in the circuit to rate at which energy is being dissipated in the form of Joule heat:

$$P = I^2R$$

or

$$I = \sqrt{\frac{P}{R}}$$

Substitute numerical values and evaluate I:

$$I = \sqrt{\frac{8\,\text{W}}{0.5\,\Omega}} = \boxed{4.00\,\text{A}}$$

(b) Apply Ohm's law to find V_R:

$$V_R = IR = (4\,\text{A})(0.5\,\Omega) = \boxed{2.00\,\text{V}}$$

(c) Apply Kirchhoff's loop rule to obtain:

$$\mathcal{E} - Ir - IR = 0$$

Solve for r:

$$r = \frac{\mathcal{E} - IR}{I} = \frac{\mathcal{E}}{I} - R$$

Substitute numerical values and evaluate r:

$$r = \frac{6\,\text{V}}{4\,\text{A}} - 0.5\,\Omega = \boxed{1.00\,\Omega}$$

*97 ••   In the circuit shown in Figure 25-59 the batteries have negligible internal resistance. Find (a) the current in each resistor, (b) the potential difference between point a and point b, and (c) the power supplied by each battery.

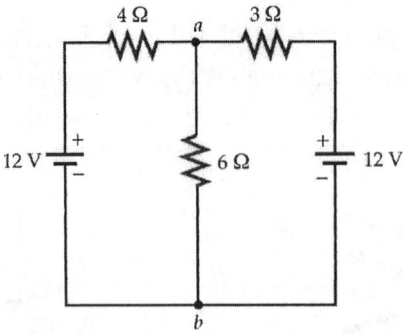

Figure 25-59 Problem 97

Picture the Problem Let I_1 be the current delivered by the left battery, I_2 the current delivered by the right battery, and I_3 the current through the 6-Ω resistor, directed down. We can apply Kirchhoff's rules to obtain three equations that we can solve simultaneously for I_1, I_2, and I_3. Knowing the currents in each branch, we can use Ohm's law to find the potential difference between points a and b and the power delivered by both the sources.

(a) Apply Kirchhoff's junction rule at junction a:

$$I_1 + I_2 = I_3$$

Apply Kirchhoff's loop rule to a loop around the outside of the circuit to obtain:

$$12\,\text{V} - (4\,\Omega)I_1 + (3\,\Omega)I_2 - 12\,\text{V} = 0$$

or

$$-(4\,\Omega)I_1 + (3\,\Omega)I_2 = 0$$

Apply Kirchhoff's loop rule to a loop around the left-hand branch of the circuit to obtain:

$$12\,\text{V} - (4\,\Omega)I_1 - (6\,\Omega)I_3 = 0$$

Solve these equations simultaneously to obtain:

$$I_1 = \boxed{0.667\,\text{A}},$$

$$I_2 = \boxed{0.889\,\text{A}},$$

and

$$I_3 = \boxed{1.56\,\text{A}}$$

(b) Apply Ohm's law to find the potential difference between points a and b:

$$V_{ab} = (6\,\Omega)I_3 = (6\,\Omega)(1.56\,\text{A})$$

$$= \boxed{9.36\,\text{V}}$$

(c) Express the power delivered by the 12-V battery in the left-hand branch of the circuit:

$$P_{\text{left}} = \mathcal{E}I_1$$

$$= (12\,\text{V})(0.667\,\text{A}) = \boxed{8.00\,\text{W}}$$

Express the power delivered by the 12-V battery in the right-hand branch of the circuit:

$$P_{\text{right}} = \mathcal{E}I_2$$

$$= (12\,\text{V})(0.889\,\text{A}) = \boxed{10.7\,\text{W}}$$

*100 •• The circuit fragment shown in Figure 25-61 is called a *voltage divider*. (a) If R_{load} is not attached, show that $V_{\text{out}} = V(R_2/(R_1 + R_2))$. (b) If $R_1 = R_2 = 10\,\text{k}\Omega$, what is the smallest value of R_{load} that can be used so that V_{out} drops by less than 10 percent from its unloaded value? (V_{out} is measured with respect to ground.)

118 Chapter 25

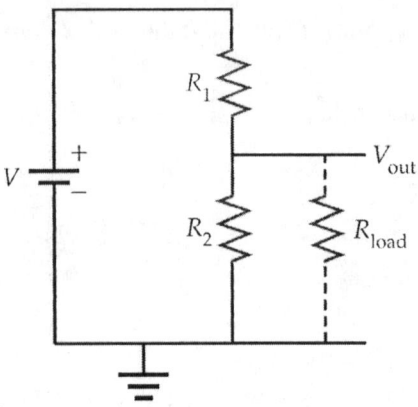

Figure 25-61 Problem 100

Picture the Problem Let the current drawn from the source be I. We can use Ohm's law in conjunction with Kirchhoff's loop rule to express the output voltage as a function of V, R_1, and R_2. In (b) we can use the result of (a) to express the condition on the output voltages in terms of the effective resistance of the loaded output and the resistances R_1 and R_2.

(a) Use Ohm's law to express V_{out} in terms of R_2 and I:

$$V_{out} = IR_2$$

Apply Kirchhoff's loop rule to the circuit to obtain:

$$V - IR_1 - IR_2 = 0$$

Solve for I:

$$I = \frac{V}{R_1 + R_2}$$

Substitute for I in the expression for V_{out} to obtain:

$$V_{out} = \left(\frac{V}{R_1 + R_2}\right) R_2 = \boxed{V\left(\frac{R_2}{R_1 + R_2}\right)}$$

(b) Relate the effective resistance of the loaded circuit R_{eff} to R_2 and R_{load}:

$$\frac{1}{R_{eff}} = \frac{1}{R_2} + \frac{1}{R_{load}}$$

Solve for R_{load}:

$$R_{load} = \frac{R_2 R_{eff}}{R_2 - R_{eff}} \quad (1)$$

Letting V'_{out} represent the output voltage under load, express the condition that V_{out} drops by less than 10 percent of its unloaded value:

$$\frac{V_{out} - V'_{out}}{V_{out}} = 1 - \frac{V'_{out}}{V_{out}} < 0.1 \quad (2)$$

Using the result from (a), express V'_{out} in terms of the effective output load R_{eff}:

$$V'_{out} = V\left(\frac{R_{eff}}{R_1 + R_{eff}}\right)$$

Electric Current and Direct-Current Circuits   119

Substitute for V_{out} and V'_{out} in equation (2) and simplify to obtain:

$$1 - \frac{\frac{R_{eff}}{R_1 + R_{eff}}}{\frac{R_2}{R_1 + R_2}} < 0.1$$

or

$$1 - \frac{R_{eff}(R_1 + R_2)}{R_2(R_1 + R_{eff})} < 0.1$$

Solve for R_{eff}:

$$R_{eff} > \frac{0.9 R_1 R_2}{R_1 + 0.1 R_2}$$

Substitute numerical values and evaluate R_{eff}:

$$R_{eff} > \frac{0.9(10\,k\Omega)(10\,k\Omega)}{10\,k\Omega + 0.1(10\,k\Omega)} = 8.18\,k\Omega$$

Finally, substitute numerical values in equation (1) and evaluate R_{load}:

$$R_{load} < \frac{(10\,k\Omega)(8.18\,k\Omega)}{10\,k\Omega - 8.18\,k\Omega} = \boxed{44.9\,k\Omega}$$

Ammeters and Voltmeters

*105   ••   A digital voltmeter can be modeled as an ideal voltmeter with an infinite internal resistance in parallel with a 10-MΩ resistor. Calculate the voltage measured by the voltmeter in the circuit shown in Figure 25-64 when
(a) $R = 1$ kΩ, (b) $R = 10$ kΩ, (c) $R = 1$ MΩ, (d) $R = 10$ MΩ, and (e) $R = 100$ MΩ. (f) What is the largest value of R possible if we wish the measured voltage to be within 10 percent of the *true* voltage (i.e., the voltage drop without the voltmeter in place)?

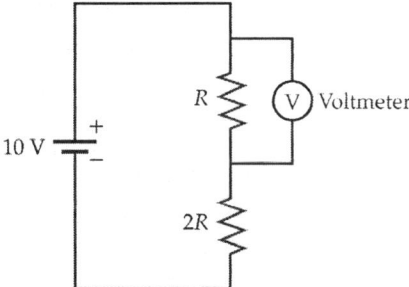

Figure 25-64 Problem 105

Picture the Problem Let I be the current drawn from source and R_{eq} the resistance equivalent to R and 10 MΩ connected in parallel and apply Kirchhoff's loop rule to express the measured voltage V across R as a function of R.

The voltage measured by the voltmeter is given by:

$$V = IR_{eq} \quad (1)$$

Apply Kirchhoff's loop rule to the circuit to obtain:

$$10\,V - IR_{eq} - I(2R) = 0$$

120 Chapter 25

Solve for I:

$$I = \frac{10\,\text{V}}{R_{eq} + 2R}$$

Express R_{eq} in terms of R and 10-MΩ resistance in parallel with it:

$$\frac{1}{R_{eq}} = \frac{1}{10\,\text{M}\Omega} + \frac{1}{R}$$

Solve for R_{eq}:

$$R_{eq} = \frac{(10\,\text{M}\Omega)R}{R + 10\,\text{M}\Omega}$$

Substitute for I in equation (1) and simplify to obtain:

$$V = \left(\frac{10\,\text{V}}{R_{eq} + 2R}\right)R_{eq} = \frac{10\,\text{V}}{1 + \frac{2R}{R_{eq}}}$$

Substitute for R_{eq} and simplify to obtain:

$$V = \frac{(10\,\text{V})(5\,\text{M}\Omega)}{R + 15\,\text{M}\Omega} \quad (2)$$

(a) Evaluate equation (2) for $R = 1\,\text{k}\Omega$:

$$V = \frac{(10\,\text{V})(5\,\text{M}\Omega)}{1\,\text{k}\Omega + 15\,\text{M}\Omega} = \boxed{3.33\,\text{V}}$$

(b) Evaluate equation (2) for $R = 10\,\text{k}\Omega$:

$$V = \frac{(10\,\text{V})(5\,\text{M}\Omega)}{10\,\text{k}\Omega + 15\,\text{M}\Omega} = \boxed{3.33\,\text{V}}$$

(c) Evaluate equation (2) for $R = 1\,\text{M}\Omega$:

$$V = \frac{(10\,\text{V})(5\,\text{M}\Omega)}{1\,\text{M}\Omega + 15\,\text{M}\Omega} = \boxed{3.13\,\text{V}}$$

(d) Evaluate equation (2) for $R = 10\,\text{M}\Omega$:

$$V = \frac{(10\,\text{V})(5\,\text{M}\Omega)}{10\,\text{M}\Omega + 15\,\text{M}\Omega} = \boxed{2.00\,\text{V}}$$

(e) Evaluate equation (2) for $R = 100\,\text{M}\Omega$:

$$V = \frac{(10\,\text{V})(5\,\text{M}\Omega)}{100\,\text{M}\Omega + 15\,\text{M}\Omega} = \boxed{0.435\,\text{V}}$$

(f) Express the condition that the measured voltage to be within 10 percent of the *true* voltage V_{true}:

$$\frac{V_{true} - V}{V_{true}} = 1 - \frac{V}{V_{true}} < 0.1$$

Substitute for V and V_{true} to obtain:

$$1 - \frac{\dfrac{(10\,\text{V})(5\,\text{M}\Omega)}{R + 15\,\text{M}\Omega}}{IR} < 0.1$$

or, because $I = 10\,\text{V}/3R$,

$$1 - \frac{\dfrac{(10\,\text{V})(5\,\text{M}\Omega)}{R + 15\,\text{M}\Omega}}{\dfrac{10}{3}\,\text{V}} < 0.1$$

Solve for R to obtain:

$$R < \frac{1.5\,M\Omega}{0.9} = \boxed{1.67\,M\Omega}$$

*109 •• Show that the meter movement in Problem 106 can be converted into a voltmeter by placing a large resistance in series with the meter movement, and find the resistance needed for a full-scale deflection when 10 V are placed across it.

Picture the Problem The circuit diagram shows a fragment of a circuit in which a resistor of resistance r is connected in series with the meter movement of Problem 106. The purpose of this resistor is to limit the current through the galvanometer movement to 50 μA and to produce a deflection of the galvanometer movement that is a measure of the potential difference V. We can apply Kirchhoff's loop rule to express r in terms of V_g, I_g, and R.

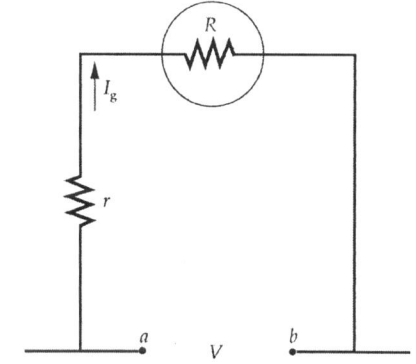

Apply Kirchhoff's loop rule to the circuit fragment to obtain:

$$V - rI_g - RI_g = 0$$

Solve for r:

$$r = \frac{V - RI_g}{I_g} = \frac{V}{I_g} - R \quad (1)$$

Use Ohm's law to relate the current I_g through the galvanometer movement to the potential difference V_g across it:

$$I_g = \frac{V_g}{R} \Rightarrow R = \frac{V_g}{I_g}$$

Use the values for V_g and I_g given in Problem 106 to evaluate R:

$$R = \frac{0.25\,V}{50\,\mu A} = 5000\,\Omega$$

Substitute numerical values in equation (1) and evaluate r:

$$r = \frac{10\,V}{50\,\mu A} - 5000\,\Omega = \boxed{195\,k\Omega}$$

Remarks: The total series resistance is the sum of r and R or 200 kΩ.

RC Circuits

*113 •• In the circuit previously shown in Figure 25-40, emf $\varepsilon = 50$ V and $C = 2.0\,\mu F$; the capacitor is initially uncharged. At 4 s after switch S is closed, the voltage drop across the resistor is 20 V. Find the resistance of the resistor.

Picture the Problem We can find the resistance of the circuit from its time constant and use Ohm's law and the expression for the current in a charging RC circuit to express τ as a function of time, V_0, and $V(t)$.

122 Chapter 25

Express the resistance of the resistor in terms of the time constant of the circuit:

$$R = \frac{\tau}{C} \qquad (1)$$

Using Ohm's law, express the voltage drop across the resistor as a function of time:

$$V(t) = I(t)R$$

Express the current in the circuit as a function of the elapsed time after the switch is closed:

$$I(t) = I_0 e^{-t/\tau}$$

Substitute to obtain:

$$V(t) = I_0 e^{-t/\tau} R = (I_0 R) e^{-t/\tau} = V_0 e^{-t/\tau}$$

Take the natural logarithm of both sides of the equation and solve for τ to obtain:

$$\tau = -\frac{t}{\ln\left[\dfrac{V(t)}{V_0}\right]}$$

Substitute in equation (1) to obtain:

$$R = -\frac{t}{C \ln\left[\dfrac{V(t)}{V_0}\right]}$$

Substitute numerical values and evaluate R using the data given for $t = 4$ s:

$$R = -\frac{4\,\text{s}}{(2\,\mu\text{F})\ln\left(\dfrac{20\,\text{V}}{50\,\text{V}}\right)} = \boxed{2.18\,\text{M}\Omega}$$

*114 •• A 0.12-μF capacitor is given a charge Q_0. After 4 s, the capacitor's charge is $\frac{1}{2}Q_0$. What is the effective resistance across this capacitor?

Picture the Problem We can find the resistance of the circuit from its time constant and use the expression for the charge on a discharging capacitor as a function of time to express τ as a function of time, Q_0, and $Q(t)$.

Express the effective resistance across the capacitor in terms of the time constant of the circuit:

$$R = \frac{\tau}{C} \qquad (1)$$

Express the charge on the capacitor as a function of the elapsed time after the switch is closed:

$$Q(t) = Q_0 e^{-t/\tau}$$

Electric Current and Direct-Current Circuits 123

Take the natural logarithm of both sides of the equation and solve for τ to obtain:
$$\tau = -\frac{t}{\ln\frac{Q(t)}{Q_0}}$$

Substitute in equation (1) to obtain:
$$R = -\frac{t}{C\ln\frac{Q(t)}{Q_0}}$$

Substitute numerical values and evaluate R:
$$R = -\frac{4\,\text{s}}{(0.12\,\mu\text{F})\ln\frac{\frac{1}{2}Q_0}{Q_0}} = \boxed{48.1\,\text{M}\Omega}$$

*120 ••• A photojournalist's flash unit uses a 9-V battery pack to charge a 0.15-μF capacitor, which is then discharged through the flash lamp of 10.5-Ω resistance when a switch is closed. The minimum voltage necessary for the flash discharge is 7 V. The capacitor is charged through an 18-kΩ resistor. (a) How much time is required to charge the capacitor to the required 7 V? (b) How much energy is released when the lamp flashes? (c) How much energy is supplied by the battery during the charging cycle and what fraction of that energy is dissipated in the resistor?

Picture the Problem We can find the time-to-discharge by expressing the voltage across the capacitor as a function of time and solving for t. We can use $U(t) = \frac{1}{2}CV^2(t)$ to find the energy released/stored in the capacitor when the lamp flashes. In part (c) we can integrate $dU_{\text{bat}} = \varepsilon dI(t)$ to find the energy supplied by the battery during the charging cycle.

(a) Express the voltage across the capacitor as a function of time:
$$V(t) = \frac{Q(t)}{C} = \frac{Q_{\text{f}}}{C}\left(1 - e^{-t/RC}\right)$$
$$= V_{\text{f}}\left(1 - e^{-t/RC}\right)$$

Solve for t to obtain:
$$t = -RC\ln\left(1 - \frac{V(t)}{V_{\text{f}}}\right)$$

Substitute numerical values and evaluate t:
$$t = -(18\,\text{k}\Omega)(0.15\,\mu\text{F})\ln\left(1 - \frac{7\,\text{V}}{9\,\text{V}}\right)$$
$$= \boxed{4.06\,\text{ms}}$$

(b) Express the energy stored in the capacitor as a function of time:
$$U(t) = \frac{1}{2}CV^2(t)$$

124   Chapter 25

Substitute for $V(t)$ to obtain:
$$U(t) = \tfrac{1}{2} C V_f^2 \left(1 - e^{-t/RC}\right)^2$$

Substitute numerical values and evaluate $U(4.06\text{ ms})$:

$$U(4.06\text{ ms}) = \tfrac{1}{2}(0.15\,\mu\text{F})(9\text{ V})^2 \left(1 - e^{-4.06\text{ ms}/(18\text{k}\Omega)(0.15\,\mu\text{F})}\right)^2 = \boxed{3.67\,\mu\text{J}}$$

(c) Relate the energy provided by the battery to its emf and the current it delivers:
$$U_{\text{bat}}(t) = \varepsilon \int_0^t I(t')\,dt' = \frac{\varepsilon^2}{R}\int_0^t e^{-t'/RC}\,dt'$$
$$= \frac{\varepsilon^2}{R}\left[RC\left(1 - e^{-t/RC}\right)\right]$$
$$= C\varepsilon^2\left(1 - e^{-t/RC}\right)$$

Substitute numerical values and evaluate $U_{\text{bat}}(4.06\text{ ms})$:

$$U_{\text{bat}}(4.06\text{ ms}) = (0.15\,\mu\text{F})(9\text{ V})^2 \left(1 - e^{-4.06\text{ ms}/(18\text{k}\Omega)(0.15\,\mu\text{F})}\right) = \boxed{9.45\,\mu\text{J}}$$

Express the fraction f of the energy supplied by the battery during the charging cycle that is dissipated in the resistor:
$$f = \frac{U_R}{U_{\text{bat}}}$$

Use conservation of energy to relate the energy supplied by the battery to the energy dissipated in the resistor and the energy released when the lamp flashes:
$$U_{\text{bat}} = U_R + U_{\text{flash}}$$
or
$$U_R = U_{\text{bat}} - U_{\text{flash}}$$

Substitute to obtain:
$$f = \frac{U_{\text{bat}} - U_{\text{flash}}}{U_{\text{bat}}} = 1 - \frac{U_{\text{flash}}}{U_{\text{bat}}}$$

Substitute numerical values and evaluate f:
$$f = 1 - \frac{3.67\,\mu\text{J}}{9.45\,\mu\text{J}} = \boxed{61.2\%}$$

*121 ••• For the circuit shown in Figure 25-68, (a) what is the initial battery current immediately after switch S is closed? (b) What is the battery current a long time after switch S is closed? (c) What is the current in the 600-Ω resistor as a function of time?

Electric Current and Direct-Current Circuits 125

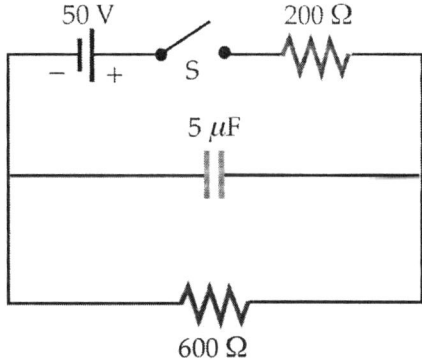

Figure 25-68 Problem 121

Picture the Problem Let $R_1 = 200\ \Omega$, $R_2 = 600\ \Omega$, I_1 and I_2 their currents, and I_3 the current into the capacitor. We can apply Kirchhoff's loop rule to find the initial battery current I_0 and the battery current I_∞ a long time after the switch is closed. In part (c) we can apply both the loop and junction rules to obtain equations that we can use to obtain a linear differential equation with constant coefficients describing the current in the 600-Ω resistor as a function of time. We can solve this differential equation by assuming a solution of a given form, differentiating this assumed solution and substituting it and its derivative in the differential equation. Equating coefficients, requiring the solution to hold for all values of the assumed constants, and invoking an initial condition will allow us to find the constants in the assumed solution.

(a) Apply Kirchhoff's loop rule to the circuit at the instant the switch is closed:

$$\mathcal{E} - (200\,\Omega)I_0 - V_{C0} = 0$$

Because the capacitor is initially uncharged:

$$V_{C0} = 0$$

Solve for and evaluate I_0:

$$I_0 = \frac{\mathcal{E}}{200\,\Omega} = \frac{50\,\text{V}}{200\,\Omega} = \boxed{0.250\,\text{A}}$$

(b) Apply Kirchhoff's loop rule to the circuit after a long time has passed:

$$50\,\text{V} - (200\,\Omega)I_\infty - (600\,\Omega)I_\infty = 0$$

Solve for I_∞ to obtain:

$$I_\infty = \frac{50\,\text{V}}{800\,\Omega} = \boxed{62.5\,\text{mA}}$$

(c) Apply the junction rule at the junction between the 200-Ω resistor

$$I_1 = I_2 + I_3 \qquad (1)$$

126 Chapter 25

and the capacitor to obtain:

Apply the loop rule to the loop containing the source, the 200-Ω resistor and the capacitor to obtain:

$$\varepsilon - R_1 I_1 - \frac{Q}{C} = 0 \qquad (2)$$

Apply the loop rule to the loop containing the 600-Ω resistor and the capacitor to obtain:

$$\frac{Q}{C} - R_2 I_2 = 0 \qquad (3)$$

Differentiate equation (2) with respect to time to obtain:

$$\frac{d}{dt}\left[\varepsilon - R_1 I_1 - \frac{Q}{C}\right] = 0 - R_1 \frac{dI_1}{dt} - \frac{1}{C}\frac{dQ}{dt}$$

$$= -R_1 \frac{dI_1}{dt} - \frac{1}{C} I_3 = 0$$

or

$$R_1 \frac{dI_1}{dt} = -\frac{1}{C} I_3 \qquad (4)$$

Differentiate equation (3) with respect to time to obtain:

$$\frac{d}{dt}\left[\frac{Q}{C} - R_2 I_2\right] = \frac{1}{C}\frac{dQ}{dt} - R_2 \frac{dI_2}{dt} = 0$$

or

$$R_2 \frac{dI_2}{dt} = \frac{1}{C} I_3 \qquad (5)$$

Using equation (1), substitute for I_3 in equation (5) to obtain:

$$\frac{dI_2}{dt} = \frac{1}{R_2 C}(I_1 - I_2) \qquad (6)$$

Solve equation (2) for I_1:

$$I_1 = \frac{\varepsilon - Q/C}{R_1} = \frac{\varepsilon - R_2 I_2}{R_1}$$

Substitute for I_1 in equation (6) and simplify to obtain the differential equation for I_2:

$$\frac{dI_2}{dt} = \frac{1}{R_2 C}\left(\frac{\varepsilon - R_2 I_2}{R_1} - I_2\right)$$

$$= \frac{\varepsilon}{R_1 R_2 C} - \left(\frac{R_1 + R_2}{R_1 R_2 C}\right) I_2$$

To solve this linear differential equation with constant coefficients we can assume a solution of the form:

$$I_2(t) = a + be^{-t/\tau} \qquad (7)$$

Differentiate $I_2(t)$ with respect to time to obtain:	$\dfrac{dI_2}{dt} = \dfrac{d}{dt}\left[a + be^{-t/\tau}\right] = -\dfrac{b}{\tau}e^{-t/\tau}$
Substitute for I_2 and dI_2/dt to obtain:	$-\dfrac{b}{\tau}e^{-t/\tau} = \dfrac{\varepsilon}{R_1 R_2 C} - \left(\dfrac{R_1 + R_2}{R_1 R_2 C}\right)\left(a + be^{-t/\tau}\right)$
Equate coefficients of $e^{-t/\tau}$ to obtain:	$\tau = \dfrac{R_1 R_2 C}{R_1 + R_2}$
Requiring the equation to hold for all values of a yields:	$a = \dfrac{\varepsilon}{R_1 + R_2}$
If I_2 is to be zero when $t = 0$:	$0 = a + b$
	or
	$b = -a = -\dfrac{\varepsilon}{R_1 + R_2}$
Substitute in equation (7) to obtain:	$I_2(t) = \dfrac{\varepsilon}{R_1 + R_2} - \dfrac{\varepsilon}{R_1 + R_2}e^{-t/\tau}$
	$= \dfrac{\varepsilon}{R_1 + R_2}\left(1 - e^{-t/\tau}\right)$
	where
	$\tau = \dfrac{R_1 R_2 C}{R_1 + R_2} = \dfrac{(200\,\Omega)(600\,\Omega)(5\,\mu F)}{200\,\Omega + 600\,\Omega}$
	$= 0.750\,\text{ms}$
Substitute numerical values and evaluate $I_2(t)$:	$I_2(t) = \dfrac{50\,\text{V}}{200\,\Omega + 600\,\Omega}\left(1 - e^{-t/0.750\,\text{ms}}\right)$
	$= \boxed{(62.5\,\text{mA})\left(1 - e^{-t/0.750\,\text{ms}}\right)}$

General Problems

*125 •• In Figure 25-71, $R_1 = 4\,\Omega$, $R_2 = 6\,\Omega$, and $R_3 = 12\,\Omega$. If we denote the currents through these resistors by I_1, I_2, and I_3, respectively, then (a) $I_1 > I_2 > I_3$. (b) $I_2 = I_3$. (c) $I_3 > I_2$. (d) none of the above is correct.

128 Chapter 25

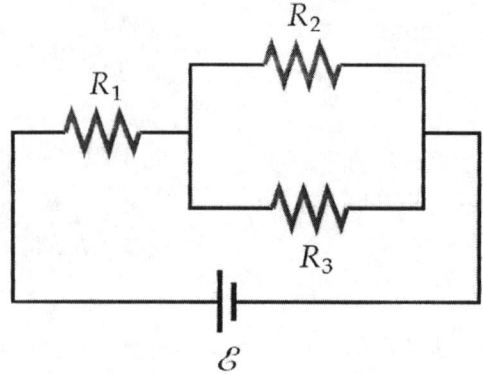

Figure 25-71 Problems 125 and 127

Determine the Concept Because all of the current drawn from the battery passes through R_1, we know that I_1 is greater than I_2 and I_3. Because $R_2 \neq R_3$, $I_2 \neq I_3$ and so (*b*) is false. Because $R_3 > R_2$, $I_3 < I_2$ and so (*c*) is false. $\boxed{(a) \text{ is correct.}}$

*131 •• A closed box has two metal terminals *a* and *b*. The inside of the box contains an unknown emf ε in series with a resistance R. When a potential difference of 21 V is maintained between terminal *a* and terminal *b*, there is a current of 1 A between the terminals *a* and *b*. If this potential difference is reversed, a current of 2 A in the reverse direction is observed. Find ε and R.

Picture the Problem We can apply Kirchhoff's loop rule to the circuit that includes the box and the 21-V source to obtain two equations in the unknowns ε and R that we can solve simultaneously.

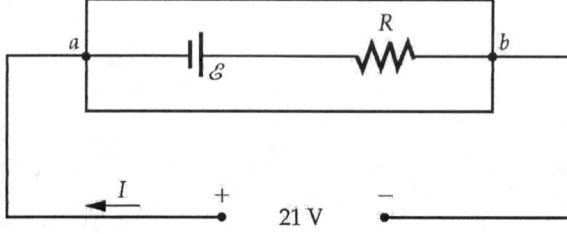

Apply Kirchhoff's loop rule to the circuit when the polarity of the 21-V source and the direction of the current are as shown in the diagram:

$$21\,\text{V} + \varepsilon - (1\,\text{A})R = 0$$

Apply Kirchhoff's loop rule to the circuit when the polarity of the source is reversed and the current is 2 A in the opposite direction:

$$-21\,\text{V} + \varepsilon + (2\,\text{A})R = 0$$

Solve these equations simultaneously to obtain:

$R = \boxed{14.0\,\Omega}$ and $\mathcal{E} = \boxed{-7.00\,\text{V}}$

*133 •• The circuit shown in Figure 25-73 is a slide-type *Wheatstone bridge*. This bridge is used to determine an unknown resistance R_x, in terms of the known resistances R_1, R_2, and R_0. The resistances R_1 and R_2 comprise a wire 1 m long. Point a is a sliding contact that is moved along the wire to vary these resistances. Resistance R_1 is proportional to the distance from the left end of the wire (labeled 0 cm) to point a, and R_2 is proportional to the distance from point a to the right end of the wire (labeled 100 cm). The sum of R_1 and R_2 remains constant. When points a and b are at the same potential, there is no current in the galvanometer and the bridge is said to be balanced. (Because the galvanometer is used to detect the absence of a current, it is called a *null detector*.) If the fixed resistance $R_0 = 200\,\Omega$, find the unknown resistance R_x if (*a*) the bridge balances at the 18-cm mark, (*b*) the bridge balances at the 60-cm mark, and (*c*) the bridge balances at the 95-cm mark.

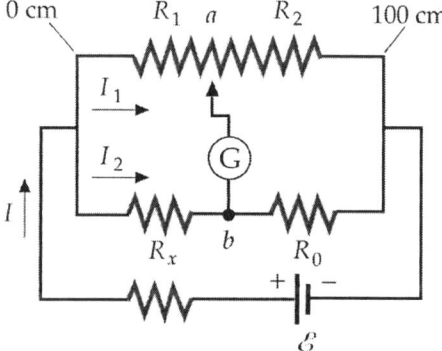

Figure 25-73 Problems 133 and 134

Picture the Problem Let the current flowing through the galvanometer by I_G. By applying Kirchhoff's rules to the loops including 1) R_1, the galvanometer, and R_x, and 2) R_2, the galvanometer, and R_0, we can obtain two equations relating the unknown resistance to R_1, R_2 and R_0. Using $R = \rho L/A$ will allow us to express R_x in terms of the length of wire L_1 that corresponds to R_1 and the length of wire L_2 that corresponds to R_2.

Apply Kirchhoff's loop rule to the loop that includes R_1, the galvanometer, and R_x to obtain:

$$-R_1 I_1 + R_x I_2 = 0 \qquad (1)$$

Apply Kirchhoff's loop rule to the loop that includes R_2, the galvanometer, and R_0 to obtain:

$$-R_2(I_1 - I_G) + R_0(I_2 + I_G) = 0 \qquad (2)$$

When the bridge is balanced,

$$R_1 I_1 = R_x I_2 \qquad (3)$$

130 Chapter 25

$I_G = 0$ and equations (1) and (2) become:	and $R_2 I_1 = R_0 I_2$	(4)
Divide equation (3) by equation (4) and solve for x to obtain:	$R_x = R_0 \dfrac{R_1}{R_2}$	(5)
Express R_1 and R_2 in terms of their lengths, cross-sectional areas, and the resistivity of their wire:	$R_1 = \rho \dfrac{L_1}{A}$ and $R_2 = \rho \dfrac{L_2}{A}$	
Substitute in equation (5) to obtain:	$R_x = R_0 \dfrac{L_1}{L_2}$	

(a) When the bridge balances at the 18-cm mark, $L_1 = 18$ cm, $L_2 = 82$ cm and:

$$R_x = (200\,\Omega)\dfrac{18\,\text{cm}}{82\,\text{cm}} = \boxed{43.9\,\Omega}$$

(b) When the bridge balances at the 60-cm mark, $L_1 = 60$ cm, $L_2 = 40$ cm and:

$$R_x = (200\,\Omega)\dfrac{60\,\text{cm}}{40\,\text{cm}} = \boxed{300\,\Omega}$$

(c) When the bridge balances at the 95-cm mark, $L_1 = 95$ cm, $L_2 = 5$ cm and:

$$R_x = (200\,\Omega)\dfrac{95\,\text{cm}}{5\,\text{cm}} = \boxed{3.80\,\text{k}\Omega}$$

*140 •• (a) Show that a leaky capacitor (one for which the resistance of the dielectric is finite) can be modeled as a capacitor with infinite resistance in parallel with a resistor. (b) Show that the time constant for discharging this capacitor is $\tau = \epsilon_0 \rho \kappa$. (c) Mica has a dielectric constant $\kappa = 5$ and a resistivity $\rho = 9 \times 10^{13}\ \Omega\cdot\text{m}$. Calculate the time it takes for the charge of a mica-filled capacitor to decrease to 10 percent of its initial value.

Picture the Problem We'll assume that the capacitor is fully charged initially and apply Kirchhoff's loop rule to the circuit fragment to obtain the differential equation describing the discharge of the leaky capacitor. We'll show that the solution to this equation is the familiar expression for an exponential decay with time constant $\tau = \epsilon_0 \rho \kappa$.

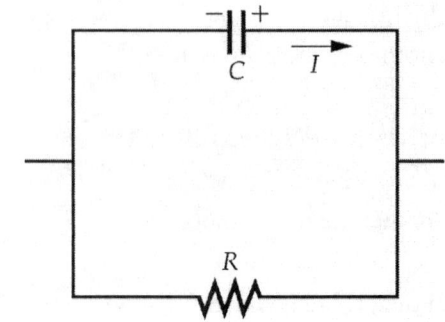

Electric Current and Direct-Current Circuits   131

(a) | If we think of the leaky capacitor as a resistor/capacitor combination, the voltage drop across the resistor must be the same as voltage drop across the capacitor. Hence, they must be in parallel.

(b) Assuming that the capacitor is initially fully charged, apply Kirchhoff's loop rule to the circuit fragment to obtain:

$$\frac{Q}{C} - RI = 0$$

or, because $I = -\frac{dQ}{dt}$,

$$\frac{Q}{C} + R\frac{dQ}{dt} = 0$$

Separate variables in this differential equation to obtain:

$$\frac{dQ}{Q} = -\frac{1}{RC}dt$$

From Problems 138 and 139 we have:

$$RC = \epsilon_0 \, \rho \kappa$$

Substitute for RC in the differential equation to obtain:

$$\frac{dQ}{Q} = -\frac{1}{\epsilon_0 \, \rho \kappa}dt$$

Integrate this equation from $Q' = Q_0$ to Q to obtain:

$$Q = Q_0 e^{-t/\tau}$$

where

$$\tau = \boxed{\epsilon_0 \, \rho \kappa}$$

(c) Because $Q/Q_0 = 0.1$:

$$e^{-t/\tau} = 0.1$$

Solve for t by taking the natural logarithm of both sides of the equation:

$$-\frac{t}{\tau} = \ln 0.1 \Rightarrow t = -\epsilon_0 \, \rho \kappa \ln 0.1$$

Substitute numerical values and evaluate t:

$$t = -(8.85 \times 10^{-12} \, C^2 / N \cdot m^2)(9 \times 10^{13} \, \Omega \cdot m)(5) \ln 0.1 = 9.17 \times 10^3 \, s = \boxed{2.55 \, h}$$

*144 ••• Capacitors C_1 and C_2 are connected in parallel by a resistor and two switches, as shown in Figure 25-76. Capacitor C_1 is initially charged to a voltage V_0, and capacitor C_2 is uncharged. The switches S_1 and S_2 are then closed. (a) What are the final charges on C_1 and C_2? (b) Compare the initial and final stored energies of the system. (c) What caused the decrease in the capacitor-stored energy?

132 Chapter 25

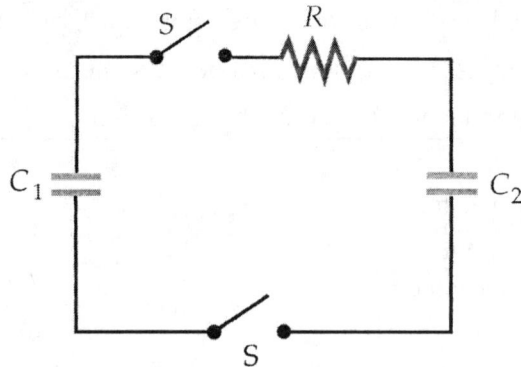

Figure 25-76 Problems 144 and 145

Picture the Problem Let Q_1 and Q_2 represent the final charges on the capacitors C_1 and C_2. Knowing that charge is conserved as it is redistributed to the two capacitors and that the final-state potential differences across the two capacitors will be the same, we can obtain two equations in the unknowns Q_1 and Q_2 that we can solve simultaneously. We can compare the initial and final energies stored in this system by expressing and simplifying their ratio. We can account for any difference between these energies by considering the role of the resistor in the circuit.

(a) Relate the total charge stored initially to the final charges Q_1 and Q_2 on C_1 and C_2:

$$Q = C_1 V_0 = Q_1 + Q_2 \qquad (1)$$

Because, in their final state, the potential differences across the two capacitors will be the same:

$$\frac{Q_1}{C_1} = \frac{Q_2}{C_2} \qquad (2)$$

Solve equation (2) for $Q2$ and substitute in equation (1) to obtain:

$$\frac{C_1}{C_2} Q_2 + Q_2 = C_1 V_0$$

Solve for Q_2 to obtain:

$$\boxed{Q_2 = \frac{C_1 C_2}{C_1 + C_2} V_0}$$

Substitute in either (1) or (2) and solve for Q_1 to obtain:

$$\boxed{Q_1 = \frac{C_1^2}{C_1 + C_2} V_0}$$

Electric Current and Direct-Current Circuits   133

(b) Express the ratio of the initial and final energies of the system:

$$\frac{U_i}{U_f} = \frac{\frac{1}{2}C_1V_0^2}{\frac{1}{2}\frac{Q_1^2}{C_1} + \frac{1}{2}\frac{Q_2^2}{C_2}}$$

$$= \frac{C_1V_0^2}{\frac{\left(\frac{C_1^2}{C_1+C_2}V_0\right)^2}{C_1} + \frac{\left(\frac{C_1C_2}{C_1+C_2}V_0\right)^2}{C_2}}$$

Simplify this expression further to obtain:

$$\frac{U_i}{U_f} = \boxed{1 + \frac{C_2}{C_1}}$$

or U_i is greater than U_f by a factor of $1 + C_2/C_1$.

(c) | The decrease in energy equals the energy dissipated as Joule heat in the resistor connecting the two capacitors. |

*147 ••• The differential resistance of a nonohmic circuit element is defined as $R_d = dV/dI$, where V is the voltage across the element and I is the current through the element. Show that for $V > 0.6$ V, the differential resistance of a diode (Problem 54) is approximately $R_d = (25 \text{ mV})/I$, and for $V < 0$, R_d increases exponentially with $|V|$. Use this result to justify the approximation given in Problem 55.

Picture the Problem We can use the definition of differential resistance and the expression for the diode current given in problem 54 to express R_d and establish the required results.

The differential resistance R_d is given by:

$$R_d = \frac{dV}{dI} = \left(\frac{dI}{dV}\right)^{-1}$$

From Problem 54, the current in the diode is given by:

$$I = I_0\left(e^{V/25\text{mV}} - 1\right) \quad (1)$$

Substitute for I to obtain:

$$R_d = \left\{\frac{d}{dV}\left[I_0\left(e^{V/25\text{mV}} - 1\right)\right]\right\}^{-1}$$

$$= \frac{25\text{ mV}}{I_0}e^{-V/25\text{mV}} \quad (2)$$

For $V > 0.6$ V, equation (1) becomes:

$$I \approx I_0 e^{V/25\text{mV}}$$

Solve for the exponential factor to obtain:

$$e^{V/25\,mV} \approx \frac{I}{I_0} \Rightarrow e^{-V/25\,mV} \approx \frac{I_0}{I}$$

Substitute in equation (2) to obtain:

$$R_d \approx \frac{25\,mV}{I_0} \frac{I_0}{I} = \boxed{\frac{25\,mV}{I}}$$

Examination of equation (2) shows that, for $V < 0$, R_d increases exponentially. This result, together with that for $V > 0.6$ V, justifies the assumptions made in Problem 55.

*151 ••• Calculate the equivalent resistance between points a and b for the infinite ladder of resistors shown in Figure 25-80, where R_1 and R_2 can take any value.

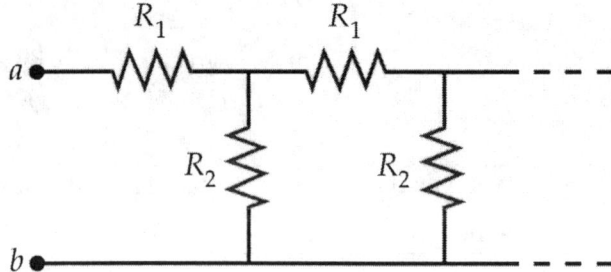

Figure 25-80 Problem 151

Picture the Problem Let R_{eq} be the equivalent resistance of the infinite ladder. If the resistance is finite and non-zero, then adding one or more stages to the ladder will not change the resistance of the network. We can apply the rules for resistance combination to the diagram shown to the right to obtain a quadratic equation in R_{eq} that we can solve for the equivalent resistance between points a and b.

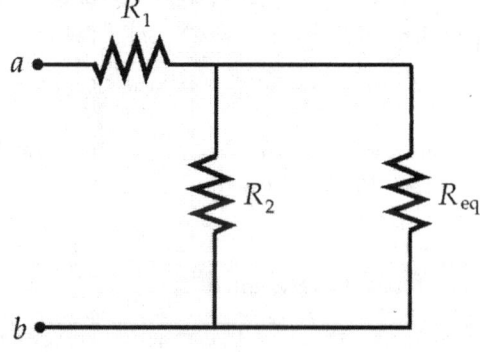

The equivalent resistance of the series combination of R_1 and $(R_2 \| R_{eq})$ is R_{eq}, so:

$$R_{eq} = R_1 + R_2 \| R_{eq} = R_1 + \frac{R_2 R_{eq}}{R_2 + R_{eq}}$$

Simplify to obtain:

$$R_{eq}^2 - R_1 R_{eq} - R_1 R_2 = 0$$

Solve for the positive value of R_{eq} to obtain:

$$R_{eq} = \boxed{\frac{R_1 + \sqrt{R_1^2 + 4R_1 R_2}}{2}}$$

Chapter 26
The Magnetic Field

Conceptual Problems

*1 • When a cathode-ray tube is placed horizontally in a magnetic field that is directed vertically upward, the electrons emitted from the cathode follow one of the dashed paths to the face of the tube in Figure 26-30. The correct path is (*a*) 1. (*b*) 2. (*c*) 3. (*d*) 4. (*e*) 5.

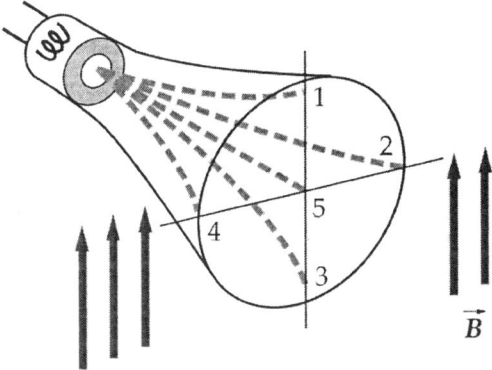

Figure 26-30 Problem 1

Determine the Concept Because the electrons are initially moving at 90° to the magnetic field, they will be deflected in the direction of the magnetic force acting on them. Use the right-hand rule based on the expression for the magnetic force acting on a moving charge $\vec{F} = q\vec{v} \times \vec{B}$, remembering that, for a negative charge, the force is in the direction opposite that indicated by the right-hand rule, to convince yourself that the particle will follow the path whose terminal point on the screen is 2. $\boxed{(b) \text{ is correct.}}$

*5 • A *flicker bulb* is a lightbulb with a long, thin filament. When it is plugged in and a magnet is brought near the lightbulb, the filament is seen to oscillate rapidly back and forth. Why does the filament oscillate, and what is the frequency of oscillation?

Determine the Concept The alternating current running through the filament is changing direction every 1/60 s, so in a magnetic field the filament experiences a force which alternates in direction at that frequency.

*8 • The north-seeking pole of a compass needle located on the magnetic equator is the end of the needle that points toward the north, and the direction of any magnetic field $\vec{B}$ is specified as the direction that the north-seeking pole of a compass needle points when the needle is aligned in the field. Suppose that the direction of the magnetic field $\vec{B}$ were instead specified as the direction of a south-seeking pole of a compass needle aligned in the field. Would the right-hand rule shown in Figure 26-2 then give the direction of the magnetic force on the moving positive charge, or would a left-hand rule be required? Explain.

135

136 Chapter 26

Determine the Concept The direction in which a particle is deflected by a magnetic field will be unchanged by any change in the definition of the direction of the magnetic field. Since we have reversed the direction of the field, we must define the direction in which particles are deflected by a "left-hand" rule instead of a "right-hand" rule.

*12 • An electron moving with speed v to the right enters a region of uniform magnetic field that points out of the paper. When the electron enters this region, it will be (*a*) deflected out of the plane of the paper. (*b*) deflected into the plane of the paper. (*c*) deflected upward. (*d*) deflected downward. (*e*) undeviated in its motion.

Determine the Concept Application of the right-hand rule indicates that a positively charged body would experience a downward force and, in the absence of other forces, be deflected downward. Because the direction of the magnetic force on an electron is opposite that of the force on a positively charged object, an electron will be deflected upward. (c) is correct.

Estimation and Approximation

*16 •• CRTs used in monitors and televisions commonly use magnetic deflection to steer the electron beams. A schematic diagram is shown in Figure 26-32. The electron beam is accelerated through a potential difference and the electron beam is then accelerated through a magnetic field that deflects the electron beam, as shown in the figure. Given the following parameters, estimate the magnitude of the magnetic field needed for maximum deflection: accelerating voltage, $V = 15$ kV; distance over which electron is in magnetic field, $d = 5$ cm; length, $L = 50$ cm; diagonal of CRT, $r = 19$ in.

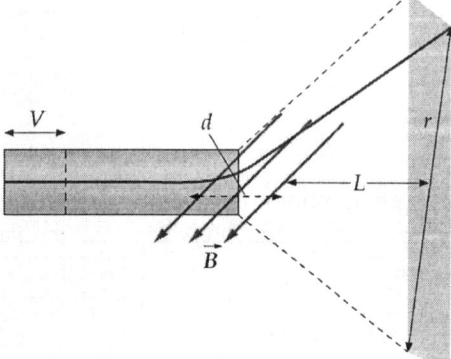

Figure 26-32 Problem 16

Picture the Problem If the electron enters the magnetic field in the coil with speed v, it will travel in a circular path under the influence of the magnetic force acting on. We can apply Newton's 2nd law to the electron in this field to obtain an expression for the magnetic field. We'll assume that the deflection of the electron is small over the distance it travels in the magnetic field, but that, once it is through the region of the magnetic field, it travels at an angle θ with respect to the direction it was originally traveling.

Apply $\sum F = ma_c$ to the electron in the magnetic field to obtain:

$$evB = m\frac{v^2}{r}$$

Solve for B:	$B = \dfrac{mv}{er}$

The kinetic energy of the electron is:	$K = eV = \tfrac{1}{2}mv^2$

Solve for v to obtain:	$v = \sqrt{\dfrac{2eV}{m}}$

Substitute for v in the expression for r:	$B = \dfrac{m}{er}\sqrt{\dfrac{2eV}{m}} = \dfrac{1}{r}\sqrt{\dfrac{2mV}{e}}$

Because $\theta \ll 1$:	$d \approx r\sin\theta \;\Rightarrow\; r \approx \dfrac{d}{\sin\theta}$

Substitute for r in the expression for B to obtain:	$B = \dfrac{\sin\theta}{d}\sqrt{\dfrac{2mV}{e}}$

For maximum deflection, $\theta \approx 45°$. Substitute numerical values and evaluate B:	$B = \dfrac{\sin 45°}{0.05\,\text{m}}\sqrt{\dfrac{2(9.11\times 10^{-31}\,\text{kg})(15\,\text{kV})}{1.60\times 10^{-19}\,\text{C}}}$ $= \boxed{5.84\,\text{mT}}$

Force Exerted by a Magnetic Field

*22 • A straight wire segment $I\vec{L} = (2.7\,\text{A})(3\,\text{cm}\,\hat{i} + 4\,\text{cm}\,\hat{j})$ is in a uniform magnetic field $\vec{B} = 1.3\,\text{T}\,\hat{i}$. Find the force on the wire.

Picture the Problem We can use $\vec{F} = I\vec{L}\times\vec{B}$ to find the force acting on the wire segment.

Express the force acting on the wire segment:	$\vec{F} = I\vec{L}\times\vec{B}$

Substitute numerical values and evaluate $\vec{F}$:	$\vec{F} = (2.7\,\text{A})[(3\,\text{cm})\hat{i} + (4\,\text{cm})\hat{j}]\times(1.3\,\text{T})\hat{i}$ $= \boxed{-(0.140\,\text{N})\hat{k}}$

*26 •• A simple gaussmeter for measuring horizontal magnetic fields consists of a stiff 50-cm wire that hangs from a conducting pivot so that its free end makes contact with a pool of mercury in a dish below. The mercury provides an electrical contact without constraining the movement of the wire. The wire has a mass of 5 g and conducts a current downward. (*a*) What is the equilibrium angular displacement of the wire from vertical if the horizontal magnetic field is 0.04 T and the current is 0.20 A? (*b*) If the

138   Chapter 26

current is 20 A and a displacement from vertical of 0.5 mm can be detected for the free end, what is the horizontal magnetic field sensitivity of this gaussmeter?

Picture the Problem The diagram shows the gaussmeter displaced from equilibrium under the influence of the gravitational and magnetic forces acting on it. We can apply the condition for translational equilibrium in the x direction to find the equilibrium angular displacement of the wire from the vertical. In part (b) we can solve the equation derived in part (a) for B and evaluate this expression for the given data to find the horizontal magnetic field sensitivity of this gaussmeter.

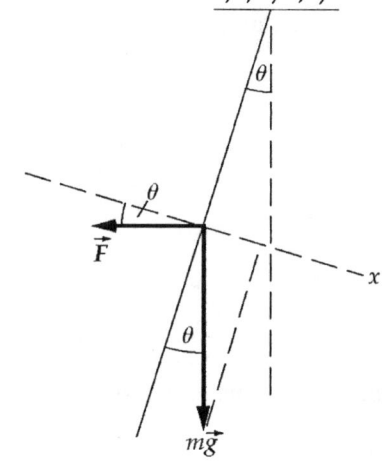

(a) Apply $\sum F_x = 0$ to the wire to obtain:

$$mg \sin\theta - F \cos\theta = 0$$

Substitute for F and solve for θ to obtain:

$$mg \sin\theta - I\ell B \cos\theta = 0 \qquad (1)$$

and

$$\theta = \tan^{-1}\left(\frac{I\ell B}{mg}\right)$$

Substitute numerical values and evaluate θ:

$$\theta = \tan^{-1}\left[\frac{(0.2\,\text{A})(0.5\,\text{m})(0.04\,\text{T})}{(0.005\,\text{kg})(9.81\,\text{m/s}^2)}\right]$$

$$= \boxed{4.66°}$$

(b) Solve equation (1) for B to obtain:

$$B = \frac{mg \tan\theta}{I\ell}$$

For a displacement from vertical of 0.5 mm:

$$\tan\theta \approx \sin\theta = \frac{0.5\,\text{mm}}{0.5\,\text{m}} = 0.001$$

and

$$\theta = 0.001\,\text{rad}$$

Substitute numerical values and evaluate B:

$$B = \frac{(0.005\,\text{kg})(9.81\,\text{m/s}^2)(0.001\,\text{rad})}{(20\,\text{A})(0.5\,\text{m})}$$

$$= \boxed{4.91\,\mu\text{T}}$$

The Magnetic Field

Motion of a Point Charge in a Magnetic Field

*31 • A proton moves in a circular orbit of radius 65 cm perpendicular to a uniform magnetic field of magnitude 0.75 T. (*a*) What is the period for this motion? (*b*) Find the speed of the proton. (*c*) Find the kinetic energy of the proton.

Picture the Problem We can apply Newton's 2nd law to the orbiting proton to relate its speed to its radius. We can then use $T = 2\pi r/v$ to find its period. In Part (*b*) we can use the relationship between T and v to determine v. In Part (*c*) we can use its definition to find the kinetic energy of the proton.

(*a*) Relate the period T of the motion of the proton to its orbital speed v:

$$T = \frac{2\pi r}{v} \quad (1)$$

Apply $\sum F_{radial} = ma_c$ to the proton to obtain:

$$qvB = m\frac{v^2}{r}$$

Solve for v/r to obtain:

$$\frac{v}{r} = \frac{qB}{m}$$

Substitute to obtain:

$$T = \frac{2\pi m}{qB}$$

Substitute numerical values and evaluate T:

$$T = \frac{2\pi(1.67 \times 10^{-27}\,\text{kg})}{(1.60 \times 10^{-19}\,\text{C})(0.75\,\text{T})} = \boxed{87.4\,\text{ns}}$$

(*b*) From equation (1) we have:

$$v = \frac{2\pi r}{T} = \frac{2\pi(0.65\,\text{m})}{87.4\,\text{ns}}$$

$$= \boxed{4.67 \times 10^7\,\text{m/s}}$$

(*c*) Using its definition, express and evaluate the kinetic energy of the proton:

$$K = \tfrac{1}{2}mv^2 = \tfrac{1}{2}(1.67 \times 10^{-27}\,\text{kg})(4.67 \times 10^7\,\text{m/s})^2 = 1.82 \times 10^{-12}\,\text{J} \times \frac{1\,\text{eV}}{1.60 \times 10^{-19}\,\text{J}}$$

$$= \boxed{11.4\,\text{MeV}}$$

*37 •• A beam of particles with velocity $\vec{v}$ enters a region of uniform magnetic field $\vec{B}$ that makes a small angle θ with $\vec{v}$. Show that after a particle moves a distance 2π

140   Chapter 26

$(m/qB)v\cos\theta$, measured along the direction of $\vec{B}$, the velocity of the particle is in the same direction as it was when the particle entered the field.

Picture the Problem The particle's velocity has a component v_1 parallel to $\vec{B}$ and a component v_2 normal to $\vec{B}$. $v_1 = v\cos\theta$ and is constant, whereas $v_2 = v\sin\theta$, being normal to $\vec{B}$, will result in a magnetic force acting on the beam of particles and circular motion perpendicular to $\vec{B}$. We can use the relationship between distance, rate, and time and Newton's 2nd law to express the distance the particle moves in the direction of the field during one period of the motion.

Express the distance moved in the direction of $\vec{B}$ by the particle during one period:
$$x = v_1 T \qquad (1)$$

Express the period of the circular motion of the particles in the beam:
$$T = \frac{2\pi r}{v_2}$$

Apply $\sum F_{radial} = ma_c$ to a particle in the beam to obtain:
$$qv_2 B = m\frac{v_2^2}{r}$$

Solve for v_2:
$$v_2 = \frac{qBr}{m}$$

Substitute to obtain:
$$T = \frac{2\pi r}{\frac{qBr}{m}} = \frac{2\pi m}{qB}$$

Because $v_1 = v\cos\theta$, equation (1) becomes:
$$x = (v\cos\theta)\left(\frac{2\pi m}{qB}\right) = \boxed{2\pi\left(\frac{m}{qB}\right)v\cos\theta}$$

The Velocity Selector

*41 • A velocity selector has a magnetic field of magnitude 0.28 T perpendicular to an electric field of magnitude 0.46 MV/m. (*a*) What must the speed of a particle be for the particle to pass through undeflected? What energy must (*b*) protons and (*c*) electrons have to pass through undeflected?

Picture the Problem Suppose that, for positively charged particles, their motion is from left to right through the velocity selector and the electric field is upward. Then the magnetic force must be downward and the magnetic field out of the page. We can apply the condition for translational equilibrium to relate v to E and B. In (*b*) and (*c*) we can use the definition of kinetic energy to find the energies of protons and electrons that pass through the velocity selector undeflected.

(a) Apply $\sum F_y = 0$ to the particle to obtain:

$F_{elec} - F_{mag} = 0$

or

$qE - qvB = 0$

Solve for v to obtain:

$v = \dfrac{E}{B}$

Substitute numerical values and evaluate v:

$v = \dfrac{0.46\,\text{MV/m}}{0.28\,\text{T}} = \boxed{1.64 \times 10^6 \text{ m/s}}$

(b) Express and evaluate the kinetic energy of protons passing through the velocity selector undeflected:

$K_p = \tfrac{1}{2} m_p v^2$
$= \tfrac{1}{2}(1.67 \times 10^{-27}\,\text{kg})(1.64 \times 10^6 \text{ m/s})^2$
$= 2.26 \times 10^{-15} \text{ J} \times \dfrac{1\,\text{eV}}{1.60 \times 10^{-19}\,\text{J}}$
$= \boxed{14.0\,\text{keV}}$

(c) The kinetic energy of electrons passing through the velocity selector undeflected is given by:

$K_e = \tfrac{1}{2} m_e v^2$
$= \tfrac{1}{2}(9.11 \times 10^{-31}\,\text{kg})(1.64 \times 10^6 \text{ m/s})^2$
$= 1.23 \times 10^{-18} \text{ J} \times \dfrac{1\,\text{eV}}{1.60 \times 10^{-19}\,\text{J}}$
$= \boxed{7.66\,\text{eV}}$

Thomson's Measurement of q/m for Electrons and the Mass Spectrometer

*43 ••   The plates of a Thomson q/m apparatus are 6.0 cm long and are separated by 1.2 cm. The end of the plates is 30.0 cm from the tube screen. The kinetic energy of the electrons is 2.8 keV. (a) If a potential of 25 V is applied across the deflection plates, by how much will the beam deflect? (b) Find the magnitude of the crossed magnetic field that will allow the beam to pass between the plates undeflected.

Picture the Problem Figure 26-18 is reproduced below. We can express the total deflection of the electron beam as the sum of the deflections while the beam is in the field between the plates and its deflection while it is in the field-free space. We can, in turn, use constant-acceleration equations to express each of these deflections. The resulting equation is in terms of v_0 and E. We can find v_0 from the kinetic energy of the beam and E from the potential difference across the plates and their separation. In part (*b*) we can equate the electric and magnetic forces acting on an electron to express B in terms of E and v_0.

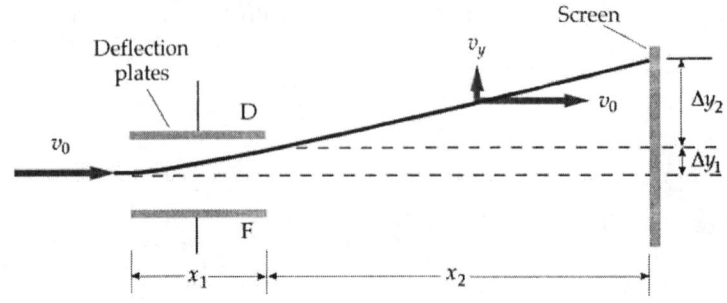

(*a*) Express the total deflection Δy of the electrons:

$$\Delta y = \Delta y_1 + \Delta y_2 \qquad (1)$$

where

Δy_1 is the deflection of the beam while it is in the electric field and Δy_2 is the deflection of the beam while it travels along a straight-line path outside the electric field.

Use a constant-acceleration equation to express Δy_1:

$$\Delta y_1 = \tfrac{1}{2} a_y (\Delta t)^2 \qquad (2)$$

where $\Delta t = x_1/v_0$ is the time an electron is in the electric field between the plates.

Apply Newton's 2nd law to an electron between the plates to obtain:

$$qE = ma_y$$

Solve for a_y and substitute into equation (2) to obtain:

$$a_y = \frac{qE}{m}$$

and

$$\Delta y_1 = \tfrac{1}{2}\left(\frac{qE}{m}\right)\left(\frac{x_1}{v_0}\right)^2 \qquad (3)$$

Express the vertical deflection Δy_2 of the electrons once they are out of the electric field:

$$\Delta y_2 = v_y \Delta t_2 \qquad (4)$$

Use a constant-acceleration equation to find the vertical speed of an electron as it leaves the electric field:

$$v_y = v_{0y} + a_y \Delta t_1$$
$$= 0 + \frac{qE}{m}\left(\frac{x_1}{v_0}\right)$$

Substitute in equation (4) to obtain:

$$\Delta y_2 = \frac{qE}{m}\left(\frac{x_1}{v_0}\right)\left(\frac{x_2}{v_0}\right) = \frac{qEx_1 x_2}{mv_0^2} \quad (5)$$

Substitute equations (3) and (5) in equation (1) to obtain:

$$\Delta y = \frac{1}{2}\left(\frac{qE}{m}\right)\left(\frac{x_1}{v_0}\right)^2 + \frac{qEx_1 x_2}{mv_0^2}$$

or

$$\Delta y = \frac{qEx_1}{mv_0^2}\left(\frac{x_1}{2} + x_2\right) \quad (6)$$

Use the definition of kinetic energy to find the speed of the electrons:

$$K = \tfrac{1}{2} mv_0^2$$

and

$$v_0 = \sqrt{\frac{2K}{m}} = \sqrt{\frac{2(2.8\,\text{keV})}{9.11\times 10^{-31}\,\text{kg}}}$$
$$= 3.14\times 10^7\,\text{m/s}$$

Express the electric field between the plates in terms of their potential difference:

$$E = \frac{V}{d}$$

Substitute numerical values and evaluate E:

$$E = \frac{V}{d} = \frac{25\,\text{V}}{1.2\,\text{cm}} = 2.08\,\text{kV/m}$$

Substitute numerical values in equation (6) and evaluate Δy:

$$\Delta y = \frac{(1.60\times 10^{-19}\,\text{C})(2.08\,\text{kV/m})(6\,\text{cm})}{(9.11\times 10^{-31}\,\text{kg})(31.4\,\text{Mm/s})^2}\left(\frac{6\,\text{cm}}{2} + 30\,\text{cm}\right) = \boxed{7.34\,\text{mm}}$$

(b) Because the electrons are deflected upward, the electric field must be downward and the magnetic field upward. Apply $\sum F_y = 0$ to an electron to obtain:

$$F_{\text{mag}} - F_{\text{elec}} = 0$$

or

$$qvB = qE$$

144   Chapter 26

Solve for B:
$$B = \frac{E}{v}$$

Substitute numerical values and evaluate B:
$$B = \frac{2.08\,\text{kV/m}}{3.14 \times 10^7\,\text{m/s}} = \boxed{66.2\,\mu\text{T}}$$

*46 ••   A beam of ^{6}Li and ^{7}Li ions passes through a velocity selector and enters a magnetic spectrometer. If the diameter of the orbit of the ^{6}Li ions is 15 cm, what is the diameter of the orbit for ^{7}Li ions?

Picture the Problem We can apply Newton's 2nd law to an ion in the magnetic field of the spectrometer to relate the diameter of its orbit to its charge, mass, velocity, and the magnetic field. If we assume that the velocity is the same for the two ions, we can then express the ratio of the two diameters as the ratio of the masses of the ions and solve for the diameter of the orbit of ^{7}Li.

Apply $\sum F_{\text{radial}} = ma_c$ to an ion in the field of the spectrometer:
$$qvB = m\frac{v^2}{r}$$

Solve for r to obtain:
$$r = \frac{mv}{qB}$$

Express the diameter of the orbit:
$$d = \frac{2mv}{qB}$$

Express the diameters of the orbits for ^{6}Li and ^{7}Li:
$$d_6 = \frac{2m_6 v}{qB} \text{ and } d_7 = \frac{2m_7 v}{qB}$$

Assume that the velocities of the two ions are the same and divide the 2nd of these diameters by the first to obtain:
$$\frac{d_7}{d_6} = \frac{\frac{2m_7 v}{qB}}{\frac{2m_6 v}{qB}} = \frac{m_7}{m_6}$$

Solve for and evaluate d_7:
$$d_7 = \frac{m_7}{m_6} d_6 = \frac{7\,\text{u}}{6\,\text{u}}(15\,\text{cm}) = \boxed{17.5\,\text{cm}}$$

The Cyclotron

*49 ••   A cyclotron for accelerating protons has a magnetic field of 1.4 T and a radius of 0.7 m. (*a*) What is the cyclotron frequency? (*b*) Find the maximum energy of the protons when they emerge. (c) How will your answers change if deuterons, which have the same charge but twice the mass, are used instead of protons?

Picture the Problem We can express the cyclotron frequency in terms of the maximum orbital radius and speed of the protons/deuterons. By applying Newton's 2nd law, we can relate the radius of the particle's orbit to its speed and, hence, express the cyclotron frequency as a function of the particle's mass and charge and the cyclotron's magnetic field. In part (b) we can use the definition of kinetic energy and their maximum speed to find the maximum energy of the emerging protons.

(a) Express the cyclotron frequency in terms of the proton's orbital speed and radius:

$$f = \frac{1}{T} = \frac{1}{2\pi r/v} = \frac{v}{2\pi r}$$

Apply $\sum F_{\text{radial}} = ma_c$ to a proton in the magnetic field of the cyclotron:

$$qvB = m\frac{v^2}{r} \quad (1)$$

Solve for r to obtain:

$$r = \frac{mv}{qB}$$

Substitute to obtain:

$$f = \frac{qBv}{2\pi mv} = \frac{qB}{2\pi m} \quad (2)$$

Substitute numerical values and evaluate f:

$$f = \frac{(1.60\times 10^{-19}\,\text{C})(1.4\,\text{T})}{2\pi(1.67\times 10^{-27}\,\text{kg})} = \boxed{21.3\,\text{MHz}}$$

(b) Express the maximum kinetic energy of a proton:

$$K_{\text{max}} = \tfrac{1}{2}mv_{\text{max}}^2$$

Solve equation (1) for v_{max} to obtain:

$$v_{\text{max}} = \frac{qBr_{\text{max}}}{m}$$

Substitute to obtain:

$$K = \tfrac{1}{2}m\left(\frac{qBr_{\text{max}}}{m}\right)^2 = \tfrac{1}{2}\left(\frac{q^2B^2}{m}\right)r_{\text{max}}^2 \quad (3)$$

146   Chapter 26

Substitute numerical values and evaluate K:

$$K = \tfrac{1}{2}\left(\frac{(1.60\times10^{-19}\,\text{C})^2(1.4\,\text{T})^2}{1.67\times10^{-27}\,\text{kg}}\right)(0.7\,\text{m})^2$$

$$= 7.36\times10^{-12}\,\text{J}\times\frac{1\,\text{eV}}{1.60\times10^{-19}\,\text{J}}$$

$$= \boxed{46.0\,\text{MeV}}$$

(c) From equation (2) we see that doubling m halves f:

$$f_{\text{deuterons}} = \tfrac{1}{2}f_{\text{protons}} = \boxed{10.7\,\text{MHz}}$$

From equation (3) we see that doubling m halves K:

$$K_{\text{deuterons}} = \tfrac{1}{2}K_{\text{protons}} = \boxed{23.0\,\text{MeV}}$$

Torques on Current Loops and Magnets

*55 •   A current-carrying wire is bent into the shape of a square of edge-length $L = 6$ cm and is placed in the xy plane. It carries a current $I = 2.5$ A. What is the magnitude of the torque on the wire if there is a uniform magnetic field of 0.3 T (a) in the z direction and (b) in the x direction?

Picture the Problem We can use $\vec{\tau} = \vec{\mu}\times\vec{B}$ to find the torque on the coil in the two orientations of the magnetic field.

Express the torque acting on the coil:

$$\vec{\tau} = \vec{\mu}\times\vec{B}$$

Express the magnetic moment of the coil:

$$\vec{\mu} = \pm IA\hat{k} = \pm IL^2\hat{k}$$

(a) Evaluate $\vec{\tau}$ for $\vec{B}$ in the z direction:

$$\vec{\tau} = \pm IL^2\hat{k}\times B\hat{k}$$
$$= \pm IL^2 B(\hat{k}\times\hat{k}) = \boxed{0}$$

(b) Evaluate $\vec{\tau}$ for $\vec{B}$ in the x direction:

$$\vec{\tau} = \pm IL^2\hat{k}\times B\hat{i} = \pm IL^2 B(\hat{k}\times\hat{i})$$
$$= \pm(2.5\,\text{A})(0.06\,\text{m})^2(0.3\,\text{T})\hat{j}$$
$$= \boxed{\pm(2.70\times10^{-3}\,\text{N}\cdot\text{m})\hat{j}}$$

Magnetic Moments

*60 •• A small magnet of length 6.8 cm is placed at an angle of 60° to the direction of a uniform magnetic field of magnitude 0.04 T. The observed torque has a magnitude of 0.10 N·m. Find the magnetic moment of the magnet.

Picture the Problem Because the small magnet can be modeled as a magnetic dipole; we can use the equation for the torque on a current loop to find its magnetic moment.

Express the magnitude of the torque acting on the magnet:

$$\tau = \mu B \sin \theta$$

Solve for μ to obtain:

$$\mu = \frac{\tau}{B \sin \theta}$$

Substitute numerical values and evaluate μ:

$$\mu = \frac{0.10\,\text{N} \cdot \text{m}}{(0.04\,\text{T})\sin 60°} = \boxed{2.89\,\text{A} \cdot \text{m}^2}$$

*64 ••• A hollow cylinder has length L and inner and outer radii R_i and R_o, respectively (Figure 26-38). The cylinder carries a uniform charge density ρ. Derive an expression for the magnetic moment as a function of ω, the angular velocity of rotation of the cylinder about its axis.

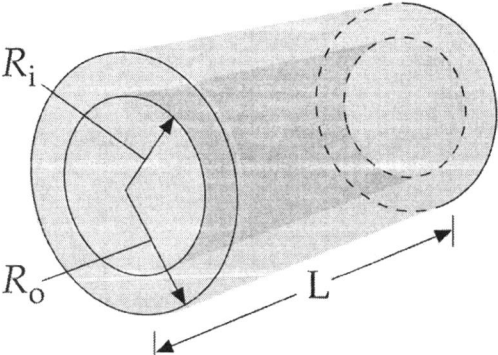

Figure 26-38 Problem 64

Picture the Problem We can express the magnetic moment of an element of charge dq in a cylinder of length L, radius r, and thickness dr, relate this charge to the length, radius, and thickness of the cylinder, express the current due to this rotating charge, substitute for A and dI in our expression for μ and then integrate to complete our derivation for the magnetic moment of the rotating cylinder as a function of its angular velocity.

Express the magnetic moment of an

$$d\mu = A\,dI$$

148   Chapter 26

element of charge dq in a cylinder of length L, radius r, and thickness dr:

where $A = \pi r^2$.

Relate the charge dq in the cylinder to the length of the cylinder, its radius, and thickness:

$$dq = 2\pi L \rho r \, dr$$

Express the current due to this rotating charge:

$$dI = \frac{\omega}{2\pi} dq = \frac{\omega}{2\pi}(2\pi L \rho r \, dr) = L\omega \rho r \, dr$$

Substitute to obtain:

$$d\mu = \pi r^2 (L\omega \rho r \, dr) = L\omega \rho \pi r^3 \, dr$$

Integrate r from R_i to R_0 to obtain:

$$\mu = L\omega \rho \pi \int_{R_i}^{R_0} r^3 \, dr = \boxed{\tfrac{1}{4} L\omega \rho \pi (R_0^4 - R_i^4)}$$

*69 •••   A uniform disk of mass m, radius R, and surface charge σ rotates about its center with angular velocity ω as shown in Figure 26-40. A uniform magnetic field of magnitude $\vec{B}$ threads the disk, making an angle θ with respect to the rotation axis of the disk. Calculate (a) the net torque acting on the disk and (b) the precession frequency of the disk in the magnetic field. (See pages 316-317 for a discussion of precession.)

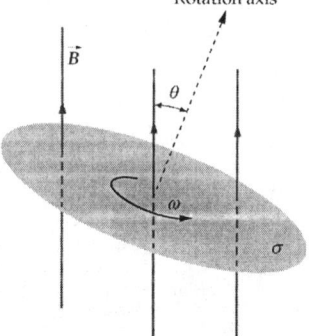

Figure 26-40 Problem 69

Picture the Problem We can use its definition to express the torque acting on the disk and the definition of the precession frequency to find the precession frequency of the disk.

(a) The magnitude of the net torque acting on the disk is:

$\tau = \mu B \sin \theta$

where μ is the magnetic moment of the disk.

From example 26-11:

$$\mu = \frac{1}{4} \pi \sigma r^4 \omega$$

Substitute for μ in the expression for τ to obtain:

$$\tau = \boxed{\frac{1}{4}\pi\sigma r^4 \omega B \sin\theta}$$

(b) The precession frequency Ω is equal to the ratio of the torque divided by the spin angular momentum:

$$\Omega = \frac{\tau}{I\omega}$$

For a solid disk, the moment of inertia is given by:

$$I = \frac{1}{2}mr^2$$

Substitute for τ and I to obtain:

$$\Omega = \frac{\frac{1}{4}\pi\sigma r^4 \omega B \sin\theta}{\frac{1}{2}mr^2\omega} = \boxed{\frac{\pi\sigma r^2 B}{2m}\sin\theta}$$

Remarks: It's interesting that the precession frequency is independent of ω.

The Hall Effect

*72 •• A copper strip ($n = 8.47 \times 10^{22}$ electrons per cubic centimeter) 2-cm wide and 0.1-cm thick is used to measure the magnitudes of unknown magnetic fields that are perpendicular to the strip. Find the magnitude of B when $I = 20$ A and the Hall voltage is (a) 2.00 μV, (b) 5.25 μV, and (c) 8.00 μV.

Picture the Problem We can use $V_H = v_d B w$ to express B in terms of V_H and $I = nqv_d A$ to eliminate the drift velocity v_d and derive an expression for B in terms of V_H, n, and t.

Relate the Hall voltage to the drift velocity and the magnetic field:

$$V_H = v_d B w$$

Solve for B to obtain:

$$B = \frac{V_H}{v_d w}$$

Express the current in the metal strip in terms of the drift velocity of the electrons:

$$I = nqv_d A$$

Solve for v_d to obtain:

$$v_d = \frac{I}{nqA}$$

150   Chapter 26

Substitute and simplify to obtain:
$$B = \frac{V_H}{\frac{I}{nqA}w} = \frac{nqAV_H}{Iw} = \frac{nqwtV_H}{Iw}$$
$$= \frac{nqt}{I}V_H$$

Substitute numerical values and simplify to obtain:

$$B = \frac{(8.47\times10^{22}\text{ cm}^{-3})(1.60\times10^{-19}\text{ C})(0.1\text{ cm})V_H}{20\text{ A}} = (6.78\times10^5\text{ s/m}^2)V_H$$

(a) Evaluate B for $V_H = 2.00$ μV:
$$B = (6.78\times10^5\text{ s/m}^2)(2.00\,\mu\text{V})$$
$$= \boxed{1.36\text{ T}}$$

(b) Evaluate B for $V_H = 5.25$ μV:
$$B = (6.78\times10^5\text{ s/m}^2)(5.25\,\mu\text{V})$$
$$= \boxed{3.56\text{ T}}$$

(c) Evaluate B for $V_H = 8.00$ μV:
$$B = (6.78\times10^5\text{ s/m}^2)(8.00\,\mu\text{V})$$
$$= \boxed{5.42\text{ T}}$$

*75 ••   Aluminum has a density of 2.7×10^3 kg/m^3 and a molar mass of 27 g/mol. The Hall coefficient of aluminum is $R = -0.3\times10^{-10}$ m^3/C. (See Problem 74 for the definition of R.) Find the number of conduction electrons per aluminum atom.

Picture the Problem We can determine the number of conduction electrons per atom from the quotient of the number density of charge carriers and the number of charge carriers per unit volume. Let the width of a slab of aluminum be w and its thickness t. We can use the definition of the Hall electric field in the slab, the expression for the Hall voltage across it, and the definition of current density to find n in terms of R and q and $n_a = \rho N_A/M$, to express n_a.

Express the number of electrons per atom N:
$$N = \frac{n}{n_a} \qquad (1)$$

where n is the number density of charge carriers and n_a is the number of atoms per unit volume.

From the definition of the Hall coefficient we have:
$$R = \frac{E_y}{J_x B_z}$$

Express the Hall electric field in the slab:	$E_y = \dfrac{V_H}{w}$

Express the current density in the slab:	$J_x = \dfrac{I}{w l} = n q v_d$

Substitute to obtain:

$$R = \dfrac{\dfrac{V_H}{w}}{n q v_d B_z} = \dfrac{V_H}{n q v_d w B_z}$$

Express the Hall voltage in terms of v_d, B, and w:

$$V_H = v_d B_z w$$

Substitute and simplify to obtain:

$$R = \dfrac{v_d B_z w}{n q v_d w B_z} = \dfrac{1}{n q}$$

Solve for and evaluate n:

$$n = \dfrac{1}{Rq} \qquad (2)$$

Express the number of atoms n_a per unit volume:

$$n_a = \rho \dfrac{N_A}{M} \qquad (3)$$

Substitute equations (2) and (3) in equation (1) to obtain:

$$N = \dfrac{M}{q R \rho N_A}$$

Substitute numerical values and evaluate N:

$$N = \dfrac{27\,\text{g/mol}}{\left(-1.60\times 10^{-19}\,\text{C}\right)\left(-0.3\times 10^{-10}\,\text{m}^3/\text{C}\right)\left(2.7\times 10^3\,\text{kg/m}^3\right)\left(6.02\times 10^{23}\,\text{atoms/mol}\right)}$$

$$= \boxed{3.46}$$

General Problems

*79 ••   A particle of mass m and charge q enters a region where there is a uniform magnetic field $\vec{B}$ along the x axis. The initial velocity of the particle is $\vec{v} = v_{0x}\hat{i} + v_{0y}\hat{j}$, so the particle moves in a helix. (a) Show that the radius of the helix is $r = mv_{0y}/qB$. (b) Show that the particle takes a time $t = 2\pi m/qB$ to make one orbit around the helix.

Picture the Problem We can use $\vec{F} = q\vec{v}\times\vec{B}$ to show that motion of the particle in the x direction is not affected by the magnetic field. The application of Newton's 2nd law to motion of the particle in yz plane will lead us to the result that $r = mv_{0y}/qB$. By expressing the period of the motion in terms of v_{0y} we can show that the time for one complete orbit around the helix is $t = 2\pi m/qB$.

(a) Express the magnetic force acting on the particle:

$$\vec{F} = q\vec{v} \times \vec{B}$$

Substitute for $\vec{v}$ and $\vec{B}$ and simplify to obtain:

$$\vec{F} = q(v_{0x}\hat{i} + v_{0y}\hat{j}) \times B\hat{i}$$
$$= qv_{0x}B(\hat{i} \times \hat{i}) + qv_{0y}B(\hat{j} \times \hat{i})$$
$$= 0 - qv_{0y}B\hat{k} = -qv_{0y}B\hat{k}$$

i.e., the motion in the direction of the magnetic field (the x direction) is not affected by the field.

Apply $\sum F_{\text{radial}} = ma_c$ to the motion of the particle in the plane perpendicular to $\hat{i}$ (i.e., the yz plane):

$$qv_{0y}B = m\frac{v_{0y}^2}{r} \quad (1)$$

Solve for r:

$$\boxed{r = \frac{mv_{0y}}{qB}}$$

(b) Relate the time for one orbit around the helix to the particle's orbital speed:

$$t = \frac{2\pi r}{v_{0y}}$$

Solve equation (1) for v_{0y}:

$$v_{0y} = \frac{qBr}{m}$$

Substitute and simplify to obtain:

$$t = \frac{2\pi r}{\frac{qBr}{m}} = \boxed{\frac{2\pi m}{qB}}$$

*80 •• A metal crossbar of mass m rides on a pair of long, horizontal conducting rails separated by a distance L and connected to a device that supplies constant current I to the circuit, as shown in Figure 26-42. A uniform magnetic field $\vec{B}$ is established, as shown. (a) If there is no friction and the bar starts from rest at $t = 0$, show that at time t the bar has velocity $v = (BIL/m)t$. (b) In which direction will the bar move? (c) If the coefficient of static friction is μ_s, find the minimum field B necessary to start the bar moving.

The Magnetic Field 153

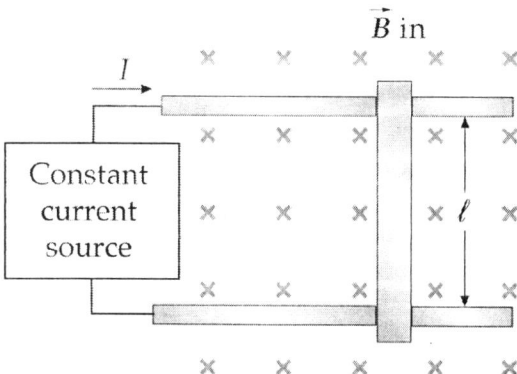

Figure 26-42 Problems 80 and 81

Picture the Problem We can use a constant-acceleration equation to relate the velocity of the crossbar to its acceleration and Newton's 2nd law to express the acceleration of the crossbar in terms of the magnetic force acting on it. We can determine the direction of motion of the crossbar using a right-hand rule or, equivalently, by applying $\vec{F} = I\vec{\ell} \times \vec{B}$. We can find the minimum field B necessary to start the bar moving by applying a condition for static equilibrium to it.

(a) Using a constant-acceleration equation, express the velocity of the bar as a function of its acceleration and the time it has been in motion:

$v = v_0 + at$

or, because $v_0 = 0$,

$v = at$

Use Newton's 2nd law to express the acceleration of the rail:

$a = \dfrac{F}{m}$

where F is the magnitude of the magnetic force acting in the direction of the crossbar's motion.

Substitute to obtain:

$v = \dfrac{F}{m}t$

Express the magnetic force acting on the current-carrying crossbar:

$F = ILB$

Substitute to obtain:

$\boxed{v = \dfrac{ILB}{m}t}$

(b) | Apply to conclude that the magnetic force is to the right and so the motion of the crossbar will also be to the right.

154   Chapter 26

(c) Apply $\sum F_x = 0$ to the crossbar:

$$ILB_{min} - f_{s,max} = 0$$

or

$$ILB_{min} - \mu_s mg = 0$$

Solve for B_{min} to obtain:

$$\boxed{B_{min} = \frac{\mu_s mg}{IL}}$$

*85   •••   A stiff, straight horizontal wire of length 25 cm and mass 20 g is supported by electrical contacts at its ends, but is otherwise free to move vertically upward. The wire is in a uniform, horizontal magnetic field of magnitude 0.4 T perpendicular to the wire. A switch connecting the wire to a battery is closed and the wire flies upward, rising to a maximum height h. The battery delivers a total charge of 2 C during the short time it makes contact with the wire. Find the height h.

Picture the Problem We can use a constant-acceleration equation to express the height to which the wire rises in terms of its initial speed and the acceleration due to gravity. We can then use the impulse-change in momentum equation to express the initial speed of the wire in terms of the impulsive magnetic force acting on it. Finally, we can use the definition of current to relate the charge delivered by the battery to the time during which the impulsive force acts.

Using a constant-acceleration equation, relate the height h to the initial and final speeds and the acceleration of the wire:

$$v^2 = v_0^2 + 2a_y h$$

or, because $v = 0$ and $a_y = g$,

$$0 = v_0^2 - 2gh$$

Solve for h:

$$h = \frac{v_0^2}{2g} \quad (1)$$

Use the impulse-momentum equation to relate the change in momentum of the wire to the impulsive force accelerating it:

$$\Delta p = F\Delta t \text{ or } p_f - p_i = F\Delta t$$

and, because $p_i = 0$, $mv_0 = F\Delta t$

Express the impulsive (magnetic) force acting on the wire:

$$F = I\ell B$$

Substitute to obtain:

$$mv_0 = I\ell B\Delta t$$

Solve for v_0 and substitute in equation (1):

$$h = \frac{\left(\dfrac{I\ell B\Delta t}{m}\right)^2}{2g} = \frac{(I\ell B\Delta t)^2}{2m^2 g}$$

Use the definition of current to relate the charge delivered by the battery to the time during which it delivers the current:

$$\Delta Q = I\Delta t$$

Substitute to obtain:

$$h = \frac{(\ell B\Delta Q)^2}{2m^2 g}$$

Substitute numerical values and evaluate h:

$$h = \frac{[(0.25\,\text{m})(0.4\,\text{T})(2\,\text{C})]^2}{2(0.02\,\text{kg})^2(9.81\,\text{m/s}^2)} = \boxed{5.10\,\text{m}}$$

*88 ••• The special theory of relativity tells us that a particle's mass depends on its speed through the formula:

$$m(v) = \frac{m}{\sqrt{1 - \dfrac{v^2}{c^2}}} = \gamma(v)m$$

where m is the particle's mass and $\gamma(v) = 1/\sqrt{1-(v^2/c^2)}$. (*a*) *Taking into account the special theory of relativity*, what is the radius and period of a particle's orbit if it has speed v and is moving in a magnetic field with magnitude B that is perpendicular to the direction of the velocity? Assume the force on the particle is given by $\vec{F} = q(\vec{v} \times \vec{B})$. The particle has mass m and charge q. (*b*) Using a spreadsheet program, make graphs of the radius and period of the orbit of an electron in a 10-T magnetic field versus $\gamma(v)$ for speeds between $v = 0.1c$ and $v = 0.999c$. Use a logarithmic scale to display $\gamma(v)$.

Picture the Problem We can apply Newton's 2nd law to the particle to derive an expression for the radius of its orbit and then express its period in terms of its orbital speed and radius.

(*a*) Because $\vec{B}$ is perpendicular to $\vec{v}$, the magnitude of force on the particle is given by:

$$F = qvB$$

Apply $\sum F = ma$ to the orbiting particle to obtain:

$$qvB = m(v)\frac{v^2}{r} = \gamma(v)m\frac{v^2}{r}$$

Solve for r:

$$r = \boxed{\frac{\gamma(v)mv}{qB}}$$

156 Chapter 26

The period T of the particle's motion is related to the radius r of its orbit and its orbital speed v:

$$T = \frac{2\pi r}{v}$$

Substitute for r and simplify to obtain:

$$\boxed{T = \frac{2\pi\gamma(v)m}{qB}}$$

(b) A spreadsheet program to calculate r and T as functions of $\ln(\gamma)$ follows. The formulas used to calculate the quantities in the columns are given in the table.

Cell	Content/Formula	Algebraic Form
B1	9.11E–31	m
B2	1.60E–19	e
B3	10	B
B4	3.00E+08	c
A7	0.100	v/c
A8	0.101	$v/c + 0.001$
B7	1/SQRT(1 – (A7)^2)	γ
C7	LN(B7)	$\ln(\gamma)$
D7	B7*$B$1*A7*$B$4/($B$2*$B$3)	$\dfrac{\gamma m v}{qB}$
E7	D7*10^8	$10^6 r$
F7	(2*PI()*A7*$B$1/($B$2*$B$3))*10^12	$\dfrac{2\pi\gamma m}{qB} \times 10^{12}$

	A	B	C	D	E	F
1	m=	9.11E–31	kg			
2	e=	1.60E–19	C			
3	B=	10	T			
4	c=	3.00E+08	m/s			
5						
6	v/c	gamma	ln(gamma)	r	r (microns)	T (ps)
7	0.100	1.0050	0.005	1.72E–05	17.2	0.358
8	0.101	1.0051	0.005	1.73E–05	17.3	0.361
9	0.102	1.0052	0.005	1.75E–05	17.5	0.365
10	0.103	1.0053	0.005	1.77E–05	17.7	0.368
11	0.104	1.0055	0.005	1.79E–05	17.9	0.372
903	0.996	11.1915	2.415	1.90E–03	1904.0	3.563
904	0.997	12.9196	2.559	2.20E–03	2200.2	3.567
905	0.998	15.8193	2.761	2.70E–03	2696.7	3.570
906	0.999	22.3663	3.108	3.82E–03	3816.6	3.574

The following graph of r as a function of $\ln(\gamma)$ was plotted using the data in columns C and E.

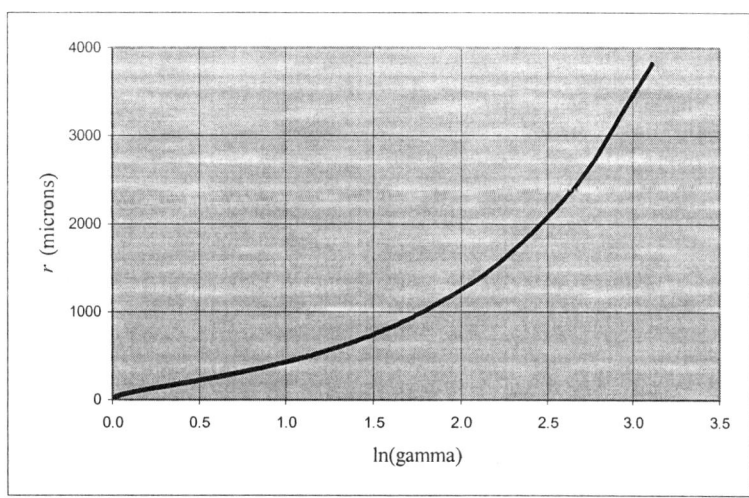

The following graph of T as a function of $\ln(\gamma)$ was plotted using the data in columns C and F.

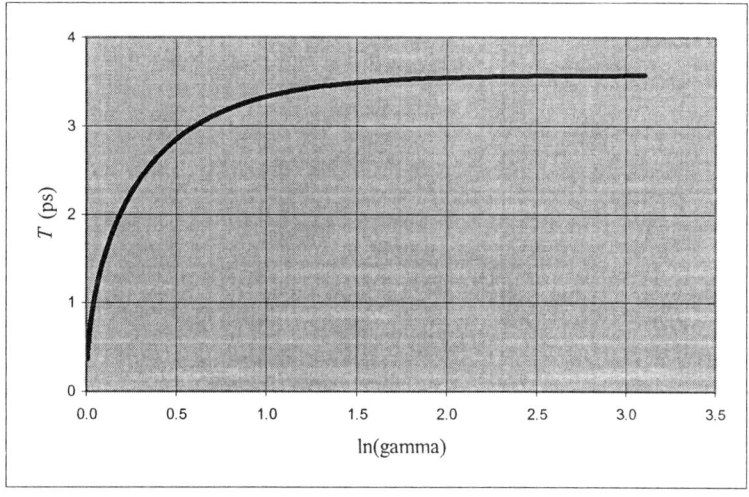

Chapter 27
Sources of the Magnetic Field

Conceptual Problems

*1 • Compare the directions of the electric force and the magnetic force between two positive charges, which move along parallel paths (*a*) in the same direction, and (*b*) in opposite directions.

Picture the Problem The electric forces are described by Coulomb's law and the laws of attraction and repulsion of charges and are independent of the fact the charges are moving. The magnetic interaction is, on the other hand, dependent on the motion of the charges. Each moving charge constitutes a current that creates a magnet field at the location of the other charge.

(*a*) The electric forces are repulsive; the magnetic forces are attractive (the two charges moving in the same direction act like two currents in the same direction).

(*b*) The electric forces are again repulsive; the magnetic forces are also repulsive.

*6 • A wire carries an electrical current straight up. What is the direction of the magnetic field due to the wire a distance of 2 m north of the wire? (*a*) North (*b*) East (*c*) West (*d*) South (*e*) Upward

Determine the Concept Applying the right-hand rule to the wire to the left we see that the magnetic field due to the current points to west at all points north of the wire. $\boxed{(c) \text{ is correct.}}$

*9 • Make a field-line sketch of the magnetic field due to the currents in the pair of coaxial coils (Figure 27-43). The currents in the coils have the same magnitude but are opposite in direction in each coil.

Picture the Problem The field-line sketch is shown below. An assumed direction for the current in the coils is shown in the diagram. Note that the field lines never begin or end and that they do not touch or cross each other. Because there are an uncountable infinity of lines, only a representative few have been shown.

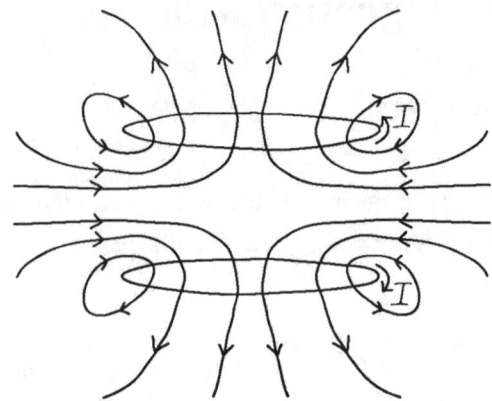

*12 • If the magnetic susceptibility is positive, (a) paramagnetic effects or ferromagnetic effects must be greater than diamagnetic effects. (b) diamagnetic effects must be greater than paramagnetic effects. (c) diamagnetic effects must be greater than ferromagnetic effects. (d) ferromagnetic effects must be greater than paramagnetic effects. (e) paramagnetic effects must be greater than ferromagnetic effects.

Determine the Concept The magnetic susceptibility χ_m is defined by the equation $\vec{M} = \chi_m \dfrac{\vec{B}_{app}}{\mu_0}$, where $\vec{M}$ is the magnetization vector and $\vec{B}_{app}$ is the applied magnetic field. For paramagnetic materials, χ_m is a small positive number that depends on temperature, whereas for diamagnetic materials, it is a small negative constant independent of temperature. $\boxed{(a) \text{ is correct.}}$

*18 • When a current is passed through the wire in Figure 27-44, will the wire tend to bunch up or form a circle?

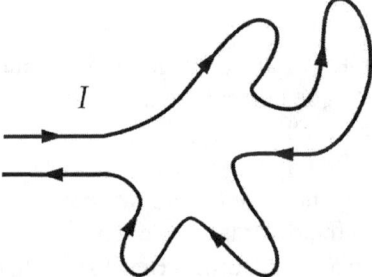

Figure 27-44 Problem 18

Determine the Concept The force per unit length experienced by each segment of the wire, due to the currents in the other segments of the wire, will be equal. These equal forces will result in the wire tending to form a circle.

Sources of the Magnetic Field   161

Estimation and Approximation

*21 ••   Estimate the transient magnetic field 100 m away from a lightning bolt if a charge of about 30 C is transferred from cloud to ground and the average velocity of the charges is 10^6 m/s.

Picture the Problem We can model the lightning bolt as a current in a long wire and use the expression for the magnetic field due to such a current to estimate the transient magnetic field 100 m from the lightning bolt.

The magnetic field due to the current in a long, straight wire is:

$$B = \frac{\mu_0}{4\pi} \frac{2I}{r}$$

where r is the distance from the wire.

Assuming that the height of the cloud is 1 km, the charge transfer will take place in roughly 10^{-3} s and the current associated with this discharge is:

$$I = \frac{\Delta Q}{\Delta t} = \frac{30\,\text{C}}{10^{-3}\,\text{s}} = 3 \times 10^4\,\text{A}$$

Substitute numerical values and evaluate B:

$$B = \frac{4\pi \times 10^{-7}\,\text{N/A}^2}{4\pi} \frac{2(3 \times 10^4\,\text{A})}{100\,\text{m}}$$

$$= \boxed{60.0\,\mu\text{T}}$$

*22 ••   The rotating disk of Problem 125 (page 896) can be used as a model for the magnetic field due to a sunspot. If the sunspot radius is approximately 10^7 m rotating at an angular velocity of about 10^{-2} rad/s, calculate the total charge Q on the sunspot needed to create a magnetic field of order 0.1 T at the center of the sunspot. What is the electrical field magnitude just above the center of the sunspot due to this charge?

Picture the Problem A rotating disk with total charge Q and surface charge density σ is shown in the diagram. We can find Q by deriving an expression for the magnetic field B at the center of the disk due to its rotation. We'll use Ampere's law to express the field dB at the center of the disk due to the element of current dI and then integrate over r to find B.

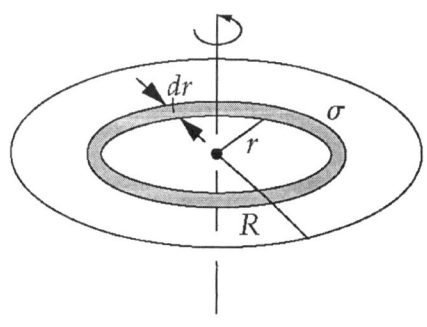

Applying Ampere's law to a circular current loop of radius r we obtain:

$$B = \frac{\mu_0 I}{2r}$$

The B field at the center of an annular ring on a rotating disk of radius r and thickness dr is:

$$dB = \frac{\mu_0}{2r} dI \qquad (1)$$

If σ represents the surface charge density, then the current in the annular ring is given by:

$$dI = \frac{\sigma(2\pi r)}{T}dr, \text{ where } \sigma = \frac{Q}{\pi R^2}$$

Because $T = \frac{2\pi}{\omega}$:

$$dI = \sigma \omega r dr$$

Substitute for dI in equation (1) to obtain:

$$dB = \frac{\mu_0}{2r}\sigma\omega r dr = \frac{\mu_0 \sigma \omega}{2}dr$$

Integrate from $r = 0$ to R to obtain:

$$B = \frac{\mu_0 \sigma \omega}{2}\int_0^R dr = \frac{\mu_0 \sigma \omega R}{2}$$

Substitution for σ yields:

$$B = \frac{\mu_0\left(\frac{Q}{\pi R^2}\right)\omega R}{2} = \frac{\mu_0 Q \omega}{2\pi R}$$

Solve for Q to obtain:

$$Q = \frac{2\pi R B}{\mu_0 \omega}$$

Substitute numerical values and evaluate Q:

$$Q = \frac{2\pi(10^7 \text{ m})(0.1\text{T})}{(4\pi \times 10^{-7} \text{ N/A}^2)(10^{-2} \text{ rad/s})} = \boxed{5.00 \times 10^{14} \text{ C}}$$

The electric field above the sunspot is given by:

$$E = \frac{\sigma}{2\epsilon_0} = \frac{Q}{2\pi \epsilon_0 R^2}$$

Substitute numerical values and evaluate E:

$$E = \frac{5.00 \times 10^{14} \text{ C}}{2\pi(8.85 \times 10^{-12} \text{ C}^2/\text{N}\cdot\text{m}^2)(10^7 \text{ m})^2}$$

$$= \boxed{90.0 \text{ GN/C}}$$

The Magnetic Field of Moving Point Charges

*27 •• Two equal charges q located at $(0, 0, 0)$ and $(0, b, 0)$ at time zero are moving with speed v in the positive x direction ($v \ll c$). Find the ratio of the magnitudes of the magnetic force and the electrostatic force on each charge.

Picture the Problem We can find the ratio of the magnitudes of the magnetic and electrostatic forces by using the expression for the magnetic field of a moving charge and Coulomb's law. Note that $\vec{v}$ and $\vec{r}$, where $\vec{r}$ is the vector from one charge to the other, are at right angles. The field $\vec{B}$ due to the charge at the origin at the location $(0, b, 0)$ is perpendicular to v and $\vec{r}$.

Express the magnitude of the magnetic force on the moving charge at $(0, b, 0)$:

$$F_B = qvB = \frac{\mu_0}{4\pi}\frac{q^2 v^2}{b^2}$$

and, applying the right hand rule, we find that the direction of the force is toward the charge at the origin; i.e., the magnetic force between the two moving charges is attractive.

Express the magnitude of the repulsive electrostatic interaction between the two charges:

$$F_E = \frac{1}{4\pi\varepsilon_0}\frac{q^2}{b^2}$$

Express the ratio of F_B to F_E and simplify to obtain:

$$\frac{F_B}{F_E} = \frac{\frac{\mu_0}{4\pi}\frac{q^2 v^2}{b^2}}{\frac{1}{4\pi\varepsilon_0}\frac{q^2}{b^2}} = \varepsilon_0 \mu_0 v^2 = \boxed{\frac{v^2}{c^2}}$$

where c is the speed of light in a vacuum.

The Magnetic Field of Currents: The Biot-Savart Law

*30 • For the current element in Problem 28, find the magnitude of $d\vec{B}$ and indicate its direction on a diagram at (a) $x = 2$ m, $y = 4$ m, $z = 0$ and (b) $x = 2$ m, $y = 0$, $z = 4$ m.

Picture the Problem We can substitute for $\vec{v}$ and q in the Biot-Savart relationship ($d\vec{B} = \frac{\mu_0}{4\pi}\frac{Id\vec{\ell}\times\hat{r}}{r^2}$), evaluate r and $\hat{r}$ for the given points, and substitute to find $d\vec{B}$.

Express the Biot-Savart law for the given current element:

$$d\vec{B} = \frac{\mu_0}{4\pi}\frac{Id\vec{\ell}\times\hat{r}}{r^2}$$

$$= (10^{-7}\,\text{N/A}^2)\frac{(2\,\text{A})(2\,\text{mm})\hat{k}\times\hat{r}}{r^2}$$

$$= (0.400\,\text{nT}\cdot\text{m}^2)\frac{\hat{k}\times\hat{r}}{r^2}$$

(a) Find r and $\hat{r}$ for the point whose coordinates are (2 m, 4 m, 0):

$$\vec{r} = (2\,\text{m})\hat{i} + (4\,\text{m})\hat{j},$$
$$r = 2\sqrt{5}\,\text{m},$$

164   Chapter 27

and
$$\hat{r} = \frac{2}{2\sqrt{5}}\hat{i} + \frac{4}{2\sqrt{5}}\hat{j} = \frac{1}{\sqrt{5}}\hat{i} + \frac{2}{\sqrt{5}}\hat{j}$$

Evaluate $d\vec{B}$ at (2 m, 4 m, 0):

$$d\vec{B}(2\,\text{m},4\,\text{m},0) = (0.400\,\text{nT}\cdot\text{m}^2)\frac{\hat{k}\times\left(\frac{1}{\sqrt{5}}\hat{i} + \frac{2}{\sqrt{5}}\hat{j}\right)}{(2\sqrt{5}\,\text{m})^2} = \boxed{-(17.9\,\text{pT})\hat{i} + (8.94\,\text{pT})\hat{j}}$$

The diagram is shown to the right:

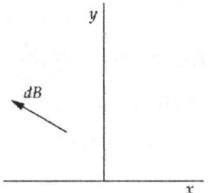

(b) Find r and $\hat{r}$ for the point whose coordinates are (2 m, 0, 4 m):

$\vec{r} = (2\,\text{m})\hat{i} + (4\,\text{m})\hat{k}$,
$r = 2\sqrt{5}\,\text{m}$,
and
$$\hat{r} = \frac{2}{2\sqrt{5}}\hat{i} + \frac{4}{2\sqrt{5}}\hat{k} = \frac{1}{\sqrt{5}}\hat{i} + \frac{2}{\sqrt{5}}\hat{k}$$

Evaluate $d\vec{B}$ at (2 m, 0, 4 m):

$$d\vec{B}(2\,\text{m},0,4\,\text{m}) = (0.400\,\text{nT}\cdot\text{m}^2)\frac{\hat{k}\times\left(\frac{1}{\sqrt{5}}\hat{i} + \frac{2}{\sqrt{5}}\hat{k}\right)}{(2\sqrt{5}\,\text{m})^2} = \boxed{(8.94\,\text{pT})\hat{j}}$$

The diagram is shown to the right:

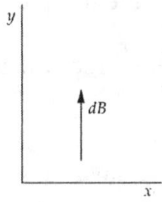

$\vec{B}$ Due to a Current Loop

*32   •   A single-turn circular loop of radius 10.0 cm is to produce a field at its center that will just cancel the earth's magnetic field at the equator, which is 0.7 G directed north. Find the current in the loop and make a sketch showing the orientation of the loop and the current.

Sources of the Magnetic Field 165

Picture the Problem We can solve $B_x = \dfrac{\mu_0}{4\pi} \dfrac{2\pi R^2 I}{(x^2 + R^2)^{3/2}}$ for I with $x = 0$ and substitute the earth's magnetic field at the equator to find the current in the loop that would produce a magnetic field equal to that of the earth.

Express B on the axis of the current loop:

$$B_x = \dfrac{\mu_0}{4\pi} \dfrac{2\pi R^2 I}{(x^2 + R^2)^{3/2}}$$

Solve for I with $x = 0$:

$$I = \dfrac{4\pi}{\mu_0} \dfrac{R}{2\pi} B_x$$

Substitute numerical values and evaluate I:

$$I = \dfrac{1}{(10^{-7}\,\text{N/A}^2)} \dfrac{(0.1\,\text{m})^3}{2\pi(0.1\,\text{m})^2} (0.7\,\text{G})\left(\dfrac{1\,\text{T}}{10^4\,\text{G}}\right) = \boxed{11.1\,\text{A}}$$

The orientation of the loop and current is shown in the sketch:

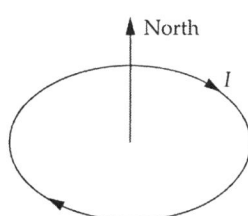

*38 •••   *Anti-Helmholtz* coils are used in many physics applications, such as laser cooling and trapping, where a spatially inhomogeneous field with a uniform gradient is desired. These coils have the same construction as a Helmholtz coil, except that the currents flow in opposite directions, so that the axial fields subtract, and the coil separation is $r\sqrt{3}$ rather than r. Graph the magnetic field as a function of x, the axial distance from the center of the coils, for an anti-Helmholtz coil using the same parameters as in Problem 36.

Picture the Problem Let the origin be midway between the coils so that one of them is centered at $x = -r\sqrt{3}/2$ and the other is centered at $x = r\sqrt{3}/2$. Let the numeral 1 denote the coil centered at $x = -r\sqrt{3}/2$ and the numeral 2 the coil centered at $x = r\sqrt{3}/2$. We can express the magnetic field in the region between the coils as the difference of the magnetic fields B_1 and B_2 due to the two coils.

Express the magnetic field on the x axis due to the coil centered at $x = -r\sqrt{3}/2$:

$$B_1(x) = \dfrac{\mu_0 N r^2 I}{2\left[\left(\dfrac{r\sqrt{3}}{2} + x\right)^2 + r^2\right]^{3/2}}$$

where N is the number of turns.

Express the magnetic field on the x axis due to the coil centered at $x = r\sqrt{3}/2$:

$$B_2(x) = \frac{\mu_0 N r^2 I}{2\left[\left(\frac{r\sqrt{3}}{2} - x\right)^2 + r^2\right]^{3/2}}$$

Subtract these equations to express the total magnetic field along the x axis:

$$B_x(x) = B_1(x) - B_2(x) = \frac{\mu_0 N r^2 I}{2\left[\left(\frac{r\sqrt{3}}{2} + x\right)^2 + r^2\right]^{3/2}} - \frac{\mu_0 N r^2 I}{2\left[\left(\frac{r\sqrt{3}}{2} - x\right)^2 + r^2\right]^{3/2}}$$

$$= \frac{\mu_0 N r^2 I}{2}\left(\left[\left(\frac{r\sqrt{3}}{2} + x\right)^2 + r^2\right]^{-3/2} - \left[\left(\frac{r\sqrt{3}}{2} - x\right)^2 + r^2\right]^{-3/2}\right)$$

The spreadsheet solution is shown below. The formulas used to calculate the quantities in the columns are as follows:

Cell	Formula/Content	Algebraic Form
B1	1.26×10^{-6}	μ_0
B2	0.30	r
B3	250	N
B3	15	I
B5	0.5*$B$1*$B$3*($B$2^2)*$B$4	$\text{Coeff} = \dfrac{\mu_0 N r^2 I}{2}$
A8	−0.30	$-r$
B8	$B$5*(($B$2*SQRT(3)/2+A8)^2 +$B$2^2)^(−3/2)	$\dfrac{\mu_0 N r^2 I}{2}\left[\left(\dfrac{r\sqrt{3}}{2} + x\right)^2 + r^2\right]^{-3/2}$
C8	$B$5* (($B$2*SQRT(3)/2−A8)^2 +$B$2^2)^(−3/2)	$\dfrac{\mu_0 N r^2 I}{2}\left[\left(\dfrac{r\sqrt{3}}{2} - x\right)^2 + r^2\right]^{-3/2}$
D8	10^4*(B8−C8)	$B_x = B_1 - B_2$

	A	B	C	D
1	mu_0=	1.26E−06	N/A^2	
2	r=	0.3	m	
3	N=	250	turns	
4	I=	15	A	
5	Coeff=	2.13E−04		
6				
7	x	B_1	B_2	B(x)
8	−0.30	5.63E−03	1.34E−03	68.4

Sources of the Magnetic Field 167

9	−0.29	5.86E−03	1.41E−03	68.9
10	−0.28	6.08E−03	1.48E−03	69.2
11	−0.27	6.30E−03	1.55E−03	69.2
12	−0.26	6.52E−03	1.62E−03	68.9
13	−0.25	6.72E−03	1.70E−03	68.4
14	−0.24	6.92E−03	1.78E−03	67.5
15	−0.23	7.10E−03	1.87E−03	66.4
61	0.23	1.87E−03	7.10E−03	−66.4
62	0.24	1.78E−03	6.92E−03	−67.5
63	0.25	1.70E−03	6.72E−03	−68.4
64	0.26	1.62E−03	6.52E−03	−68.9
65	0.27	1.55E−03	6.30E−03	−69.2
66	0.28	1.48E−03	6.08E−03	−69.2
67	0.29	1.41E−03	5.86E−03	−68.9
68	0.30	1.34E−03	5.63E−03	−68.4

The following graph of B_x as a function of x was plotted using the data in the above table.

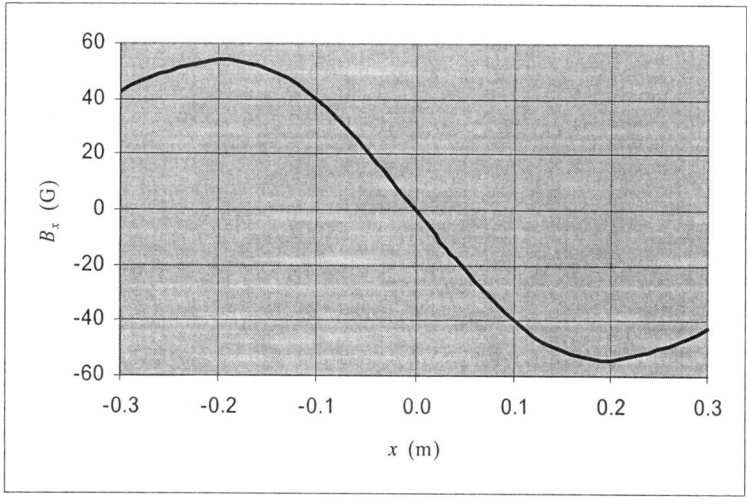

Straight-Line Current Segments

*43 • If the currents in Figure 27-45 are in the negative x direction, find $\vec{B}$ at the points on the y axis at (a) $y = -3$ cm, (b) $y = 0$, (c) $y = +3$ cm, and (d) $y = +9$ cm.

Picture the Problem Let + denote the wire (and current) at $y = +6$ cm and − the wire (and current) at $y = -6$ cm. We can use $B = \dfrac{\mu_0}{4\pi}\dfrac{2I}{R}$ to find the magnetic field due to each of the current carrying wires and superimpose the magnetic fields due to the currents in the wires to find B at the given points on the y axis. We can apply the right-hand rule to find the direction of each of the fields and, hence, of $\vec{B}$.

168   Chapter 27

(a) Express the resultant magnetic field at $y = -3$ cm:

$$\vec{B}(-3\,\text{cm}) = \vec{B}_+(-3\,\text{cm}) + \vec{B}_-(-3\,\text{cm})$$

Find the magnitudes of the magnetic fields at $y = -3$ cm due to each wire:

$$B_+(-3\,\text{cm}) = (10^{-7}\,\text{T}\cdot\text{m/A})\frac{2(20\,\text{A})}{0.09\,\text{m}}$$
$$= 44.4\,\mu\text{T}$$

and

$$B_-(-3\,\text{cm}) = (10^{-7}\,\text{T}\cdot\text{m/A})\frac{2(20\,\text{A})}{0.03\,\text{m}}$$
$$= 133\,\mu\text{T}$$

Apply the right-hand rule to find the directions of $\vec{B}_+$ and $\vec{B}_-$:

$$\vec{B}_+(-3\,\text{cm}) = (44.4\,\mu\text{T})\hat{k}$$
and
$$\vec{B}_-(-3\,\text{cm}) = -(133\,\mu\text{T})\hat{k}$$

Substitute to obtain:

$$\vec{B}(-3\,\text{cm}) = (44.4\,\mu\text{T})\hat{k} - (133\,\mu\text{T})\hat{k}$$
$$= \boxed{-(88.6\,\mu\text{T})\hat{k}}$$

(b) Express the resultant magnetic field at $y = 0$:

$$\vec{B}(0) = \vec{B}_+(0) + \vec{B}_-(0)$$

Because $\vec{B}_+(0) = -\vec{B}_-(0)$:

$$\vec{B}(0) = \boxed{0}$$

(c) Proceed as in (a) to obtain:

$$\vec{B}_+(3\,\text{cm}) = (133\,\mu\text{T})\hat{k},$$
$$\vec{B}_-(3\,\text{cm}) = -(44.4\,\mu\text{T})\hat{k},$$
and
$$\vec{B}(3\,\text{cm}) = (133\,\mu\text{T})\hat{k} - (44.4\,\mu\text{T})\hat{k}$$
$$= \boxed{-(88.6\,\mu\text{T})\hat{k}}$$

(d) Proceed as in (a) with $y = 9$ cm to obtain:

$$\vec{B}_+(9\,\text{cm}) = -(133\,\mu\text{T})\hat{k},$$
$$\vec{B}_-(9\,\text{cm}) = -(26.7\,\mu\text{T})\hat{k},$$
and
$$\vec{B}(9\,\text{cm}) = -(133\,\mu\text{T})\hat{k} - (26.7\,\mu\text{T})\hat{k}$$
$$= \boxed{-(160\,\mu\text{T})\hat{k}}$$

*52 •• Three long, parallel straight wires pass through the corners of an equilateral triangle of sides 10 cm, as shown in Figure 27-47, where a dot means that the current is out of the paper and a cross means that the current is into the paper. If each current is 15 A, find (a) the force per unit length on the upper wire and (b) the magnetic field B at the upper wire due to the two lower wires

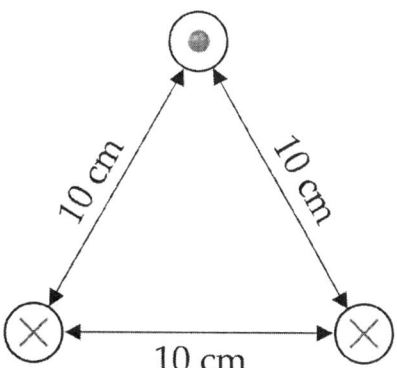

Figure 27-47 Problems 52 and 53

Picture the Problem Note that the forces on the upper wire are away from and directed along the lines to the lower wire and that their horizontal components cancel. We can use $\dfrac{F}{\ell} = 2\dfrac{\mu_0}{4\pi}\dfrac{I^2}{R}$ to find the resultant force in the upward direction (the y direction) acting on the top wire. In part (b) we can use the right-hand rule to determine the directions of the magnetic fields at the upper wire due to the currents in the two lower wires and use $B = \dfrac{\mu_0}{4\pi}\dfrac{2I}{R}$ to find the magnitude of the resultant field due to these currents.

(a) Express the force per unit length each of the lower wires exerts on the upper wire:

$$\dfrac{F}{\ell} = 2\dfrac{\mu_0}{4\pi}\dfrac{I^2}{R}$$

Noting that the horizontal components add up to zero, express the net upward force per unit length on the upper wire:

$$\sum \dfrac{F_y}{\ell} = 2\dfrac{\mu_0}{4\pi}\dfrac{I^2}{R}\cos 30°$$
$$+ 2\dfrac{\mu_0}{4\pi}\dfrac{I^2}{R}\cos 30°$$
$$= 4\dfrac{\mu_0}{4\pi}\dfrac{I^2}{R}\cos 30°$$

Substitute numerical values and evaluate $\sum \dfrac{F_y}{\ell}$:

$$\sum \dfrac{F_y}{\ell} = 4(10^{-7}\,\text{T}\cdot\text{m/A})\dfrac{(15\,\text{A})^2}{0.1\,\text{m}}\cos 30°$$
$$= \boxed{7.79\times 10^{-4}\,\text{N/m}}$$

170 Chapter 27

(b) Noting, from the geometry of the wires, the magnetic field vectors both are at an angle of 30° with the horizontal and that their y components cancel, express the resultant magnetic field:

$$\vec{B} = 2\frac{\mu_0}{4\pi}\frac{2I}{R}\cos 30°\hat{i}$$

Substitute numerical values and evaluate B:

$$B = 2(10^{-7}\,\text{T}\cdot\text{m/A})\frac{2(15\,\text{A})}{0.1\,\text{m}}\cos 30°$$

$$= \boxed{52.0\,\mu\text{T}}$$

*57 •• Four long, straight parallel wires each carry current I. In a plane perpendicular to the wires, the wires are at the corners of a square of side a. Find the force per unit length on one of the wires if (a) all the currents are in the same direction and (b) the currents in the wires at adjacent corners are oppositely directed.

Picture the Problem Choose a coordinate system with its origin at the lower left-hand corner of the square, the positive x axis to the right and the positive y axis upward. Let the numeral 1 denote the wire and current in the upper left-hand corner of the square, the numeral 2 the wire and current in the lower left-hand corner (at the origin) of the square, and the numeral 3 the wire and current in the lower right-hand corner of the square. We can use $B = \frac{\mu_0}{4\pi}\frac{2I}{R}$ and the right-hand rule to find the magnitude and direction of the magnetic field at, say, the upper right-hand corner due to each of the currents, superimpose these fields to find the resultant field, and then use $F = I\ell B$ to find the force per unit length on the wire.

(a) Express the resultant magnetic field at the upper right-hand corner:

$$\vec{B} = \vec{B}_1 + \vec{B}_2 + \vec{B}_3 \qquad (1)$$

When all the currents are into the paper their magnetic fields at the upper right-hand corner are as shown to the right:

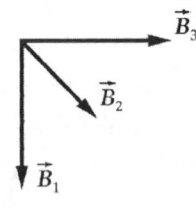

Express the magnetic field due to the current I_1:

$$\vec{B}_1 = -\frac{\mu_0}{4\pi}\frac{2I}{a}\hat{j}$$

Express the magnetic field due to the current I_2:

$$\vec{B}_2 = \frac{\mu_0}{4\pi} \frac{2I}{a\sqrt{2}} \cos 45°(\hat{i} - \hat{j})$$

$$= \frac{\mu_0}{4\pi} \frac{2I}{2a}(\hat{i} - \hat{j})$$

Express the magnetic field due to the current I_3:

$$\vec{B}_3 = \frac{\mu_0}{4\pi} \frac{2I}{a} \hat{i}$$

Substitute in equation (1) and simplify to obtain:

$$\vec{B} = -\frac{\mu_0}{4\pi} \frac{2I}{a} \hat{j} + \frac{\mu_0}{4\pi} \frac{2I}{2a}(\hat{i} - \hat{j}) + \frac{\mu_0}{4\pi} \frac{2I}{a} \hat{i} = \frac{\mu_0}{4\pi} \frac{2I}{a}\left(-\hat{j} + \frac{1}{2}(\hat{i} - \hat{j}) + \hat{i}\right)$$

$$= \frac{\mu_0}{4\pi} \frac{2I}{a}\left[\left(1 + \frac{1}{2}\right)\hat{i} + \left(-1 - \frac{1}{2}\right)\hat{j}\right] = \frac{3\mu_0 I}{4\pi a}[\hat{i} - \hat{j}]$$

Using the expression for the magnetic force on a current-carrying wire, express the force per unit length on the wire at the upper right-hand corner:

$$\frac{F}{\ell} = BI \qquad (2)$$

Substitute to obtain:

$$\frac{\vec{F}}{\ell} = \frac{3\mu_0 I^2}{4\pi a}[\hat{i} - \hat{j}]$$

and

$$\frac{F}{\ell} = \sqrt{\left(\frac{3\mu_0 I^2}{4\pi a}\right)^2 + \left(\frac{3\mu_0 I^2}{4\pi a}\right)^2}$$

$$= \boxed{\frac{3\sqrt{2}\mu_0 I^2}{4\pi a}}$$

(b) When the current in the upper right-hand corner of the square is out of the page, and the currents in the wires at adjacent corners are oppositely directed, the magnetic fields at the upper right-hand are as shown to the right:

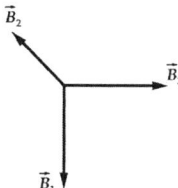

172 Chapter 27

Express the magnetic field at the upper right-hand corner due to the current I_2:

$$\vec{B}_2 = \frac{\mu_0}{4\pi} \frac{2I}{a\sqrt{2}} \cos 45°\left(-\hat{i} + \hat{j}\right)$$

$$= \frac{\mu_0}{4\pi} \frac{2I}{2a}\left(-\hat{i} + \hat{j}\right)$$

Using $\vec{B}_1$ and $\vec{B}_3$ from (a), substitute in equation (1) and simplify to obtain:

$$\vec{B} = -\frac{\mu_0}{4\pi} \frac{2I}{a} \hat{j} + \frac{\mu_0}{4\pi} \frac{2I}{2a}\left(-\hat{i} + \hat{j}\right) + \frac{\mu_0}{4\pi} \frac{2I}{a} \hat{i} = \frac{\mu_0}{4\pi} \frac{2I}{a}\left(-\hat{j} + \frac{1}{2}\left(-\hat{i} + \hat{j}\right) + \hat{i}\right)$$

$$= \frac{\mu_0}{4\pi} \frac{2I}{a}\left[\left(1-\frac{1}{2}\right)\hat{i} + \left(-1+\frac{1}{2}\right)\hat{j}\right] = \frac{\mu_0}{4\pi} \frac{2I}{a}\left[\frac{1}{2}\hat{i} - \frac{1}{2}\hat{j}\right] = \frac{\mu_0 I}{4\pi a}\left[\hat{i} - \hat{j}\right]$$

Substitute in equation (2) to obtain:

$$\frac{\vec{F}}{\ell} = \frac{\mu_0 I^2}{4\pi a}\left[\hat{i} - \hat{j}\right]$$

and

$$\frac{F}{\ell} = \sqrt{\left(\frac{\mu_0 I^2}{4\pi a}\right)^2 + \left(\frac{\mu_0 I^2}{4\pi a}\right)^2}$$

$$= \boxed{\frac{\sqrt{2}\mu_0 I^2}{4\pi a}}$$

$\vec{B}$ Due to a Current in a Solenoid

*60 • A solenoid 2.7-m long has a radius of 0.85 cm and 600 turns. It carries a current I of 2.5 A. What is the approximate magnetic field B on the axis of the solenoid?

Picture the Problem We can use $B_x = \mu_0 n I$ to find the approximate magnetic field on the axis and inside the solenoid.

Express B_x as a function of n and I:

$$B_x = \mu_0 n I$$

Substitute numerical values and evaluate B_x:

$$B_x = \left(4\pi \times 10^{-7} \text{ N/A}^2\right)\left(\frac{600}{2.7 \text{ m}}\right)(2.5 \text{ A})$$

$$= \boxed{0.698 \text{ mT}}$$

Ampère's Law

*63 • A long, straight, thin-walled cylindrical shell of radius R carries a current I. Find B inside the cylinder and outside the cylinder.

Sources of the Magnetic Field   173

Picture the Problem We can apply Ampère's law to a circle centered on the axis of the cylinder and evaluate this expression for $r < R$ and $r > R$ to find B inside and outside the cylinder.

Apply Ampère's law to a circle centered on the axis of the cylinder:

$$\oint_C \vec{B} \cdot d\vec{\ell} = \mu_0 I_C$$

Note that, by symmetry, the field is the same everywhere on this circle.

Evaluate this expression for $r < R$:

$$\oint_C \vec{B}_{inside} \cdot d\vec{\ell} = \mu_0(0) = 0$$

Solve for B_{inside} to obtain:

$$B_{inside} = \boxed{0}$$

Evaluate this expression for $r > R$:

$$\oint_C \vec{B}_{outside} \cdot d\vec{\ell} = B(2\pi R) = \mu_0 I$$

Solve for $B_{outside}$ to obtain:

$$B_{outside} = \boxed{\frac{\mu_0 I}{2\pi R}}$$

*67 •• Show that a uniform magnetic field with no fringing field, such as that shown in Figure 27-50, is impossible because it violates Ampère's law. Do this by applying Ampère's law to the rectangular curve shown by the dashed lines.

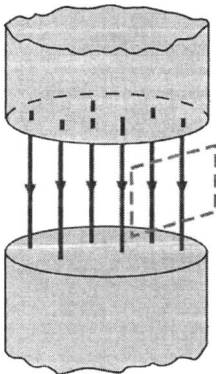

Figure 27-50 Problem 68

Determine the Concept The contour integral consists of four portions, two horizontal portions for which $\oint_C \vec{B} \cdot d\vec{\ell} = 0$, and two vertical portions. The portion within the magnetic field gives a nonvanishing contribution, whereas the portion outside the field gives no contribution to the contour integral. Hence, the contour integral has a finite value. However, it encloses no current; thus, it appears that Ampère's law is violated. What this demonstrates is that there must be a fringing field so that the contour integral does vanish.

174   Chapter 27

*72 •• The *xz* plane contains an infinite sheet of current in the positive *z* direction. The current per unit length (along the *x* direction) is λ. Figure 27-52a shows a point *P* above the sheet ($y > 0$) and two portions of the current sheet labeled I_1 and I_2. (*a*) What is the direction of the magnetic field $\vec{B}$ at *P* due to the two portions of the current shown? (*b*) What is the direction of the magnetic field $\vec{B}$ at point *P* due to the entire sheet? (*c*) What is the direction of $\vec{B}$ at a point below the sheet ($y < 0$)? (*d*) Apply Ampère's law to the rectangular curve shown in Figure 27-52b to show that the magnetic field at any point above the sheet is given by $\vec{B} = -\tfrac{1}{2}\mu_0\lambda\hat{i}$.

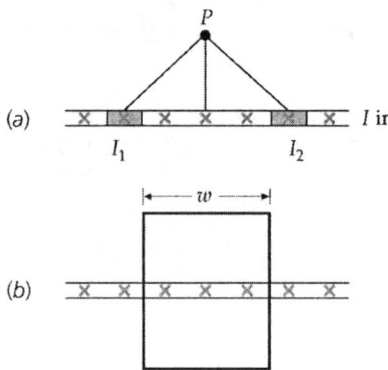

Figure 27-52 Problem 72

Picture the Problem In parts (*a*), (*b*), and (*c*) we can use a right-hand rule to determine the direction of the magnetic field at points above and below the infinite sheet of current. In part (*d*) we can evaluate $\oint_C \vec{B}\cdot d\vec{\ell}$ around the specified path and equate it to $\mu_0 I_C$ and solve for *B*.

(*a*) At *P* the magnetic field points to the right (i.e., in the $-\hat{i}$ direction) since its vertical components cancel.

(*b*) Because the sheet is infinite, the same argument used in (*a*) applies; B is in the $-\hat{i}$ direction.

(*c*) Below the sheet the magnetic field points to the left, i.e., in the $\hat{i}$ direction. The vertical components cancel.

(*d*) Express $\oint_C \vec{B}\cdot d\vec{\ell}$, in the counterclockwise direction, for the given path:

$$\oint_C \vec{B}\cdot d\vec{\ell} = 2\int_{\text{parallel}} \vec{B}\cdot d\vec{\ell} + 2\int_{\perp} \vec{B}\cdot d\vec{\ell}$$

Sources of the Magnetic Field 175

For the paths perpendicular to the sheet, $\vec{B}$ and $d\vec{\ell}$ are perpendicular to each other and:

$$\int_{\perp} \vec{B} \cdot d\vec{\ell} = 0$$

For the paths parallel to the sheet, $\vec{B}$ and $d\vec{\ell}$ are in the same direction and:

$$\int_{parallel} \vec{B} \cdot d\vec{\ell} = Bw$$

Substitute to obtain:

$$\oint_C \vec{B} \cdot d\vec{\ell} = 2 \int_{parallel} \vec{B} \cdot d\vec{\ell} = 2Bw$$
$$= \mu_0 I_C = \mu_0 (\lambda w)$$

Solve for B:

$$B = \tfrac{1}{2}\mu_0 \lambda$$

and

$$\boxed{\vec{B}_{above} = -\tfrac{1}{2}\mu_0 \lambda \hat{i}}$$

Magnetization and Magnetic Susceptibility

*79 •• A cylinder of magnetic material is placed in a long solenoid of n turns per unit length and current I. The values for magnetic field B within the material versus nI are given below. Use these values to plot B versus B_{app} and K_m versus nI.

nI, A/m	0	50	100	150	200	500	1000	10,000
B, T	0	0.04	0.67	1.00	1.2	1.4	1.6	1.7

Picture the Problem We can use the data in the table and $B_{app} = \mu_0 nI$ to plot B versus B_{app}. We can find K_m using $B = K_m B_{app}$.

We can find the applied field B_{app} for a long solenoid using:

$$B_{app} = \mu_0 nI$$

K_m can be found from B_{app} and B using:

$$K_m = \frac{B}{B_{app}}$$

The following graph was plotted using a spreadsheet program. The abscissa values for the graph were obtained by multiplying nI by μ_0. B initially rises rapidly, and then becomes nearly flat. This is characteristic of a ferromagnetic material.

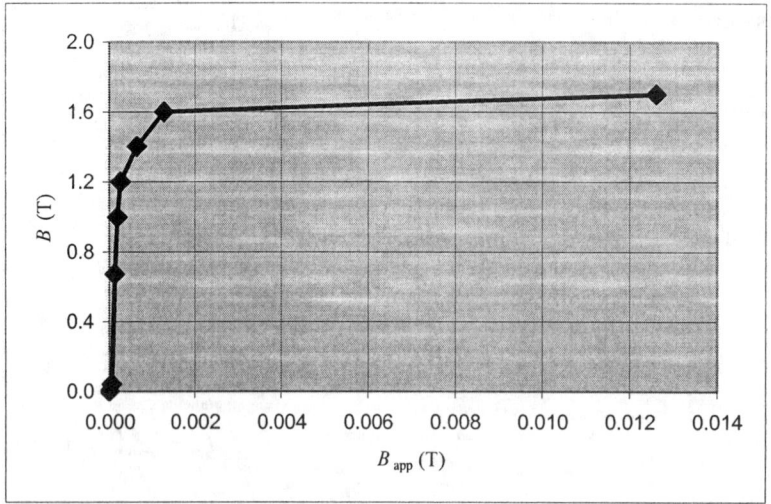

The graph of K_m versus nI shown below was also plotted using a spreadsheet program. Note that K_m becomes quite large for small values of nI but then diminishes. A more revealing graph would be to plot $B/(nI)$, which would be quite large for small values of nI and then drop to nearly zero at $nI = 10{,}000$ A/m, corresponding to saturation of the magnetization.

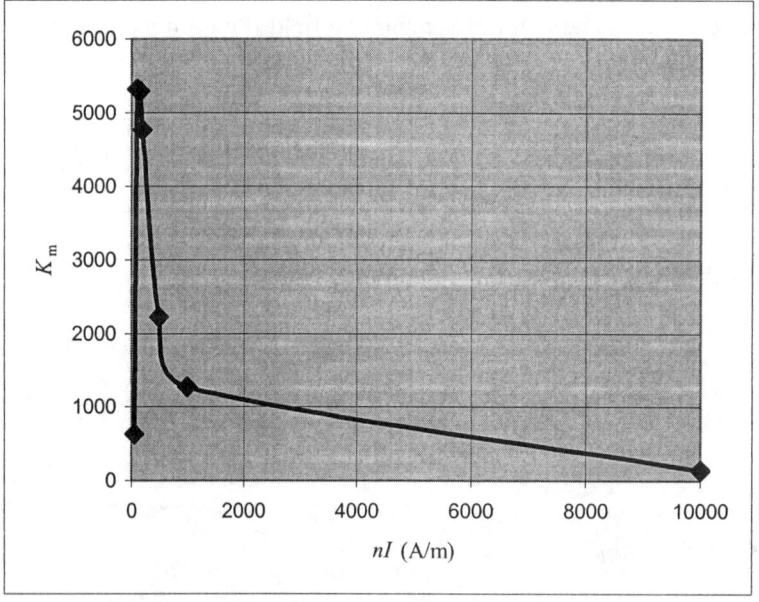

Atomic Magnetic Moments

*82 •• Nickel has a density of 8.7 g/cm^3 and a molecular mass of 58.7 g/mol. Nickel's saturation magnetization is given by $\mu_0 M_s = 0.61$ T. Calculate the magnetic moment of a nickel atom in Bohr magnetons.

Sources of the Magnetic Field 177

Picture the Problem We can find the magnetic moment of a nickel atom μ from its relationship the saturation magnetization M_S using $M_S = n\mu$ where n is the number of molecules. n, in turn, can be found from Avogadro's number, the density of nickel, and its molar mass using $n = \dfrac{N_A \rho}{M}$.

Express the saturation magnetic field in terms of the number of molecules per unit volume and the magnetic moment of each molecule:

$$M_S = n\mu$$
or
$$\mu = \dfrac{M_S}{n}$$

Express the number of molecules per unit volume in terms of Avogadro's number N_A, the molecular mass M, and the density ρ:

$$n = \dfrac{N_A \rho}{M}$$

Substitute and simplify to obtain:

$$\mu = \dfrac{M_S}{\dfrac{N_A \rho}{M}} = \dfrac{\mu_0 M_S}{\dfrac{\mu_0 N_A \rho}{M}} = \dfrac{\mu_0 M_S M}{\mu_0 N_A \rho}$$

Substitute numerical values and evaluate μ:

$$\mu = \dfrac{(0.61\,\text{T})(58.7\times 10^{-3}\,\text{kg/mol})}{(4\pi\times 10^{-7}\,\text{N/A}^2)(6.02\times 10^{23}\,\text{atoms/mol})(8.7\,\text{g/cm}^3)} = 5.44\times 10^{-24}\,\text{A}\cdot\text{m}^2$$

Express the value of 1 Bohr magneton:

$$\mu_B = 9.27\times 10^{-24}\,\text{A}\cdot\text{m}^2$$

Divide μ by μ_B to obtain:

$$\dfrac{\mu}{\mu_B} = \dfrac{5.44\times 10^{-24}\,\text{A}\cdot\text{m}^2}{9.27\times 10^{-24}\,\text{A}\cdot\text{m}^2} = 0.587$$

or
$$\mu = \boxed{0.587\,\mu_B}$$

Paramagnetism

*86 •• Assume that the magnetic moment of an aluminum atom is 1 Bohr magneton. The density of aluminum is 2.7 g/cm^3, and its molecular mass is 27 g/mol. (*a*) Calculate M_s and $\mu_0 M_s$ for aluminum. (*b*) Use the results of Problem 84 to calculate χ_m at T = 300 K. (*c*) Explain why the result for Part (*b*) is larger than the value listed in Table 27-1.

178 Chapter 27

Picture the Problem In (*a*) we can express the saturation magnetic field in terms of the number of molecules per unit volume and the magnetic moment of each molecule and use $n = N_A \rho / M$ to express the number of molecules per unit volume in terms of Avogadro's number N_A, the molecular mass M, and the density ρ. We can use $\chi_m = \mu_0 \mu M_S / 3kT$ from Problem 84 to calculate χ_m.

(*a*) Express the saturation magnetic field in terms of the number of molecules per unit volume and the magnetic moment of each molecule:

$$M_S = n\mu_B$$

Express the number of molecules per unit volume in terms of Avogadro's number N_A, the molecular mass M, and the density ρ:

$$n = \frac{N_A \rho}{M}$$

Substitute to obtain:

$$M_S = \frac{N_A \rho}{M} \mu_B$$

Substitute numerical values and evaluate M_S:

$$M_S = \frac{(6.02 \times 10^{23} \text{ atoms/mol})(2.7 \times 10^3 \text{ kg/m}^3)(9.27 \times 10^{-24} \text{ A} \cdot \text{m}^2)}{27 \text{ g/mol}}$$

$$= \boxed{5.58 \times 10^5 \text{ A/m}}$$

and

$$B_S = \mu_0 M_S = (4\pi \times 10^{-7} \text{ N/A}^2)(5.58 \times 10^5 \text{ A/m}) = \boxed{0.701 \text{ T}}$$

(*b*) From Problem 84 we have:

$$\chi_m = \frac{\mu_0 \mu M_S}{3kT}$$

Substitute numerical values and evaluate χ_m:

$$\chi_m = \frac{(4\pi \times 10^{-7} \text{ N/A}^2)(9.27 \times 10^{-24} \text{ A} \cdot \text{m}^2)(5.58 \times 10^5 \text{ A/m})}{3(1.381 \times 10^{-23} \text{ J/K})(300 \text{ K})} = \boxed{5.23 \times 10^{-4}}$$

(*c*) $\boxed{\text{In calculating } \chi_m \text{ in } (b) \text{ we neglected any diamagnetic effects.}}$

Sources of the Magnetic Field

Ferromagnetism

*90 •   For annealed iron, the relative permeability K_m has its maximum value of approximately 5500 at $B_{app} = 1.57 \times 10^{-4}$ T. Find M and B when K_m is maximum.

Picture the Problem We can use $B = K_m B_{app}$ to find B and $M = (K_m - 1)B_{app}/\mu_0$ to find M.

Express B in terms of M and K_m:	$B = K_m B_{app}$
Substitute numerical values and evaluate B:	$B = (5500)(1.57 \times 10^{-4}\text{ T})$ $= \boxed{0.864\text{ T}}$
Relate M to K_m and B_{app}:	$M = (K_m - 1)\dfrac{B_{app}}{\mu_0} \approx \dfrac{K_m B_{app}}{\mu_0}$
Substitute numerical values and evaluate M:	$M = \dfrac{(5500)(1.57 \times 10^{-4}\text{ T})}{4\pi \times 10^{-7}\text{ N/A}^2}$ $= \boxed{6.87 \times 10^5 \text{ A/m}}$

*96 ••   Two long straight wires 4-cm apart are embedded in a uniform insulator that has a relative permeability of $K_m = 120$. The wires carry 40 A in opposite directions. (*a*) What is the magnetic field at the midpoint of the plane of the wires? (*b*) What is the force per unit length on the wires?

Picture the Problem Because the wires carry equal currents in opposite directions, the magnetic field midway between them will be twice that due to either current alone and will be greater, by a factor of K_m, than it would be in the absence of the insulator. We can use Ampère's law to find the field, due to either current, at the midpoint of the plane of the wires and $d\vec{F} = I d\vec{\ell} \times \vec{B}$ to find the force per unit length on either wire.

(*a*) Relate the magnetic field in the insulator to the magnetic field in its absence:	$B = K_m B_{app}$
Apply Ampère's law to a closed circular path a distance r from a current-carrying wire to obtain:	$\oint_C \vec{B} \cdot d\vec{\ell} = B_{app}(2\pi r) = \mu_0 I_C = \mu_0 I$

180   Chapter 27

Solve for B_{app} to obtain:

$$B_{app} = \frac{\mu_0 I}{2\pi r}$$

Because there are two current carrying wires, with their currents in opposite directions, the fields are additive and:

$$B = 2K_m \frac{\mu_0 I}{2\pi r} = \frac{K_m \mu_0 I}{\pi r}$$

Substitute numerical values and evaluate B:

$$B = \frac{120(4\pi \times 10^{-7} \text{ N/A}^2)(40 \text{ A})}{\pi(0.02 \text{ m})}$$

$$= \boxed{96.0 \text{ mT}}$$

(b) Express the force per unit length experienced by either wire due to the current in the other:

$$\frac{F}{\ell} = BI$$

Apply Ampère's law to obtain:

$$\oint_C \vec{B} \cdot d\vec{\ell} = B(2\pi r) = \mu_0 I_C = \mu_0 I$$

where r is the separation of the wires.

Solve for B:

$$B = \frac{\mu_0 I}{2\pi r} \text{ and } B_{app} = \frac{K_m \mu_0 I}{2\pi r}$$

Substitute to obtain:

$$\frac{F}{\ell} = \frac{K_m \mu_0 I^2}{2\pi r}$$

Substitute numerical values and evaluate $\frac{F}{\ell}$:

$$\frac{F}{\ell} = \frac{120(4\pi \times 10^{-7} \text{ N/A}^2)(40 \text{ A})^2}{2\pi(0.04 \text{ m})}$$

$$= \boxed{0.960 \text{ N/m}}$$

General Problems

*101 • In Figure 27-55, find the magnetic field at point P, which is at the common center of the two semicircular arcs.

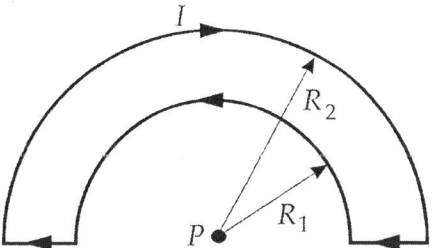

Figure 27-55 Problem 101

Picture the Problem Let out of the page be the positive x direction. Because point P is on the line connecting the straight segments of the conductor, these segments do not contribute to the magnetic field at P. Hence, the resultant magnetic field at P will be the sum of the magnetic fields due to the current in the two semicircles, and we can use the expression for the magnetic field at the center of a current loop to find $\vec{B}_P$.

Express the resultant magnetic field at P:

$$\vec{B}_P = \vec{B}_1 + \vec{B}_2$$

Express the magnetic field at the center of a current loop:

$$B = \frac{\mu_0 I}{2R}$$

where R is the radius of the loop.

Express the magnetic field at the center of half a current loop:

$$B = \frac{1}{2}\frac{\mu_0 I}{2R} = \frac{\mu_0 I}{4R}$$

Express $\vec{B}_1$ and $\vec{B}_2$:

$$\vec{B}_1 = \frac{\mu_0 I}{4R_1}\hat{i}$$

and

$$\vec{B}_2 = -\frac{\mu_0 I}{4R_2}\hat{i}$$

Substitute to obtain:

$$\vec{B}_P = \frac{\mu_0 I}{4R_1}\hat{i} - \frac{\mu_0 I}{4R_2}\hat{i} = \boxed{\frac{\mu_0 I}{4}\left(\frac{1}{R_1} - \frac{1}{R_2}\right)\hat{i}}$$

*104 •• A power cable carrying 50 A is 2 m below the earth's surface, but the cable's direction and precise position are unknown. Show how you could locate the cable using a compass. Assume that you are at the equator, where the earth's magnetic field is 0.7 G north.

Picture the Problem Depending on the direction of the wire, the magnetic field due to its current (provided this field is a large enough fraction of the earth's magnetic field) will either add to or subtract from the earth's field and moving the compass over the ground in the vicinity of the wire will indicate the direction of the current.

Apply Ampère's law to a circle of radius r and concentric with the center of the wire:

$$\oint \vec{B} \cdot d\vec{\ell} = B_{\text{wire}}(2\pi r) = \mu_0 I_C = \mu_0 I$$

Solve for B to obtain:

$$B_{\text{wire}} = \frac{\mu_0 I}{2\pi r}$$

Substitute numerical values and evaluate B_{wire}:

$$B_{\text{wire}} = \frac{(4\pi \times 10^{-7}\ \text{N/A}^2)(50\ \text{A})}{2\pi(2\ \text{m})}$$
$$= 0.0500\ \text{G}$$

Express the ratio of B_{wire} to B_{earth}:

$$\frac{B_{\text{wire}}}{B_{\text{earth}}} = \frac{0.05\ \text{G}}{0.7\ \text{G}} \approx 7\%$$

Thus, the field of the current-carrying wire should be detectable with a good compass.

If the cable runs east-west, its magnetic field is in the north-south direction and thus either adds to or subtracts from the earth's field, depending on the current direction and location of the compass. Moving the compass over the region one should be able to detect the change.

If the cable runs north-south, its magnetic field is perpendicular to that of the earth, and moving the compass about one should observe a change in the direction of the compass needle.

*108 •• A very long straight wire carries a current of 20 A. An electron 1 cm from the center of the wire is moving with a speed of 5.0×10^6 m/s. Find the force on the electron when it moves (*a*) directly away from the wire, (*b*) parallel to the wire in the direction of the current, and (*c*) perpendicular to the wire and tangent to a circle around the wire.

Picture the Problem Chose the coordinate system shown to the right. Then the current is in the positive z direction. Assume that the electron is at (1 cm, 0, 0). We can use $\vec{F} = q\vec{v} \times \vec{B}$ to relate the magnetic force on the electron to $\vec{v}$ and $\vec{B}$ and $\vec{B} = \frac{\mu_0}{4\pi}\frac{2I}{r}\hat{j}$ to express the magnetic field at the location of the electron. We'll need to express $\vec{v}$ for each of the three situations described in the problem in order to evaluate $\vec{F} = q\vec{v} \times \vec{B}$.

Sources of the Magnetic Field 183

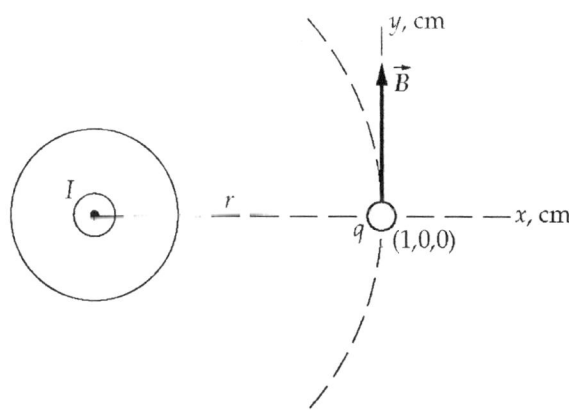

| Express the magnetic force acting on the electron: | $\vec{F} = q\vec{v} \times \vec{B}$ |

| Express the magnetic field due to the current in the wire as a function of distance from the wire: | $\vec{B} = \dfrac{\mu_0}{4\pi} \dfrac{2I}{r} \hat{j}$ |

| Substitute to obtain: | $\vec{F} = q\vec{v} \times \dfrac{\mu_0}{4\pi} \dfrac{2I}{r} \hat{j} = \dfrac{2q\mu_0 I}{4\pi r} (\vec{v} \times \hat{j})$ (1) |

| (a) Express the velocity of the electron when it moves directly away from the wire: | $\vec{v} = v\hat{i}$ |

| Substitute to obtain: | $\vec{F} = \dfrac{2q\mu_0 I}{4\pi r}(v\hat{i} \times \hat{j}) = \dfrac{2q\mu_0 I v}{4\pi r}\hat{k}$ |

Substitute numerical values and evaluate $\vec{F}$:

$$\vec{F} = \dfrac{2(4\pi \times 10^{-7}\,\text{N/A}^2)(-1.6 \times 10^{-19}\,\text{C})(5 \times 10^6\,\text{m/s})(20\,\text{A})\hat{k}}{4\pi(0.01\,\text{m})}$$

$$= \boxed{(-3.20 \times 10^{-16}\,\text{N})\hat{k}}$$

| (b) Express $\vec{v}$ when the electron is traveling parallel to the wire in the direction of the current: | $\vec{v} = v\hat{k}$ |

| Substitute in equation (1) to obtain: | $\vec{F} = \dfrac{2q\mu_0 I}{4\pi r}(v\hat{k} \times \hat{j}) = -\dfrac{2q\mu_0 I v}{4\pi r}\hat{i}$ |

184   Chapter 27

Substitute numerical values and evaluate $\vec{F}$:

$$\vec{F} = -\frac{2(4\pi \times 10^{-7} \text{ N/A}^2)(-1.6 \times 10^{-19} \text{ C})(5 \times 10^6 \text{ m/s})(20 \text{ A})\hat{i}}{4\pi(0.01\text{ m})} = \boxed{(3.20 \times 10^{-16} \text{ N})\hat{i}}$$

(c) Express $\vec{v}$ when the electron is traveling perpendicular to the wire and tangent to a circle around the wire:

$$\vec{v} = v\hat{j}$$

Substitute in equation (1) to obtain:

$$\vec{F} = \frac{2q\mu_0 I}{4\pi r}(v\hat{j} \times \hat{j}) = \boxed{0}$$

*111   ••   Figure 27-60 shows a bar magnet suspended by a thin wire that provides a restoring torque $-\kappa\theta$. The magnet is 16 cm long, has a mass of 0.8 kg, a dipole moment of $\mu = 0.12$ A·m^2, and it is located in a region where a uniform magnetic field B can be established. When the external magnetic field is 0.2 T and the magnet is given a small angular displacement $\Delta\theta$, the bar magnet oscillates about its equilibrium position with a period of 0.500 s. Determine the constant κ and the period of this torsional pendulum when $B = 0$.

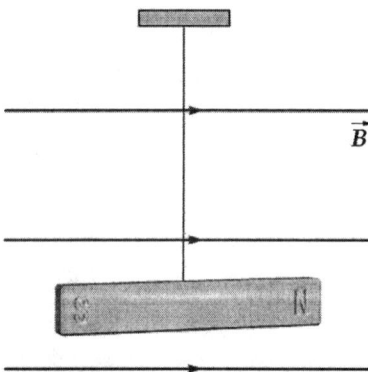

Figure 27-60 Problem 111

Picture the Problem We can apply Newton's 2nd law for rotational motion to obtain the differential equation of motion of the bar magnet. While this equation is not linear, we can use a small-angle approximation to render it linear and obtain an expression for the square of the angular frequency that we can solve for κ when there is an external field and for the period T in the absence of an external field.

Apply $\sum \tau = I\alpha$ to the bar magnet when $B \neq 0$ to obtain the differential equation of motion for the magnet:

$$-\kappa\theta - \mu B \sin\theta = I\frac{d^2\theta}{dt^2}$$

where I is the moment of inertia of the magnet about an axis through its point of suspension.

For small displacements from equilibrium ($\theta \ll 1$):	$-\kappa\theta - \mu B\theta \approx I\dfrac{d^2\theta}{dt^2}$
Rewrite the differential equation as:	$I\dfrac{d^2\theta}{dt^2} + (\kappa + \mu B)\theta = 0$
	or
	$\dfrac{d^2\theta}{dt^2} + \left(\dfrac{\kappa + \mu B}{I}\right)\theta = 0$
Because the coefficient of the linear term is the square of the angular frequency, we have:	$\omega^2 = \dfrac{\kappa + \mu B}{I}$ (1)
Express the moment of inertia (see Table 9-1) of the bar magnet about an axis through its center:	$I = \tfrac{1}{12} mL^2$
Substitute to obtain:	$\omega^2 = \dfrac{\kappa + \mu B}{\tfrac{1}{12} mL^2}$
Solve for κ to obtain:	$\kappa = \tfrac{1}{12}mL^2\omega^2 - \mu B = \tfrac{1}{12}mL^2\left(\dfrac{4\pi^2}{T^2}\right) - \mu B$
	$= \dfrac{\pi^2 mL^2}{3T^2} - \mu B$

Substitute numerical values and evaluate κ:

$$\kappa = \frac{\pi^2 (0.8\,\text{kg})(0.16\,\text{m})^2}{3(0.5\,\text{s})^2} - (0.12\,\text{A}\cdot\text{m}^2)(0.2\,\text{T}) = \boxed{0.246\,\text{N}\cdot\text{m/rad}}$$

Substitute $B = 0$ and $\omega = 2\pi/T$ in equation (1) to obtain:	$\dfrac{4\pi^2}{T^2} = \dfrac{\kappa}{I}$
Solve for T:	$T = 2\pi\sqrt{\dfrac{I}{\kappa}} = 2\pi\sqrt{\dfrac{mL^2}{12\kappa}} = \pi L\sqrt{\dfrac{m}{3\kappa}}$
Substitute numerical values and evaluate T:	$T = \pi(0.16\,\text{m})\sqrt{\dfrac{0.8\,\text{kg}}{3(0.246\,\text{N}\cdot\text{m/rad})}}$
	$= \boxed{0.523\,\text{s}}$

186   Chapter 27

*116 •• An iron bar of length 1.4 m has a diameter of 2 cm and a uniform magnetization of 1.72×10^6 A/m directed along the bar's length. The bar is stationary in space and is suddenly demagnetized so that its magnetization disappears. What is the rotational angular velocity of the bar if its angular momentum is conserved? (Assume that Equation 27-27 holds where m is the mass of an electron and $q = -e$.)

Picture the Problem We can use the definition of angular momentum and Equation 27-27, together with the definition of the magnetization M of the iron bar, to derive an expression for the rotational angular velocity of the bar just after it has been demagnetized.

Assuming its angular momentum to be conserved, use the definition of L to express the angular momentum of the iron bar just after it has been demagnetized:	$L = I\omega$
Solve for the angular velocity ω:	$\omega = \dfrac{L}{I}$
Assuming that Equation 27-27 holds yields:	$L = \dfrac{2m}{q}\mu = \dfrac{2m_e}{e}MV = \dfrac{2m_e}{e}M\pi r^2 \ell$ where r is the radius of the bar and ℓ its length.
Modeling the bar as a cylinder, express its moment of inertia with respect to its axis:	$I = \tfrac{1}{2}mr^2 = \tfrac{1}{2}\rho V r^2 = \tfrac{1}{2}\rho \pi r^4 \ell$
Substitute to obtain:	$\omega = \dfrac{\dfrac{2m_e}{e}M\pi r^2 \ell}{\tfrac{1}{2}\rho\pi r^4 \ell} = \dfrac{4m_e M}{e\rho r^2}$

Substitute numerical values (see Table 13-1 for the density of iron) and evaluate ω:

$$\omega = \dfrac{4(9.11\times 10^{-31}\text{ kg})(1.72\times 10^6 \text{ A/m})}{(1.6\times 10^{-19}\text{ C})(7.96\times 10^3\text{ kg/m}^3)(0.01\text{ m})^2} = \boxed{4.92\times 10^{-5}\text{ rad/s}}$$

*118 •• A relatively inexpensive ammeter, called a *tangent galvanometer*, can be made using the earth's field. A plane circular coil of N turns and radius R is oriented such that the field B_c it produces in the center of the coil is either east or west. A compass is placed at the center of the coil. When there is no current in the coil, the compass needle points north. When there is a current I, the compass needle points in the direction of the resultant magnetic field $\vec{B}$ at an angle θ to the north. Show that the current I is related to θ and to the horizontal component of the earth's field B_e by

$$I = \frac{2RB_e}{\mu_0 N} \tan\theta$$

Picture the Problem Note that B_e and B_c are perpendicular to each other and that the resultant magnetic field is at an angle θ with north. We can use trigonometry to relate B_c and B_e and express B_c in terms of the geometry of the coil and the current flowing in it.

Express B_c in terms of B_e:

$$B_c = B_e \tan\theta$$

where θ is the angle of the resultant field from north.

Express the field B_c due to the current in the coil:

$$B_c = \frac{N\mu_0 I}{2R}$$

where N is the number of turns.

Substitute to obtain:

$$\frac{N\mu_0 I}{2R} = B_e \tan\theta$$

Solve for I:

$$\boxed{I = \frac{2RB_e}{\mu_0 N} \tan\theta}$$

*125 ••• A disk of radius R carries a fixed charge density σ and rotates with angular velocity ω. (a) Consider a circular strip of radius r and width dr with charge dq. Show that the current produced by this strip $dI = (\omega/2\pi) dq = \omega\sigma r\, dr$. (b) Use your result from Part (a) to show that the magnetic field at the center of the disk is $B = \tfrac{1}{2}\mu_0 \sigma \omega R$. (c) Use your result from Part (a) to find the magnetic field at a point on the axis of the disk a distance x from the center.

188 Chapter 27

Picture the Problem The diagram shows the rotating disk and the circular strip of radius r and width dr with charge dq. We can use the definition of surface charge density to express dq in terms of r and dr and the definition of current to show that $dI = \omega \sigma r \, dr$. We can then use this current and expression for the magnetic field on the axis of a current loop to obtain the results called for in (b) and (c).

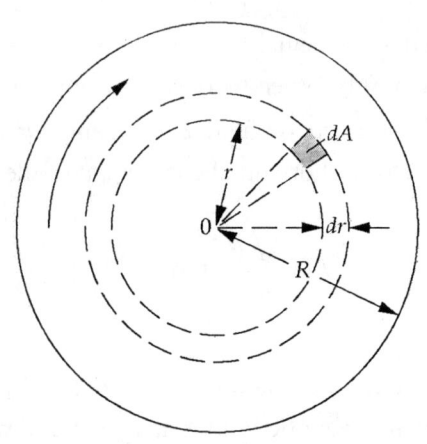

(a) Express the total charge dq that passes a given point on the circular strip once each period:

$$dq = \sigma dA = 2\pi \sigma r \, dr$$

Using its definition, express the current in the element of width dr:

$$dI = \frac{dq}{dt} = \frac{2\pi \sigma r \, dr}{\frac{2\pi}{\omega}} = \boxed{\omega \sigma r \, dr}$$

(c) Express the magnetic field dB_x at a distance x along the axis of the disk due to the current loop of radius r and width dr:

$$dB_x = \frac{\mu_0}{4\pi} \frac{2\pi r^2 \, dI}{\left(x^2 + r^2\right)^{3/2}}$$

$$= \frac{\mu_0 \omega \sigma r^3}{2\left(x^2 + r^2\right)^{3/2}} dr$$

Integrate from $r = 0$ to $r = R$ to obtain:

$$B_x = \frac{\mu_0 \omega \sigma}{2} \int_0^R \frac{r^3}{\left(x^2 + r^2\right)^{3/2}} dr$$

$$= \boxed{\frac{\mu_0 \omega \sigma}{2}\left(\frac{R^2 + 2x^2}{\sqrt{R^2 + x^2}} - 2x\right)}$$

(b) Evaluate B_x for $x = 0$:

$$B_x(0) = \frac{\mu_0 \omega \sigma}{2}\left(\frac{R^2}{\sqrt{R^2}}\right) = \boxed{\tfrac{1}{2}\mu_0 \sigma \omega R}$$

Chapter 28
Magnetic Induction

Conceptual Problems

*1 • A conducting loop lies in the plane of this page and carries a clockwise induced current. Which of the following statements could be true? (*a*) A constant magnetic field is directed into the page. (*b*) A constant magnetic field is directed out of the page. (*c*) An increasing magnetic field is directed into the page. (*d*) A decreasing magnetic field is directed into the page. (*e*) A decreasing magnetic field is directed out of the page.

Determine the Concept We know that the magnetic flux (in this case the magnetic field because the area of the conducting loop is constant and its orientation is fixed) must be changing so the only issues are whether the field is increasing or decreasing and in which direction. Because the direction of the magnetic field associated with the clockwise current is into the page, the changing field that is responsible for it must be either increasing out of the page (not included in the list of possible answers) or a decreasing field directed into the page. (d) is correct.

*6 • If the current through an inductor were doubled, the energy stored in the inductor would be (*a*) the same. (*b*) doubled. (*c*) quadrupled. (*d*) halved. (*e*) quartered.

Determine the Concept The magnetic energy stored in an inductor is given by $U_\mathrm{m} = \tfrac{1}{2}LI^2$. Doubling I quadruples U_m. (c) is correct.

*10 • A pendulum is fabricated from a thin, flat piece of metal. At the bottom of its arc, it passes between the poles of a strong permanent magnet. In Figure 28-42 *a*, the metal sheet is continuous, whereas in Figure 28-42*b*, there are slots in it. The pendulum with slots swings back and forth many times, but the pendulum without slots comes to a stop in no more than one complete oscillation. Explain why.

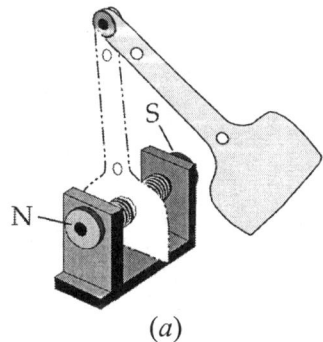
(*a*)

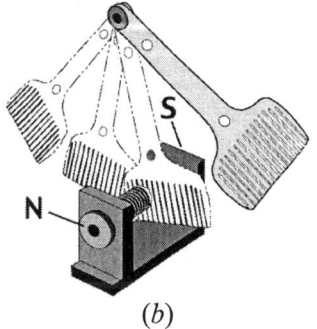
(*b*)

Figure 28-42 Problem 10

Chapter 28

Determine the Concept In the configuration shown in (a), energy is dissipated by eddy currents from the emf induced by the pendulum movement. In the configuration shown in (b), the slits inhibit the eddy currents and the braking effect is greatly reduced.

Estimation and Approximation

*13 •• A physics teacher attempts the following emf demonstration. She has two of her students hold a long wire connected to a voltmeter. The wire is held slack, so that there is a large arc in it. When she says "start", the students being rotating the wire in a large vertical arc, as if they were playing jump-rope. The students stand 3.0 m apart, and the sag in the wire is about 1.5 m. (You may idealize the shape of wire as a semi-circular arc of diameter $d = 1.5$ m.) The induced emf from the jump rope is then measured on the voltmeter. (a) Estimate a reasonable value for the maximum angular velocity which the students can rotate the wire. (b) From this, estimate the maximum emf induced in the wire. The magnitude of the earth's magnetic field is approximately 0.7 G. (c) Can the students rotate the jump-rope fast enough to generate an emf of 1 V? (d) Suggest modifications to the demonstration that would allow higher emfs to be generated.

Picture the Problem We can use Faraday's law to relate the induced emf to the angular velocity with which the students turn the jump rope.

(a) It seems unlikely that the students could turn the "jump rope" wire faster than 5 revolutions per second. This corresponds to a maximum angular velocity of:

$$\omega = 5\frac{\text{rev}}{\text{s}} \times \frac{2\pi \text{ rad}}{\text{rev}} = \boxed{31.4 \text{ rad/s.}}$$

(b) The magnetic flux ϕ_m through the rotating circular loop of wire varies sinusoidally with time according to:

$$\phi_m = BA \sin \omega t$$
and
$$\frac{d\phi_m}{dt} = BA\omega \cos \omega t$$

Because the average value of the cosine function, over one revolution, is ½, the average rate at which the flux changes through the circular loop is:

$$\left.\frac{d\phi_m}{dt}\right|_{av} = \tfrac{1}{2} BA\omega = \tfrac{1}{2}\pi r^2 B\omega$$

From Faraday's law, the magnitude of the induced emf in the loop is:

$$\mathcal{E} = \frac{d\phi_m}{dt} = \tfrac{1}{2}\pi r^2 B\omega$$

Substitute numerical values and evaluate $\mathcal{E}$:

$$\mathcal{E} = \tfrac{1}{2}\pi \left(\frac{1.5 \text{ m}}{2}\right)^2 \left(0.7 \text{ G} \times \frac{1 \text{ T}}{10^4 \text{ G}}\right)(31.4 \text{ rad/s}) = \boxed{1.94 \text{ mV}}$$

(c) No. To generate an emf of 1 V, the students would have to rotate the jump rope about 500 times faster.

(d) The use of multiple strands of lighter wire (so that the composite wire could be rotated at the same angular speed) looped several times around would increase the induced emf.

Magnetic Flux

*17 • A circular coil has 25 turns and a radius of 5 cm. It is at the equator, where the earth's magnetic field is 0.7 G north. Find the magnetic flux through the coil when its plane is (a) horizontal, (b) vertical with its axis pointing north, (c) vertical with its axis pointing east, and (d) vertical with its axis making an angle of 30° with north.

Picture the Problem Because the coil defines a plane with area A and $\vec{B}$ is constant in magnitude and direction over the surface and makes an angle θ with the unit normal vector, we can use $\phi_m = NBA\cos\theta$ to find the magnetic flux through the coil.

Substitute for N, B, and A to obtain:

$$\phi_m = NB\pi r^2 \cos\theta = 25\left(0.7\,\text{G} \cdot \frac{1\,\text{T}}{10^4\,\text{G}}\right)\pi(5\times 10^{-2}\,\text{m})^2 \cos\theta$$

$$= (1.37\times 10^{-5}\,\text{Wb})\cos\theta$$

(a) When the plane of the coil is horizontal, $\theta = 90°$:

$\phi_m = (1.37\times 10^{-5}\,\text{Wb})\cos 90°$
$= \boxed{0}$

(b) When the plane of the coil is vertical with its axis pointing north, $\theta = 0°$:

$\phi_m = (1.37\times 10^{-5}\,\text{Wb})\cos 0°$
$= \boxed{1.37\times 10^{-5}\,\text{Wb}}$

(c) When the plane of the coil is vertical with its axis pointing east, $\theta = 90°$:

$\phi_m = (1.37\times 10^{-5}\,\text{Wb})\cos 90°$
$= \boxed{0}$

(d) When the plane of the coil is vertical with its axis making an angle of 30° with north, $\theta = 30°$:

$\phi_m = (1.37\times 10^{-5}\,\text{Wb})\cos 30°$
$= \boxed{1.19\times 10^{-5}\,\text{Wb}}$

*24 •• A long straight wire carries a current I. A rectangular loop with two sides parallel to the straight wire has sides a and b, with its near side a distance d from the straight wire, as shown in Figure 28-45. (a) Compute the magnetic flux through the rectangular loop. (*Hint:* Calculate the flux through a strip of area $dA = b\,dx$ and integrate from $x = d$ to $x = d + a$.) (b) Evaluate your answer for $a = 5$ cm, $b = 10$ cm, $d = 2$ cm, and $I = 20$ A.

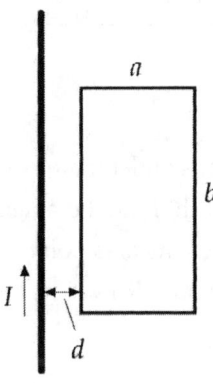

Figure 28-45 Problem 24

Picture the Problem We can use the hint to set up the element of area dA and express the flux $d\phi_m$ through it and then carry out the details of the integration to express ϕ_m.

(a) Express the flux through the strip of area dA:

$$d\phi_m = B\,dA$$

where $dA = b\,dx$.

Express B at a distance x from a long, straight wire:

$$B = \frac{\mu_0}{4\pi}\frac{2I}{x} = \frac{\mu_0}{2\pi}\frac{I}{x}$$

Substitute to obtain:

$$d\phi_m = \frac{\mu_0}{2\pi}\frac{I}{x}b\,dx = \frac{\mu_0 Ib}{2\pi}\frac{dx}{x}$$

Integrate from $x = d$ to $x = d + a$:

$$\phi_m = \frac{\mu_0 Ib}{2\pi}\int_d^{d+a}\frac{dx}{x} = \boxed{\frac{\mu_0 Ib}{2\pi}\ln\frac{d+a}{d}}$$

(b) Substitute numerical values and evaluate ϕ_m:

$$\phi_m = \frac{(4\pi \times 10^{-7}\text{ N/A}^2)(20\text{ A})(0.1\text{m})}{2\pi}\ln\left(\frac{7\text{ cm}}{2\text{ cm}}\right) = \boxed{5.01 \times 10^{-7}\text{ Wb}}$$

Magnetic Induction

Induced EMF and Faraday's Law

*27 • A uniform magnetic field $\vec{B}$ is established perpendicular to the plane of a loop of radius 5 cm, resistance 0.4 Ω, and negligible self-inductance. The magnitude of $\vec{B}$ is increasing at a rate of 40 mT/s. Find (*a*) the induced emf ε in the loop, (*b*) the induced current in the loop, and (*c*) the rate of joule heating in the loop.

Picture the Problem We can find the induced emf by applying Faraday's law to the loop. The application of Ohm's law will yield the induced current in the loop and we can find the rate of joule heating using $P = I^2 R$.

(*a*) Apply Faraday's law to express the induced emf in the loop in terms of the rate of change of the magnetic field:

$$|\varepsilon| = \frac{d\phi_m}{dt} = \frac{d}{dt}(AB) = A\frac{dB}{dt} = \pi R^2 \frac{dB}{dt}$$

Substitute numerical values and evaluate $|\varepsilon|$:

$$|\varepsilon| = \pi(0.05\,\text{m})^2 (40\,\text{mT/s}) = \boxed{0.314\,\text{mV}}$$

(*b*) Using Ohm's law, relate the induced current to the induced voltage and the resistance of the loop and evaluate *I*:

$$I = \frac{\varepsilon}{R} = \frac{0.314\,\text{mV}}{0.4\,\Omega} = \boxed{0.785\,\text{mA}}$$

(*c*) Express the rate at which power is dissipated in a conductor in terms of the induced current and the resistance of the loop and evaluate *P*:

$$P = I^2 R = (0.785\,\text{mA})^2 (0.4\,\Omega) = \boxed{0.247\,\mu\text{W}}$$

*31 •• A 100-turn circular coil has a diameter of 2 cm and resistance of 50 Ω. The plane of the coil is perpendicular to a uniform magnetic field of magnitude 1 T. The direction of the field is suddenly reversed. (*a*) Find the total charge that passes through the coil. If the reversal takes 0.1 s, find (*b*) the average current in the coil and (*c*) the average emf in the coil.

Picture the Problem We can use the definition of average current to express the total charge passing through the coil as a function of I_{av}. Because the induced current is proportional to the induced emf and the induced emf, in turn, is given by Faraday's law, we can express ΔQ as a function of the number of turns of the coil, the magnetic field, the resistance of the coil, and the area of the coil. Knowing the reversal time, we can find the average current from its definition and the average emf in the coil from Ohm's law.

194 Chapter 28

(a) Express the total charge that passes through the coil in terms of the induced current:

$$\Delta Q = I_{av}\Delta t$$

Relate the induced current to the induced emf:

$$I = I_{av} = \frac{\mathcal{E}}{R}$$

Using Faraday's law, express the induced emf in terms of ϕ_m:

$$\mathcal{E} = -\frac{\Delta\phi_m}{\Delta t}$$

Substitute and simplify to obtain:

$$\Delta Q = \frac{\mathcal{E}}{R}\Delta t = -\frac{\frac{\Delta\phi_m}{\Delta t}}{R}\Delta t = -\frac{2\phi_m}{R}$$

$$= -\frac{2NBA}{R} = -\frac{2NB\left(\frac{\pi}{4}d^2\right)}{R}$$

$$= -\frac{NB\pi d^2}{2R}$$

where d is the diameter of the coil.

Substitute numerical values and evaluate ΔQ:

$$\Delta Q = -\frac{(100)(1\,\text{T})\pi(0.02\,\text{m})^2}{2(50\,\Omega)}$$

$$= \boxed{-1.26\,\text{mC}}$$

(b) Apply the definition of average current to obtain:

$$I_{av} = \frac{\Delta Q}{\Delta t} = \frac{1.26\,\text{mC}}{0.1\,\text{s}} = \boxed{12.6\,\text{mA}}$$

(c) Using Ohm's law, relate the average emf in the coil to the average current:

$$\mathcal{E}_{av} = I_{av}R = (12.6\,\text{mA})(50\,\Omega)$$

$$= \boxed{630\,\text{mV}}$$

Motional EMF

*36 • A rod 30 cm long moves at 8 m/s in a plane perpendicular to a magnetic field of 500 G. The velocity of the rod is perpendicular to its length. Find (a) the magnetic force on an electron in the rod, (b) the electrostatic field $\vec{E}$ in the rod, and (c) the potential difference V between the ends of the rod.

Picture the Problem We can apply the equation for the force on a charged particle

Magnetic Induction   195

moving in a magnetic field to find the magnetic force acting on an electron in the rod. We can use $\vec{E} = \vec{v} \times \vec{B}$ to find E and $V = E\ell$, where ℓ is the length of the rod, to find the potential difference between its ends.

(a) Relate the magnetic force on an electron in the rod to the speed of the rod, the electronic charge, and the magnetic field in which the rod is moving:

$\vec{F} = q\vec{v} \times \vec{B}$
and
$F = qvB\sin\theta$

Substitute numerical values and evaluate F:

$F = (1.6 \times 10^{-19}\,\text{C})(8\,\text{m/s})(0.05\,\text{T})\sin 90°$
$= \boxed{6.40 \times 10^{-20}\,\text{N}}$

(b) Express the electrostatic field $\vec{E}$ in the rod in terms of the magnetic field $\vec{B}$:

$\vec{E} = \vec{v} \times \vec{B}$
and
$E = vB\sin\theta$

Substitute numerical values and evaluate B:

$E = (8\,\text{m/s})(0.05\,\text{T})\sin 90°$
$= \boxed{0.400\,\text{V/m}}$

(c) Relate the potential difference between the ends of the rod to its length ℓ and the electric field E:

$V = E\ell$

Substitute numerical values and evaluate V:

$V = (0.4\,\text{V/m})(0.3\,\text{m}) = \boxed{0.120\,\text{V}}$

*42 •• In Example 28-9, find the total energy dissipated in the resistance and show that it is equal to $\tfrac{1}{2}mv_0^2$.

Picture the Problem In Example 28-9 it is shown that the speed of the rod is given by $v = v_0 e^{-(B^2\ell^2/mR)t}$. We can use the definition of power and the expression for a motional emf to express the power dissipated in the resistance in terms of B, ℓ, v, and R. We can then separate the variables and integrate over all time to show that the total energy dissipated is equal to the initial kinetic energy of the rod.

Express the power dissipated in terms of $\mathcal{E}$ and R:

$P = \dfrac{\mathcal{E}^2}{R}$

Express $\mathcal{E}$ as a function of B, ℓ, and

$\mathcal{E} = B\ell v$

v:

where
$$v = v_0 e^{-(B^2\ell^2/mR)t}$$

Substitute to obtain:
$$P = \frac{(B\ell v)^2}{R}$$

The total energy dissipated as the rod comes to rest is obtained by integrating $dE = P\,dt$:

$$E = \int_0^\infty \frac{(B\ell v)^2}{R}\,dt$$

$$= \int_0^\infty \frac{\left(B\ell v_0 e^{-(B^2\ell^2/mR)t}\right)^2}{R}\,dt$$

$$= \frac{B^2\ell^2 v_0^2}{R}\int_0^\infty e^{-2(B^2\ell^2/mR)t}\,dt$$

Evaluate the integral (by changing variables to $u = -\dfrac{2B^2\ell^2}{mR}$) to obtain:

$$E = \frac{B^2\ell^2 v_0^2}{R}\left(\frac{mR}{2B^2\ell^2}\right) = \boxed{\tfrac{1}{2}mv_0^2}$$

*45 •• In Figure 28-50, a conducting rod of mass m and negligible resistance is free to slide without friction along two parallel rails of negligible resistance separated by a distance ℓ and connected by a resistance R. The rails are attached to a long inclined plane that makes an angle θ with the horizontal. There is a magnetic field B directed upward. (*a*) Show that there is a retarding force directed up the incline given by $F = (B^2\ell^2 v\cos^2\theta)/R$. (*b*) Show that the terminal speed of the rod is $v_t = (mgR\sin\theta)/(B^2\ell^2\cos^2\theta)$.

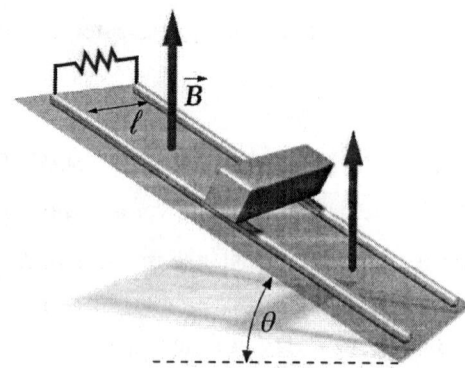

Figure 28-50 Problem 45

Picture the Problem The free-body diagram shows the forces acting on the rod as it slides down the inclined plane. The retarding force is the component of F_m acting up the incline, i.e., in the $-x$ direction. We can express F_m using the expression for the force acting on a conductor moving in a magnetic field. Recognizing that only the horizontal component of the rod's velocity $\vec{v}$ produces an induced emf, we can apply the expression for a motional emf in conjunction with Ohm's law to find the induced current in the rod. In part (b) we can apply Newton's 2nd law to obtain an expression for dv/dt and set this expression equal to zero to obtain v_t.

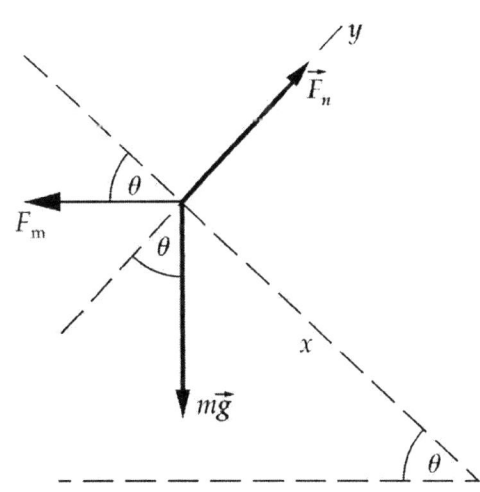

(a) Express the retarding force acting on the rod:

$F = F_m \cos\theta$ (1)

where

$F_m = I\ell B$

and I is the current induced in the rod as a consequence of its motion in the magnetic field.

Express the induced emf due to the motion of the rod in the magnetic field:

$\mathcal{E} = B\ell v \cos\theta$

Using Ohm's law, relate the current I in the circuit to the induced emf:

$I = \dfrac{\mathcal{E}}{R} = \dfrac{B\ell v \cos\theta}{R}$

Substitute in equation (1) to obtain:

$F = \left(\dfrac{B\ell v \cos\theta}{R}\right)\ell B \cos\theta$

$= \boxed{\dfrac{B^2\ell^2 v}{R}\cos^2\theta}$

(b) Apply $\sum F_x = ma_x$ to the rod:

$mg\sin\theta - \dfrac{B^2\ell^2 v}{R}\cos^2\theta = m\dfrac{dv}{dt}$

and

198   Chapter 28

$$\frac{dv}{dt} = g\sin\theta - \frac{B^2\ell^2 v}{mR}\cos^2\theta$$

When the rod reaches its terminal velocity v_t, $dv/dt = 0$ and:

$$0 = g\sin\theta - \frac{B^2\ell^2 v_t}{mR}\cos^2\theta$$

Solve for v_t to obtain:

$$\boxed{v_t = \frac{mgR\sin\theta}{B^2\ell^2\cos^2\theta}}$$

*51 ••• The loop in Problem 24 moves away from the wire with a constant speed v. At time $t = 0$, the left side of the loop is a distance d from the long straight wire. (a) Compute the emf in the loop by computing the motional emf in each segment of the loop that is parallel to the long wire. Explain why you can neglect the emf in the segments that are perpendicular to the wire. (b) Compute the emf in the loop by first computing the flux through the loop as a function of time and then using $\varepsilon = -d\phi_m/dt$ and compare your answer with that obtained in Part (a).

Picture the Problem We can use the expression for a motional emf and Ampere's law to express the net emf induced in the moving loop. We can also use express the magnetic flux through the loop and apply Faraday's law to obtain the same result.

(a) Express the motional emf induced in the segments parallel to the current-carrying wire:

$$\varepsilon = B(x)vb$$

Using Ampere's law, express $B(d + vt)$ and $B(d + a + vt)$:

$$B(d+vt) = \frac{\mu_0 I}{2\pi(d+vt)}$$

and

$$B(d+a+vt) = \frac{\mu_0 I}{2\pi(d+a+vt)}$$

Substitute to express ε_1 for the near wire and ε_2 for the far wire:

$$\varepsilon_1 = \frac{\mu_0 Ivb}{2\pi(d+vt)}$$

and

$$\varepsilon_1 = \frac{\mu_0 Ivb}{2\pi(d+a+vt)}$$

Noting that the emfs both point upward and hence oppose one another, express the net emf induced in the loop:

$$\mathcal{E} = \mathcal{E}_1 - \mathcal{E}_2$$
$$= \frac{\mu_0 I v b}{2\pi(d+vt)} - \frac{\mu_0 I v b}{2\pi(d+a+vt)}$$
$$= \boxed{\frac{\mu_0 I v b}{2\pi}\left(\frac{1}{d+vt} - \frac{1}{d+a+vt}\right)}$$

> The motion of the segments perpendicular to the long wire does not change the flux through the rectangular loop. Consequently, these segments do not contribute to the the induced emf.

(b) From Faraday's law we have:

$$\mathcal{E} = -\frac{d\phi_m}{dt}$$

Express the magnetic flux in an area of length b and width vdt:

$$d\phi_m = B(x)dA = B(x)b\,dx$$

where, from Ampere's law,

$$B(x) = \frac{\mu_0 I}{2\pi x}$$

Substitute and integrate from $x = d + vt$ to $d + a + vt$:

$$\phi_m = \int_{d+vt}^{d+a+vt} B(x)dx = \frac{\mu_0 I b}{2\pi}\int_{d+vt}^{d+a+vt}\frac{dx}{x}$$
$$= \frac{\mu_0 I b}{2\pi}\ln\left[\frac{d+a+vt}{d+vt}\right]$$

Differentiate with respect to time and simplify to obtain:

$$\mathcal{E} = -\frac{d}{dt}\left[\frac{\mu_0 I b}{2\pi}\ln\frac{d+a+vt}{d+vt}\right] = -\frac{\mu_0 I b}{2\pi}\frac{d}{dt}\left[\ln\frac{d+a+vt}{d+vt}\right]$$
$$= -\frac{\mu_0 I b}{2\pi}\left[\left(\frac{d+vt}{d+a+vt}\right)\left(\frac{(d+vt)v-(d+a+vt)v}{(d+vt)^2}\right)\right]$$
$$= -\frac{\mu_0 I b v}{2\pi}\left[\frac{(d+vt)-(d+a+vt)}{(d+vt)(d+a+vt)}\right] = -\frac{\mu_0 I b v}{2\pi}\left[\frac{1}{d+a+vt} - \frac{1}{d+vt}\right]$$
$$= \boxed{\frac{\mu_0 I b v}{2\pi}\left[\frac{1}{d+vt} - \frac{1}{d+a+vt}\right]}$$

Inductance

*54 • A coil with self-inductance L carries a current I, given by $I = I_0 \sin 2\pi f t$. Find and graph the flux ϕ_m and the self-induced emf as functions of time.

Picture the Problem We can apply $\phi_m = LI$ to find ϕ_m and Faraday's law to find the self-induced emf as functions of time.

Use the definition of self-inductance to express ϕ_m:
$$\phi_m = LI = \boxed{LI_0 \sin 2\pi ft}$$

The graph of the flux ϕ_m as a function of time shown below was plotted using a spreadsheet program. The maximum value of the flux is LI_0 and we have chosen $2\pi f = 1$ rad/s.

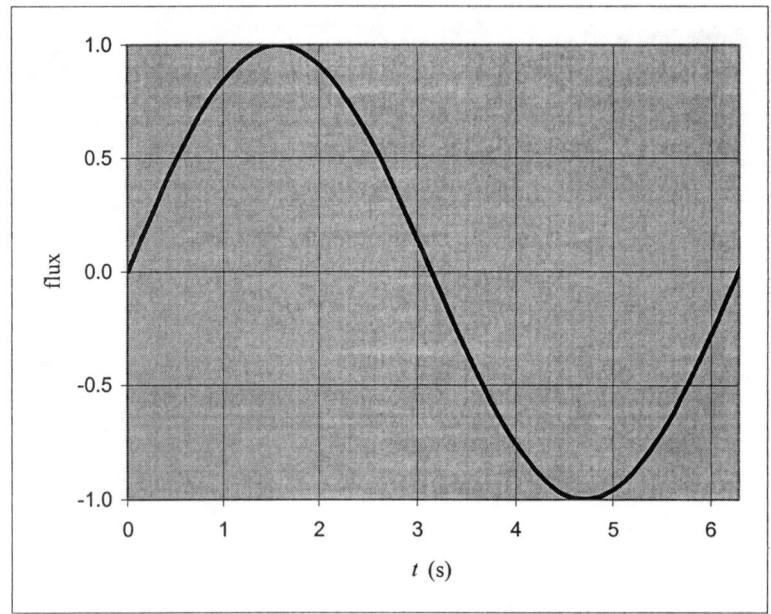

Apply Faraday's law to relate ε, L, and dI/dt:
$$\varepsilon = -L\frac{dI}{dt} = -L\frac{d}{dt}[I_0 \sin 2\pi ft]$$
$$= \boxed{-2\pi f LI_0 \cos 2\pi ft}$$

The graph of the emf ε as a function of time shown below was plotted using a spreadsheet program. The maximum value of the induced emf is $2\pi f LI_0$ and we have chosen $2\pi f = 1$ rad/s.

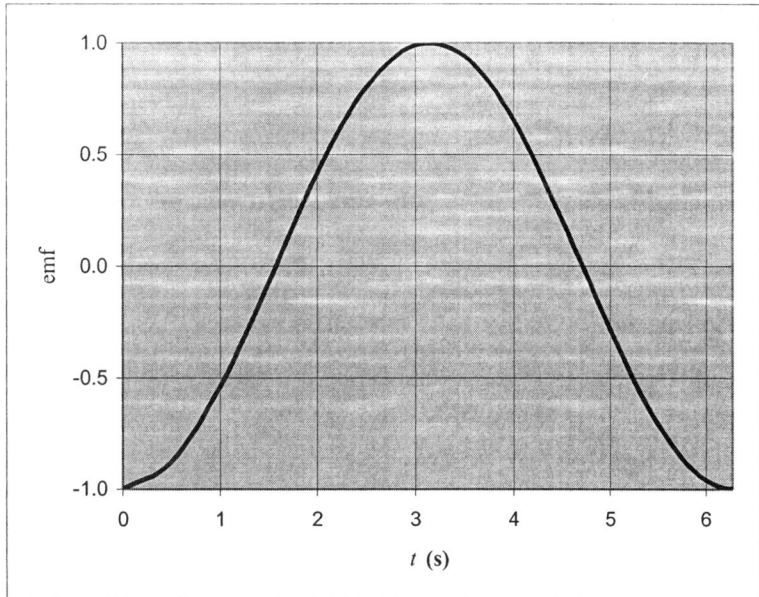

*57 •• A long insulated wire with a resistance of 18 Ω/m is to be used to construct a resistor. First, the wire is bent in half, and then the doubled wire is wound in a cylindrical form as shown in Figure 28-53. The diameter of the cylindrical form is 2 cm, its length is 25 cm, and the total length of wire is 9 m. Find the resistance and inductance of this wire-wound resistor.

Figure 28-53 Problem 57

Picture the Problem Note that the current in the two parts of the wire is in opposite directions. Consequently, the total flux in the coil is zero. We can find the resistance of the wire-wound resistor from the length of wire used and the resistance per unit length.

Because the total flux in the coil is zero:

$$L = \boxed{0}$$

Express the total resistance of the wire:

$$R = \left(18\frac{\Omega}{m}\right)L = \left(18\frac{\Omega}{m}\right)(9\,m) = \boxed{162\,\Omega}$$

202   Chapter 28

Magnetic Energy

*61 ••   In a plane electromagnetic wave, such as a light wave, the magnitudes of the electric fields and magnetic fields are related by $E = cB$, where $c = 1/\sqrt{\varepsilon_0 \mu_0}$ is the speed of light. Show that in this case the electric energy and the magnetic energy densities are equal.

Picture the Problem We can examine the ratio of u_m to u_E with $E = cB$ and $c = 1/\sqrt{\varepsilon_0 \mu_0}$ to show that the electric and magnetic energy densities are equal.

Express the ratio of the energy density in the magnetic field to the energy density in the electric field:

$$\frac{u_m}{u_E} = \frac{\frac{B^2}{2\mu_0}}{\frac{1}{2}\varepsilon_0 E^2} = \frac{B^2}{\mu_0 \varepsilon_0 E^2}$$

Substitute $E = cB$:

$$\frac{u_m}{u_E} = \frac{B^2}{\mu_0 \varepsilon_0 c^2 B^2} = \frac{1}{\mu_0 \varepsilon_0 c^2}$$

Substitute for c:

$$\frac{u_m}{u_E} = \frac{\mu_0 \varepsilon_0}{\mu_0 \varepsilon_0} = 1 \Rightarrow \boxed{u_m = u_E}$$

*64 ••   You are given a length d of wire which has radius a, and told to wind it into an inductor in the shape of a cylinder with a circular cross-section of radius r. The windings are to be as close together as possible without overlapping. Show that the self-inductance of this inductor is $L = \mu_0 \left(\dfrac{rd}{4a} \right)$.

Picture the Problem The wire of length d and radius a is shown in the diagram, as is the inductor constructed with this wire and whose inductance L is to be found. We can use the equation for the self-inductance of a cylindrical inductor to derive an expression for L.

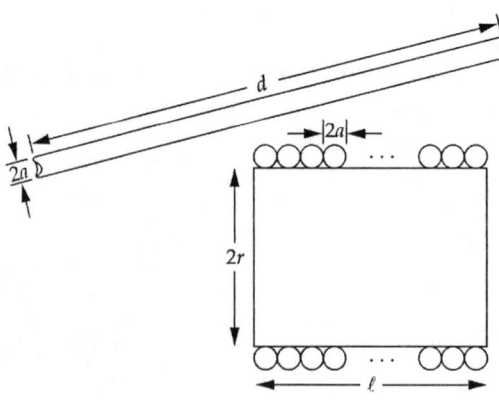

The self-inductance of an inductor with length ℓ, cross-sectional area A,

$$L = \mu_0 n^2 A \ell \qquad (1)$$

and number of turns per unit length n is:

The number of turns N is given by:
$$N = \frac{\ell}{2a}$$

The number of turns per unit length n is:
$$n = \frac{N}{\ell} = \frac{1}{2a}$$

Assuming that $a \ll r$, the length of the wire d is related to n and r:
$$d = N(2\pi r) = \left(\frac{\ell}{2a}\right) 2\pi r = \frac{\pi r}{a}\ell$$

Solve for ℓ to obtain:
$$\ell = \frac{ad}{\pi r}$$

Substitute for ℓ, A, and n in equation (1) to obtain:
$$L = \mu_0 \left(\frac{1}{2a}\right)^2 (\pi r^2)\left(\frac{ad}{\pi r}\right) = \boxed{\mu_0\left(\frac{rd}{4a}\right)}$$

RL Circuits

*69 •• In the circuit of Figure 28-29, let $\varepsilon_0 = 12$ V, $R = 3$ Ω, and $L = 0.6$ H. The switch is closed at time $t = 0$. At time $t = 0.5$ s, find (*a*) the rate at which the battery supplies power, (*b*) the rate of joule heating, and (*c*) the rate at which energy is being stored in the inductor.

Picture the Problem We can find the current using $I = I_f(1 - e^{-t/\tau})$, where $I_f = \varepsilon_0/R$, and $\tau = L/R$, and its rate of change by differentiating this expression with respect to time.

Express the dependence of the current on I_f and τ:
$$I = I_f(1 - e^{-t/\tau})$$

Evaluate I_f and τ:
$$I_f = \frac{\varepsilon_0}{R} = \frac{12\,\text{V}}{3\,\Omega} = 4\,\text{A}$$
and
$$\tau = \frac{L}{R} = \frac{0.6\,\text{H}}{3\,\Omega} = 0.2\,\text{s}$$

Substitute to obtain:
$$I = (4\,\text{A})(1 - e^{-t/0.2\,\text{s}}) = (4\,\text{A})(1 - e^{-5t\,\text{s}^{-1}})$$

204   Chapter 28

Express dI/dt:
$$\frac{dI}{dt} = (4\,\text{A})\left(-e^{-5t\text{s}^{-1}}\right)\left(-5\,\text{s}^{-1}\right)$$
$$= (20\,\text{A/s})e^{-5t\text{s}^{-1}}$$

(a) Find the current at $t = 0.5$ s:
$$I(0.5\,\text{s}) = (4\,\text{A})\left(1 - e^{-5(0.5\,\text{s})\text{s}^{-1}}\right)$$
$$= 3.67\,\text{A}$$

The rate at which the battery supplies power at $t = 0.5$ s is:
$$P(0.5\,\text{s}) = I(0.5\,\text{s})\mathcal{E}$$
$$= (3.67\,\text{A})(12\,\text{V})$$
$$= \boxed{44.0\,\text{W}}$$

(b) The rate of joule heating is:
$$P_{\text{J}}(0.5\,\text{s}) = [I(0.5\,\text{s})]^2 R$$
$$= (3.67\,\text{A})^2(3\,\Omega)$$
$$= \boxed{40.4\,\text{W}}$$

(c) Using the expression for the magnetic energy stored in an inductor, express the rate at which energy is being stored:
$$\frac{dU_{\text{L}}}{dt} = \frac{d}{dt}\left[\tfrac{1}{2}LI^2\right] = LI\frac{dI}{dt}$$

Substitute for L, I, and dI/dt to obtain:
$$\frac{dU_{\text{L}}}{dt} = \frac{d}{dt}\left[\tfrac{1}{2}LI^2\right] = LI\frac{dI}{dt}$$

Substitute numerical values and evaluate $\dfrac{dU_{\text{L}}}{dt}$:

$$\frac{dU_{\text{L}}}{dt} = (0.6\,\text{H})(4\,\text{A})\left(1 - e^{-5t\text{s}^{-1}}\right)(20\,\text{A/s})e^{-5t\text{s}^{-1}} = (48\,\text{W})\left(1 - e^{-5t\text{s}^{-1}}\right)e^{-5t\text{s}^{-1}}$$

Evaluate this expression for $t = 0.5$ s:
$$\frac{dU_{\text{L}}}{dt} = (48\,\text{W})\left(1 - e^{-5(0.5\,\text{s})\text{s}^{-1}}\right)e^{-5(0.5\,\text{s})\text{s}^{-1}}$$
$$= (48\,\text{W})\left(1 - e^{-2.5}\right)e^{-2.5}$$
$$= \boxed{3.62\,\text{W}}$$

Remarks: Note that, to a good approximation, $dU_{\text{L}}/dt = P - P_{\text{J}}$.

*75 •••   Given the circuit shown in Figure 28-56, assume that the switch S has been closed for a long time so that steady currents exist in the inductor, and that the inductor L has negligible resistance. (a) Find the battery current, the current in the 100 Ω resistor and the current through the inductor. (*b*) Find the initial voltage across the inductor when the switch S is opened. (*c*) Using a spreadsheet program, make graphs of the current and voltage across the inductor as a function of time.

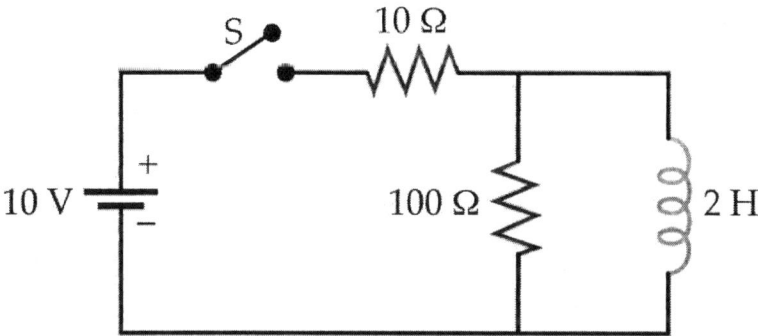

Figure 28-56 Problem 75

Picture the Problem The self-induced emf in the inductor is proportional to the rate at which the current through it is changing. Under steady-state conditions, $dI/dt = 0$ and so the self-induced emf in the inductor is zero. We can use Kirchhoff's loop rule to obtain the current through and the voltage across the inductor as a function of time.

(*a*) Because, under steady-state conditions, the self-induced emf in the inductor is zero and because the inductor has negligible resistance, we can apply Kirchhoff's loop rule to the loop that includes the source, the 10-Ω resistor, and the inductor to find the current drawn from the battery and flowing through the inductor and the 10-Ω resistor:

$$10\,\text{V} - (10\,\Omega)I = 0$$
and
$$I = \frac{10\,\text{V}}{10\,\Omega} = \boxed{1.00\,\text{A}}$$

By applying Kirchhoff's junction rule at the junction between the resistors, we can conclude that:

$$I_{100\text{-}\Omega\,\text{resistor}} = I_{\text{battery}} - I_{\text{inductor}} = \boxed{0}$$

(*b*) When the switch is closed, the current cannot immediately go to zero in the circuit because of the inductor. For a time, a current will circulate in the circuit loop between the inductor and the 100-Ω resistor. Because the current flowing through this circuit is initially 1 A, the voltage drop across the 100-Ω resistor is initially $\boxed{100\,\text{V}.}$ Conservation of energy (Kirchhoff's loop rule) requires that the voltage drop across the inductor is also $\boxed{100\,\text{V}.}$

(c) Apply Kirchhoff's loop rule to the RL circuit to obtain:

$$L\frac{dI}{dt} + IR = 0$$

The solution to this differential equation is:

$$I(t) = I_0 e^{-\frac{R}{L}t} = I_0 e^{-\frac{t}{\tau}}$$

where $\tau = \dfrac{L}{R} = \dfrac{2\,\text{H}}{100\,\Omega} = 0.02\,\text{s}$

A spreadsheet program to generate the data for graphs of the current and the voltage across the inductor as functions of time is shown below. The formulas used to calculate the quantities in the columns are as follows:

Cell	Formula/Content	Algebraic Form
B1	2	L
B2	100	R
B3	1	I_0
A6	0	t_0
B6	$B$3*EXP((-$B$2/$B$1)*A6)	$I_0 e^{-\frac{R}{L}t}$

	A	B	C
1	L=	2	H
2	R=	100	ohms
3	I_0=	1	A
4			
5	t	I(t)	V(t)
6	0.000	1.00E+00	100.00
7	0.005	7.79E–01	77.88
8	0.010	6.07E–01	60.65
9	0.015	4.72E–01	47.24
10	0.020	3.68E–01	36.79
11	0.025	2.87E–01	28.65
12	0.030	2.23E–01	22.31
32	0.130	1.50E–03	0.15
33	0.135	1.17E–03	0.12
34	0.140	9.12E–04	0.09
35	0.145	7.10E–04	0.07
36	0.150	5.53E–04	0.06

The following graph of the current in the inductor as a function of time was plotted using the data in columns A and B of the spreadsheet program.

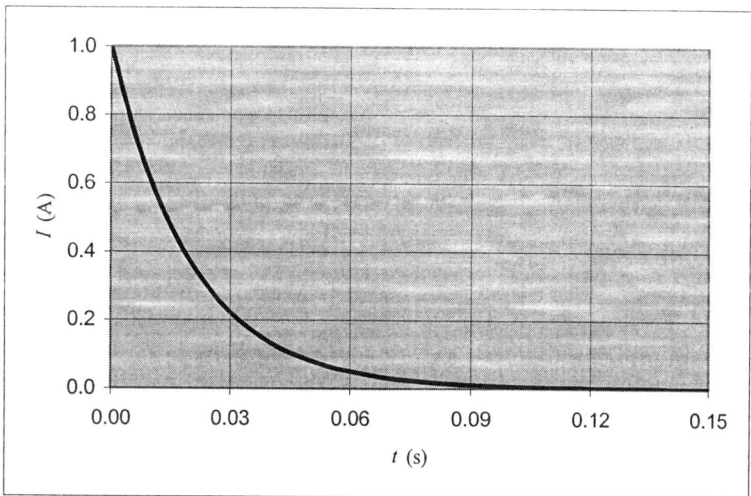

The following graph of the voltage across the inductor as a function of time was plotted using the data in columns A and C of the spreadsheet program.

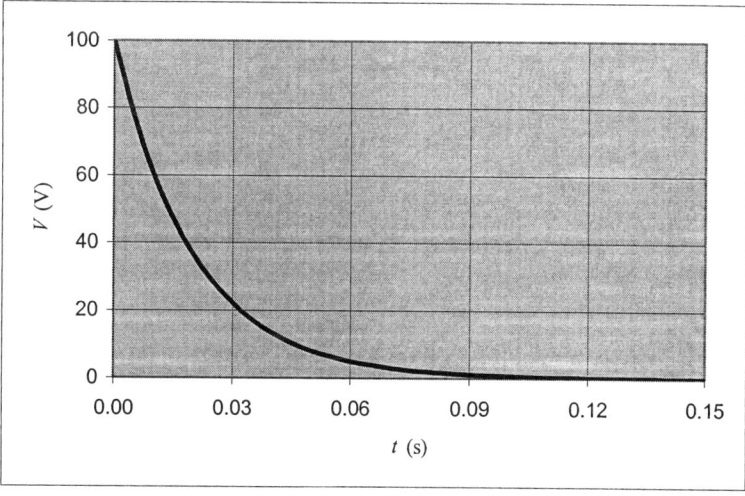

*80 •• For the circuit of Example 28-13, find the time at which the power dissipation in the resistor equals the rate at which magnetic energy is stored in the inductor.

Picture the Problem If the current is initially zero in an LR circuit, its value at some later time t is given by $I = I_f\left(1 - e^{-t/\tau}\right)$, where $I_f = \varepsilon_0/R$ and $\tau = L/R$ is the time constant for the circuit. We can find the time at which the power dissipation in the resistor equals the rate at which magnetic energy is stored in the inductor by equating expressions for these rates and using the expression for I and its rate of change.

Express the rate at which magnetic energy is stored in the inductor:
$$\frac{dU_L}{dt} = \frac{d}{dt}\left[\tfrac{1}{2}LI^2\right] = LI\frac{dI}{dt}$$

Express the rate at which power is
$$P = I^2 R$$

208   Chapter 28

dissipated in the resistor:

Equate these expressions to obtain:
$$I^2 R = LI \frac{dI}{dt}$$

Simplify to obtain:
$$I = \tau \frac{dI}{dt} \qquad (1)$$

Express the current and its rate of change:
$$I = I_f \left(1 - e^{-t/\tau}\right)$$
and
$$\frac{dI}{dt} = I_f \frac{d}{dt}\left(1 - e^{-t/\tau}\right) = -I_f e^{-t/\tau}\left(-\frac{1}{\tau}\right)$$
$$= \frac{I_f}{\tau} e^{-t/\tau}$$

Substitute in equation (1) to obtain:
$$I_f \left(1 - e^{-t/\tau}\right) = \tau \left(\frac{I_f}{\tau} e^{-t/\tau}\right)$$

or
$$1 - e^{-t/\tau} = e^{-t/\tau} \Rightarrow 1 = 2e^{-t/\tau}$$

Solve for t:
$$t = -\tau \ln \tfrac{1}{2}$$

Using $\tau = 333$ μs from Example 28-11, evaluate t to obtain:
$$t = -(333\,\mu s)\ln\tfrac{1}{2} = \boxed{231\,\mu s}$$

General Problems

*85 •• Figure 28-58 shows an ac generator. It consists of a rectangular loop of dimensions a and b with N turns connected to slip rings. The loop rotates with an angular velocity ω in a uniform magnetic field $\vec{B}$. (a) Show that the potential difference between the two slip rings is $\varepsilon = NBab\omega \sin \omega t$. (b) If $a = 1$ cm, $b = 2$ cm, $N = 1000$, and $B = 2$ T, at what angular frequency ω must the coil rotate to generate an emf whose maximum value is 110 V?

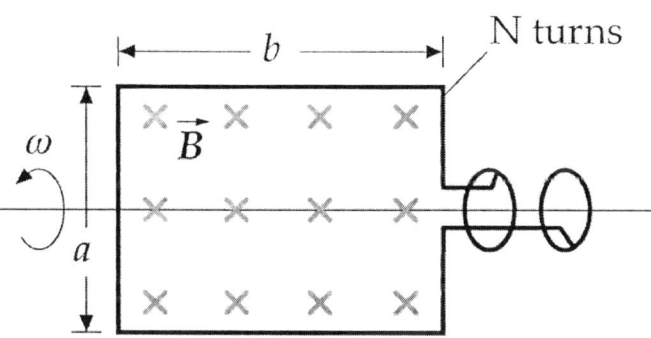

Figure 28-58 Problems 85 and 86

Picture the Problem We can apply Faraday's law and the definition of magnetic flux to derive an expression for the induced emf in the coil (potential difference between the slip rings). In part (b) we can solve this equation for ω under the given conditions.

(a) Use Faraday's law to express the induced emf:

$$\varepsilon = -\frac{d\phi_m}{dt}$$

Using the definition of magnetic flux, relate the magnetic flux through the loop to its angular velocity:

$$\phi_m(t) = NBA\cos\omega t$$

Substitute to obtain:

$$\varepsilon = -\frac{d}{dt}[NBA\cos\omega t]$$
$$= -NBab\omega(-\sin\omega t)$$
$$= \boxed{NBab\omega\sin\omega t}$$

(b) Express the condition under which $\varepsilon = \varepsilon_{max}$:

$$\sin\omega t = 1$$

Solve for and evaluate ω under this condition:

$$\omega = \frac{\varepsilon_{max}}{NBab}$$
$$= \frac{110\,\text{V}}{(1000)(2\,\text{T})(0.01\,\text{m})(0.02\,\text{m})}$$
$$= \boxed{275\,\text{rad/s}}$$

***88** ••   Show that the effective inductance for two inductors L_1 and L_2 connected in parallel, so that none of the flux from either passes through the other, is given by

$$\frac{1}{L_{eff}} = \frac{1}{L_1} + \frac{1}{L_2}.$$

Picture the Problem We can use the common potential difference across the parallel combination of inductors and the fact that the current into the parallel combination is the sum of the currents through each inductor to find an expression of the equivalent inductance.

Define L_{eff} by:
$$L_{eff} = \frac{\varepsilon}{dI/dt}$$

or

$$\frac{dI}{dt} = \varepsilon \frac{1}{L_{eff}} \quad (1)$$

Relate the common potential difference across the inductors to their inductances and the rate at which the current is changing in each:

$$\varepsilon_1 = L_1 \frac{dI_1}{dt} \quad (2)$$

and

$$\varepsilon_2 = L_2 \frac{dI_2}{dt} \quad (3)$$

Because the current divides at the parallel junction:

$$I = I_1 + I_2$$

and

$$\frac{dI}{dt} = \frac{dI_1}{dt} + \frac{dI_2}{dt}$$

Solve equations (2) and (3) for dI_1/dt and dI_2/dt and substitute to obtain:

$$\frac{dI}{dt} = \frac{\varepsilon_1}{L_1} + \frac{\varepsilon_2}{L_2}$$

Express the relationship between an emf ε applied across the parallel combination of inductors and the emfs ε_1 and ε_2 across the individual inductors:

$$\varepsilon = \varepsilon_1 = \varepsilon_2$$

Substitute to obtain:

$$\frac{dI}{dt} = \frac{\varepsilon}{L_1} + \frac{\varepsilon}{L_2} = \varepsilon\left(\frac{1}{L_1} + \frac{1}{L_2}\right)$$

Substitute in equation (1) and solve for $1/L_{eff}$:

$$\frac{1}{L_{eff}} = \boxed{\frac{1}{L_1} + \frac{1}{L_2}}$$

*89 •• Figure 28-59a shows an experiment designed to measure the acceleration of gravity. A large plastic tube is encircled by a wire, which is arranged in single loops separated by a distance of 10 cm. A strong magnet is dropped through the top of the loop. As the magnet falls through each loop, the voltage rises, then rapidly falls through 0 to a large negative value as the magnet passes through the loop, and the returns to 0. The shape of the voltage signal is shown in Figure 28-59(b). (a) Explain how this experiment works. (b) Explain why the tube cannot be made of a conductive material. (c) Qualitatively explain the shape of the voltage signal in Figure 28-59b. (d) The times at which the voltage crosses 0 as the magnet falls through each loop in succession are given in the table in the next column. Use these data to calculate a value for g.

(a)

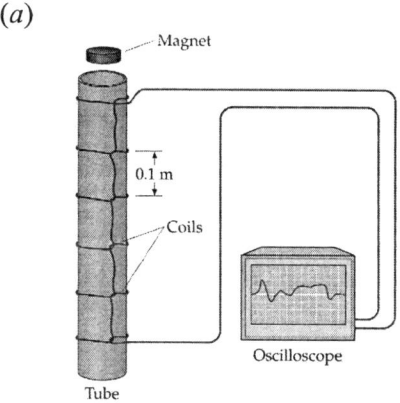

(b)

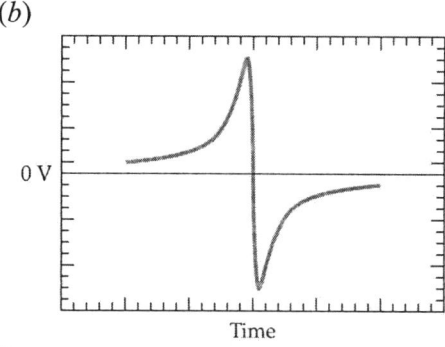

Figure 28-59 Problem 89

Loop Number	Zero crossing time (s)
1	0.011189
2	0.063133
3	0.10874
4	0.14703
5	0.18052
6	0.21025
7	0.23851
8	0.26363
9	0.28853
10	0.31144
11	0.33494
12	0.35476
13	0.37592
14	0.39107

Picture the Problem

(a) As the magnet passes through the coil, it induces an emf because of the changing flux through the coil. This allows the coil to "sense" when the magnet is passing through it.

(b) One cannot use a cylinder made of conductive material because eddy currents induced in it by a falling magnet would slow the magnet.

(c) As the magnet approaches the loop, the flux increases, resulting in the increasing voltage signal. When the magnet is passing the coil, the flux goes from increasing to decreasing, so the induced emf becomes zero and then negative. The time at which the induced emf is zero is the time at which the magnet is at the center of the coil.

(d) Each time represents a point when the distance has increased by 10 cm. The following graph of distance versus time was plotted using a spreadsheet program. The regression curve, obtained using Excel's "Add Trendline" feature, is shown as a dashed line.

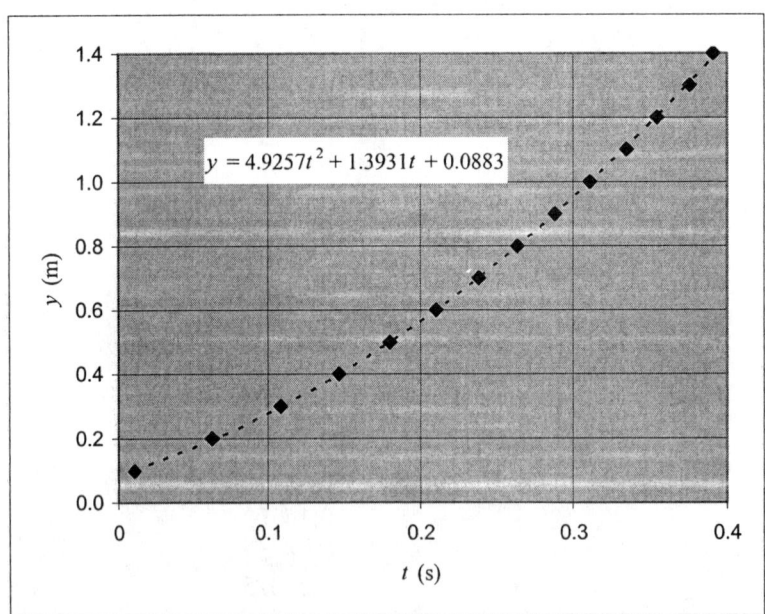

The coefficient of the second-degree term is $\frac{1}{2}g$. Consequently,

$$g = 2(4.9257 \text{ m/s}^2) = \boxed{9.85 \text{ m/s}^2}$$

*91 •• Suppose the coil of Problem 90 is rotated about its vertical centerline at constant angular velocity of 2 rad/s. Find the induced current as a function of time.

Picture the Problem We can apply Faraday's law and the definition of magnetic flux to derive an expression for the induced emf in the coil. We can then apply Ohm's law to find the induced current as a function of time. Note that only half of the loop is in the magnetic field.

Apply Ohm's law to relate the induced current to the induced emf:
$$I(t) = \frac{\varepsilon(t)}{R} \quad (1)$$

Use Faraday's law to express the induced emf:
$$\varepsilon(t) = -\frac{d\phi_m(t)}{dt}$$

Using the definition of magnetic flux, relate the magnetic flux through the loop to its angular velocity:
$$\phi_m(t) = NBA\cos\omega t$$

Substitute to obtain:
$$\varepsilon(t) = -\frac{d}{dt}[NBA\cos\omega t]$$
$$= -NBA\omega(-\sin\omega t)$$
$$= NBA\omega\sin\omega t$$

Substitute in equation (1) to obtain:
$$I(t) = \frac{NBA\omega}{R}\sin\omega t$$

Substitute numerical values and evaluate $I(t)$:

$$I(t) = \frac{(80)(1.4\,\text{T})(0.25\,\text{m})(0.15\,\text{m})(2\,\text{rad/s})}{24\,\Omega}\sin(2\,\text{rad/s})t$$
$$= \boxed{(0.350\,\text{A})\sin(2\,\text{rad/s})t}$$

*96 ••• Figure 28-62 shows a rectangular loop of wire, 0.30 m wide and 1.50 m long, in the vertical plane and perpendicular to a uniform magnetic field $B = 0.40$ T, directed inward as shown. The portion of the loop not in the magnetic field is 0.10 m long. The resistance of the loop is 0.20 Ω and its mass is 0.05 kg. The loop is released from rest at $t = 0$. (*a*) What is the magnitude and direction of the induced current when the loop has a downward velocity *v*? (*b*) What is the force that acts on the loop as a result of this current? (*c*) What is the net force acting on the loop? (*d*) Write the equation of motion of the loop. (*e*) Obtain an expression for the velocity of the loop as a function of time. (*f*) Integrate the expression obtained in Part (*e*) to find the displacement *y* as a function of time. (*g*) Using a spreadsheet program, make a graph of the position *y* of the loop as a function of time for values of *y* between 0 m and 1.4 m (i.e., when the loop leaves the magnetic field). At what time *t* does *y* = 1.4 m? Compare this to the time it would have taken if *B* = 0.

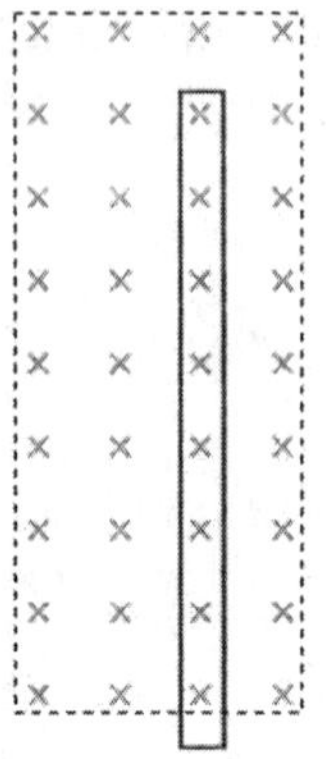

Figure 28-62 Problems 96 and 97

Picture the Problem We can use $I = \varepsilon/R$ and $\varepsilon = Bv\ell$ to find the current induced in the loop and Lenz's law to determine its direction. We can apply the equation for the force on a current-carrying wire to find the net magnetic force acting on the loop and then sum the forces to find the net force on the loop. Separating the variables in the differential equation and integrating will lead us to an expression for $v(t)$ and a second integration to an expression for $y(t)$. We can solve the latter equation for $y = 1.40$ m to find the time it takes the loop to exit the magnetic field and our expression for $v(t)$ to find its exit speed. Finally, we can use a constant-acceleration equation to find its exit speed in the absence of the magnetic field.

(*a*) Relate the magnitude of the induced current to the induced emf and the resistance of the loop:

$$I = \frac{\varepsilon}{R}$$

Relate the induced emf to the motion of the loop:

$$\varepsilon = Bv\ell$$

Substitute for ε to obtain:

$$\boxed{I = \frac{B\ell}{R}v}$$

As the loop falls, the flux into the page decreases. The direction of the induced current is such that its magnetic field opposes this decrease, i.e., clockwise.

(*b*) Express the velocity-dependent force that acts on the loop in terms

$$F_v = BI\ell$$

of the current in the loop:

Substitute for I to obtain:
$$F_v = B\left(\frac{B\ell}{R}\right)v\ell = \boxed{\frac{B^2\ell^2}{R}v}$$

> Apply $d\vec{F} = Id\vec{\ell} \times \vec{B}$ to the horizontal portion of the loop that is in the magnetic field to conclude that the net magnetic force is upward.

Note that the magnetic force on the left side of the loop is to the left and the magnetic force on the right side of the loop is to the right.

(c) The net force acting on the loop is the difference between the downward gravitational force and the upward magnetic force:
$$F_{net} = mg - F_v$$
$$= \boxed{mg - \frac{B^2\ell^2}{R}v}$$

(d) Apply Newton's 2nd law of motion to the loop to obtain its equation of motion:
$$mg - \frac{B^2\ell^2}{R}v = m\frac{dv}{dt}$$
or
$$\boxed{\frac{dv}{dt} = g - \frac{B^2\ell^2}{mR}v}$$

Factor g to obtain an alternate form of the equation of motion:
$$\frac{dv}{dt} = g\left(1 - \frac{B^2\ell^2}{mgR}v\right) = \boxed{g\left(1 - \frac{v}{v_t}\right)}$$

where $v_t = \frac{mgR}{B^2\ell^2}$

(e) Separate the variables to obtain:
$$\frac{dv}{g - \frac{B^2\ell^2}{mR}v} = dt$$

or
$$\frac{dv}{a - bv} = dt$$

where $a = g$ and $b = \frac{B^2\ell^2}{mR}$

Integrate v' from 0 to v and t' from 0 to t:
$$\int_0^v \frac{dv'}{a - bv'} = \int_0^t dt' \Rightarrow -\frac{1}{b}\ln\left(\frac{a - bv}{a}\right) = t$$

Transform from logarithmic to
exponential form and solve for v to
obtain:
$$v(t) = \frac{a}{b}\left(1 - e^{-bt}\right)$$

Noting that $v_t = \dfrac{a}{b}$, we have:
$$v(t) = \boxed{v_t\left(1 - e^{-t/\tau}\right)}$$
where $\tau = \dfrac{v_t}{a} = \dfrac{v_t}{g}$.

(f) Write v as dy/dt and separate variables to obtain:
$$dy = v_t\left(1 - e^{-t/\tau}\right)dt$$

Integrate y' from 0 to y and t' from 0 to t:

$$\int_0^y dy' = v_t \int_0^t \left(1 - e^{-t'/\tau}\right)dt'$$

and

$$y(t) = \boxed{v_t\left[t - \tau\left(1 - e^{-t/\tau}\right)\right]}$$

(g) A spreadsheet program to generate the data for graphs of position y as a function of time t is shown below. The formulas used to calculate the quantities in the columns are as follows:

Cell	Formula/Content	Algebraic Form
B1	0.05	m
B2	0.2	R
B3	0.4	B
B4	0.3	L
B5	$B$1*$B$7*$B$2/($B$3^2*$B$4^2)	v_t
B6	$B$5/$B$7	τ
B7	9.81	g
A10	0.00	t
B10	$B$5*(A10−$B$6*(1−EXP(−A10/$B$6)))	y
C10	0.5*$B$7*A10^2	$\tfrac{1}{2}gt^2$

	A	B	C
1	m=	0.05	kg
2	R=	0.2	ohms
3	B=	0.4	T
4	L=	0.3	m
5	vt=	6.813	m/s
6	tau=	0.694	s
7	g=	9.81	m/s^2
8			

9	t	y	y (no B)
10	0.00	0.000	0.000
11	0.05	0.012	0.012
12	0.10	0.047	0.049
13	0.15	0.103	0.110
14	0.20	0.179	0.196
15	0.25	0.273	0.307
16	0.30	0.384	0.441
17	0.35	0.511	0.601
18	0.40	0.654	0.785
19	0.45	0.809	0.993
20	0.50	0.978	1.226
21	0.55	1.159	1.484
22	0.60	1.351	1.766
23	0.65	1.553	2.072
24	0.70	1.764	2.403
25	0.75	1.985	2.759
26	0.80	2.214	3.139
27	0.85	2.451	3.544
28	0.90	2.695	3.973
29	0.95	2.946	4.427
30	1.00	3.202	4.905

Examining the table, we see that $y = 1.4$ m when $t \approx$ $\boxed{0.60 \text{ s}}$.

The following graph shows y as a function of t for $B \neq 0$ (solid curve) and $B = 0$ (dashed curve).

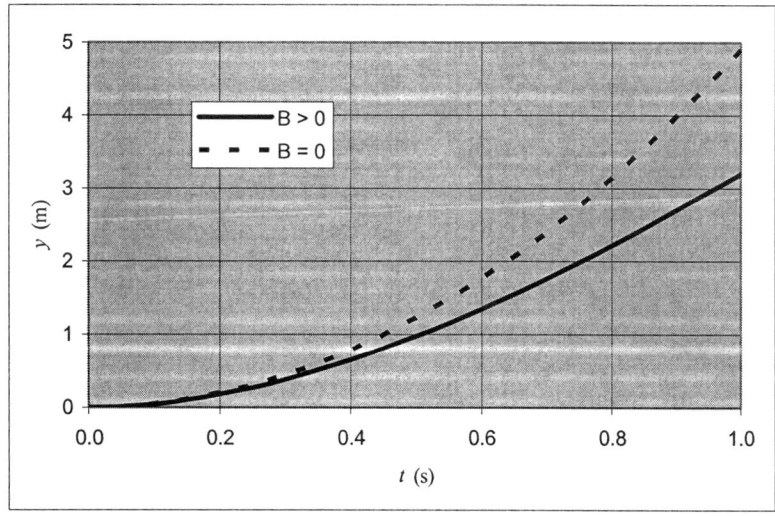

Chapter 29
Alternating-Current Circuits

Conceptual Problems

*1 • As the frequency in the simple ac circuit in Figure 29-27 increases, the rms current through the resistor (a) increases. (b) does not change. (c) may increase or decrease depending on the magnitude of the original frequency.
(d) may increase or decrease depending on the magnitude of the resistance.
(e) decreases.

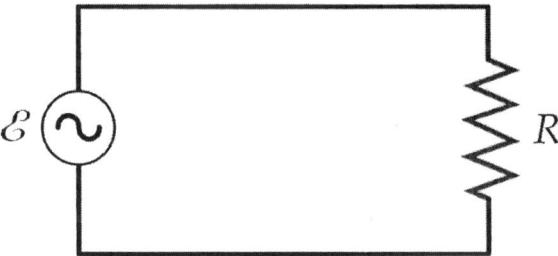

Figure 29-27 Problem 1

Determine the Concept Because the rms current through the resistor is given by $I_{rms} = \mathcal{E}_{rms}/R$ and both $\mathcal{E}_{rms}$ and R are independent of frequency, $\boxed{(b) \text{ is correct.}}$

*5 • If the frequency in the circuit in Figure 29-29 is doubled, the capacitive reactance of the circuit will (a) increase by a factor of 2. (b) not change. (c) decrease by a factor of 2. (d) increase by a factor of 4. (e) decrease by a factor of 4.

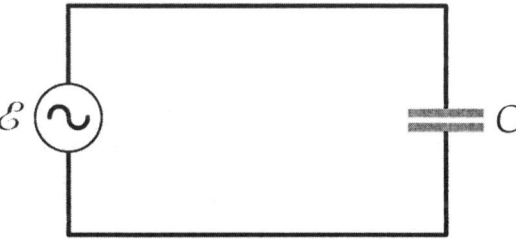

Figure 29-29 Problem 5

Determine the Concept The capacitive reactance of an capacitor varies with the frequency according to $X_C = 1/\omega C$. Hence, doubling ω will halve X_C. $\boxed{(c) \text{ is correct.}}$

*9 •• Making LC circuits with oscillation frequencies of thousands of hertz or more is easy, but making LC circuits that have small frequencies is difficult. Why?

Determine the Concept To make an *LC* circuit with a small resonance frequency requires a large inductance and large capacitance. Neither is easy to construct.

*12 • Are there any disadvantages to having a radio tuning circuit with an extremely large *Q* factor?

Determine the Concept Yes; the bandwidth must be wide enough to accommodate the modulation frequency.

Estimation and Approximation

*18 •• The impedances of motors, transformers, and electromagnets have inductive reactance. Suppose that the phase angle of the total impedance of a large industrial plant is 25° when the plant is under full operation and using 2.3 MW of power. The power is supplied to the plant from a substation 4.5 km from the plant; the 60 Hz rms line voltage at the plant is 40,000 V. The resistance of the transmission line from the substation to the plant is 5.2 Ω. The cost per kilowatt-hour is 0.07 dollars. The plant pays only for the actual energy used. (*a*) What are the resistance and inductive reactance of the plant's total load? (*b*) What is the current in the power lines and what must be the rms voltage at the substation to maintain the voltage at the plant at 40,000 V? (*c*) How much power is lost in transmission? (*d*) Suppose that the phase angle of the plant's impedance were reduced to 18° by adding a bank of capacitors in series with the load. How much money would be saved by the electric utility during one month of operation, assuming the plant operates at full capacity for 16 h each day? (*e*) What must be the capacitance of this bank of capacitors?

Picture the Problem We can find the resistance and inductive reactance of the plant's total load from the impedance of the load and the phase constant. The current in the power lines can be found from the total impedance of the load the potential difference across it and the rms voltage at the substation by applying Kirchhoff's loop rule to the substation-transmission wires-load circuit. The power lost in transmission can be found from $P_{trans} = I_{rms}^2 R_{trans}$. We can find the cost savings by finding the difference in the power lost in transmission when the phase angle is reduced to 18°. Finally, we can find the capacitance that is required to reduce the phase angle to 18° by first finding the capacitive reactance using the definition of $\tan\delta$ and then applying the definition of capacitive reactance to find *C*.

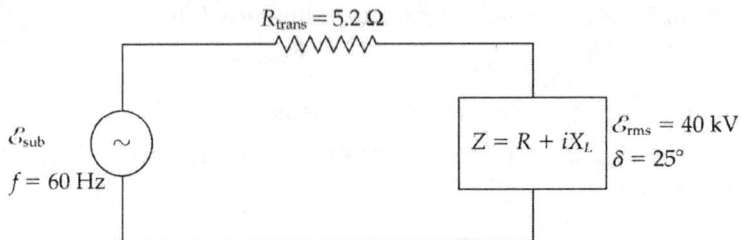

Alternating-Current Circuits   221

(a) Relate the resistance and inductive reactance of the plant's total load to Z and δ:

$R = Z \cos\delta$
and
$X_L = Z \sin\delta$

Express Z in terms of the current I in the power lines and voltage ε_{rms} at the plant:

$Z = \dfrac{\varepsilon_{rms}}{I}$

Express the power delivered to the plant in terms of ε_{rms}, I_{rms}, and δ and solve for I_{rms}:

$P_{av} = \varepsilon_{rms} I_{rms} \cos\delta$
and
$I_{rms} = \dfrac{P_{av}}{\varepsilon_{rms} \cos\delta}$ (1)

Substitute to obtain:

$Z = \dfrac{\varepsilon_{rms}^2 \cos\delta}{P_{av}}$

Substitute numerical values and evaluate Z:

$Z = \dfrac{(40\,\text{kV})^2 \cos 25°}{2.3\,\text{MW}} = 630\,\Omega$

Substitute numerical values and evaluate R and X_L:

$R = (630\,\Omega)\cos 25° = \boxed{571\,\Omega}$
and
$X_L = (630\,\Omega)\sin 25° = \boxed{266\,\Omega}$

(b) Use equation (1) to find the current in the power lines:

$I_{rms} = \dfrac{2.3\,\text{MW}}{(40\,\text{kV})\cos 25°} = \boxed{63.4\,\text{A}}$

Apply Kirchhoff's loop rule to the circuit:

$\varepsilon_{sub} - I_{rms} R_{trans} - I Z_{tot} = 0$

Solve for ε_{sub}:

$\varepsilon_{sub} = I_{rms}(R_{trans} + Z_{tot})$

Evaluate Z_{tot}:

$Z_{tot} = \sqrt{R^2 + X_L^2}$
$= \sqrt{(571\,\Omega)^2 + (266\,\Omega)^2} = 630\,\Omega$

Substitute numerical values and evaluate ε_{sub}:

$\varepsilon_{sub} = (63.4\,\text{A})(5.2\,\Omega + 630\,\Omega)$
$= \boxed{40.3\,\text{kV}}$

222 Chapter 29

(c) The power lost in transmission is:

$$P_{trans} = I_{rms}^2 R_{trans} = (63.4\,A)^2(5.2\,\Omega)$$
$$= \boxed{20.9\,kW}$$

(d) Express the cost savings ΔC in terms of the difference in energy consumption $(P_{25°} - P_{18°})\Delta t$ and the per-unit cost u of the energy:

$$\Delta C = (P_{25°} - P_{18°})\Delta t u$$

Express the power list in transmission when $\delta = 18°$:

$$P_{18°} = I_{18°}^2 R_{trans}$$

Find the current in the transmission lines when $\delta = 18°$:

$$I_{18°} = \frac{2.3\,MW}{(40\,kV)\cos 18°} = 60.5\,A$$

Evaluate $P_{18°}$:

$$P_{18°} = (60.5\,A)^2(5.2\,\Omega) = 19.0\,kW$$

Substitute numerical values and evaluate ΔC:

$$\Delta C = (20.9\,kW - 19.0\,kW)(16\,h/d)(30\,d/month)(\$0.07/kW\cdot h) = \boxed{\$63.84}$$

Relate the new phase angle δ to the inductive reactance X_L, the reactance due to the added capacitance X_C, and the resistance of the load R:

$$\tan\delta = \frac{X_L - X_C}{R}$$

Solve for and evaluate X_C:

$$X_C = X_L - R\tan\delta$$
$$= 266\,\Omega - (571\,\Omega)\tan 18° = 80.5\,\Omega$$

Substitute numerical values and evaluate C:

$$C = \frac{1}{2\pi(60\,s^{-1})(80.5\,\Omega)} = \boxed{33.0\,\mu F}$$

Alternating Current Generators

*21 • A 2-cm by 1.5-cm rectangular coil has 300 turns and rotates in a magnetic field of 4000 G. (a) What is the maximum emf generated when the coil rotates at 60 Hz? (b) What must its frequency be to generate a maximum emf of 110 V?

Picture the Problem We can use the relationship $\mathcal{E}_{max} = 2\pi NBAf$ to relate the maximum emf generated to the area of the coil, the number of turns of the coil, the magnetic field in which the coil is rotating, and the frequency at which it rotates.

(a) Relate the induced emf to the magnetic field in which the coil is rotating:

$$\mathcal{E}_{max} = NBA\omega = 2\pi NBAf \quad (1)$$

Substitute numerical values and evaluate $\mathcal{E}_{max}$:

$$\mathcal{E}_{max} = 2\pi(300)(0.4\,\text{T})(2\times10^{-2}\,\text{m})(1.5\times10^{-2}\,\text{m})(60\,\text{s}^{-1}) = \boxed{13.6\,\text{V}}$$

(b) Solve equation (1) for f:

$$f = \frac{\mathcal{E}_{max}}{2\pi NBA}$$

Substitute numerical values and evaluate f:

$$f = \frac{110\,\text{V}}{2\pi(300)(0.4\,\text{T})(2\times10^{-2}\,\text{m})(1.5\times10^{-2}\,\text{m})} = \boxed{486\,\text{Hz}}$$

Alternating Current in a Resistor

***23 •** A 100-W light bulb is plugged into a standard 120-V (rms) outlet. Find (a) I_{rms}, (b) I_{max}, and (c) the maximum power.

Picture the Problem We can use $P_{av} = \mathcal{E}_{rms} I_{rms}$ to find I_{rms}, $I_{max} = \sqrt{2} I_{rms}$ to find I_{max}, and $P_{max} = I_{max} \mathcal{E}_{max}$ to find P_{max}.

(a) Relate the average power delivered by the source to the rms voltage across the bulb and the rms current through it:

$$P_{av} = \mathcal{E}_{rms} I_{rms}$$

Solve for and evaluate I_{rms}:

$$I_{rms} = \frac{P_{av}}{\mathcal{E}_{rms}} = \frac{100\,\text{W}}{120\,\text{V}} = \boxed{0.833\,\text{A}}$$

(b) Express I_{max} in terms of I_{rms}:

$$I_{max} = \sqrt{2} I_{rms}$$

Substitute for I_{rms} and evaluate I_{max}:

$$I_{max} = \sqrt{2}(0.833\,\text{A}) = \boxed{1.18\,\text{A}}$$

(c) Express the maximum power in terms of the maximum voltage and

$$P_{max} = I_{max} \mathcal{E}_{max}$$

224   Chapter 29

maximum current:

Substitute numerical values and evaluate P_{max}: $\quad P_{max} = (1.18\,\text{A})\sqrt{2}(120\,\text{V}) = \boxed{200\,\text{W}}$

Alternating Current in Inductors and Capacitors

*29 •   An emf of 10 V maximum and frequency 20 Hz is applied to a 20-μF capacitor. Find (*a*) I_{max} and (*b*) I_{rms}.

Picture the Problem We can use $I_{max} = \varepsilon_{max}/X_C$ and $X_C = 1/\omega C$ to express I_{max} as a function of ε_{max}, f, and C. Once we've evaluate I_{max}, we can use $I_{rms} = I_{max}/\sqrt{2}$ to find I_{rms}.

Express I_{max} in terms of ε_{max} and X_C: $\quad I_{max} = \dfrac{\varepsilon_{max}}{X_C}$

Express the capacitive reactance: $\quad X_C = \dfrac{1}{\omega C} = \dfrac{1}{2\pi f C}$

Substitute to obtain: $\quad I_{max} = 2\pi f C \varepsilon_{max}$

(*a*) Substitute numerical values and evaluate I_{max}:
$$I_{max} = 2\pi (20\,\text{s}^{-1})(20\,\mu\text{F})(10\,\text{V})$$
$$= \boxed{25.1\,\text{mA}}$$

(*b*) Express I_{rms} in terms of I_{max}: $\quad I_{rms} = \dfrac{I_{max}}{\sqrt{2}} = \dfrac{25.1\,\text{mA}}{\sqrt{2}} = \boxed{17.8\,\text{mA}}$

LC and *RLC* Circuits without a Generator

*32 •   Show from the definitions of the henry and the farad that $1/\sqrt{LC}$ has the unit s^{-1}.

Picture the Problem We can use $X_L = \omega L$ and $X_C = 1/\omega C$ to show the $1/\sqrt{LC}$ has the unit s^{-1}. Alternatively, we can use the dimensions of C and L to establish this result.

Substitute the units for L and C in the expression $1/\sqrt{LC}$ to obtain:
$$\dfrac{1}{\sqrt{\text{H}\cdot\text{F}}} = \dfrac{1}{\sqrt{(\Omega\cdot\text{s})\left(\dfrac{\text{s}}{\Omega}\right)}} = \dfrac{1}{\sqrt{\text{s}^2}} = \boxed{\text{s}^{-1}}$$

Alternatively, use the defining equation $(C = Q/V)$ for capacitance to obtain the dimension of C:

$$[C] = \frac{[Q]}{[V]}$$

Solve the defining equation $(V = L\, dI/dt)$ for inductance to obtain the dimension of L:

$$[L] = \frac{[V]}{\left[\frac{dI}{dt}\right]} = \frac{[V]}{\frac{[Q]}{[T]^2}} = \frac{[V][T]^2}{[Q]}$$

Express the dimension of $1/\sqrt{LC}$:

$$\left[\frac{1}{\sqrt{LC}}\right] = \left[\frac{1}{\sqrt{[L][C]}}\right] = \left[\frac{1}{\sqrt{\frac{[V][T]^2}{[Q]}\frac{[Q]}{[V]}}}\right]$$

$$= \left[\frac{1}{\sqrt{[T]^2}}\right] = \frac{1}{[T]}$$

Because the SI unit of time is the second, we've shown that $1/\sqrt{LC}$ has units of $\boxed{s^{-1}}$.

*37 •• An inductor and a capacitor are connected, as shown in Figure 29-30. With the switch open, the left plate of the capacitor has charge Q_0. The switch is closed and the charge and current vary sinusoidally with time. (*a*) Plot both Q versus t and I versus t and explain how to interpret these two plots to illustrate that the current leads the charge by 90°. (*b*) Using a trig identity, show the expression for the current (Equation 29-38) leads the expression for the charge (Equation 29-39) by 90°. That is, show that

$$I = -I_{peak}\sin\omega t = I_{peak}\cos(\omega t + \frac{\pi}{2}).$$

Figure 29-30 Problem 37

Picture the Problem Let Q represent the instantaneous charge on the capacitor and apply Kirchhoff's loop rule to obtain the differential equation for the circuit. We can then solve this equation to obtain an expression for the charge on the capacitor as a function of time and, by differentiating this expression with respect to time, an expression for the current as a function of time. We'll use a spreadsheet program to plot the graphs.

226   Chapter 29

Apply Kirchhoff's loop rule to a clockwise loop just after the switch is closed:

$$\frac{Q}{C} + L\frac{dI}{dt} = 0$$

Because $I = dQ/dt$:

$$L\frac{d^2Q}{dt^2} + \frac{Q}{C} = 0 \text{ or } \frac{d^2Q}{dt^2} + \frac{1}{LC}Q = 0$$

The solution to this equation is:

$$Q(t) = Q_0 \cos(\omega t - \delta)$$

where $\omega = \sqrt{\dfrac{1}{LC}}$

Because $Q(0) = Q_0$, $\delta = 0$ and:

$$Q(t) = Q_0 \cos \omega t$$

The current in the circuit is the derivative of Q with respect to t:

$$I = \frac{dQ}{dt} = \frac{d}{dt}[Q_0 \cos \omega t] = -\omega Q_0 \sin \omega t$$

(a) A spreadsheet program was used to plot the following graph showing both the charge on the capacitor and the current in the circuit as functions of time. L, C, and Q_0 were all arbitrarily set equal to one to obtain these graphs. Note that the current leads the charge by one-fourth of a cycle or 90°.

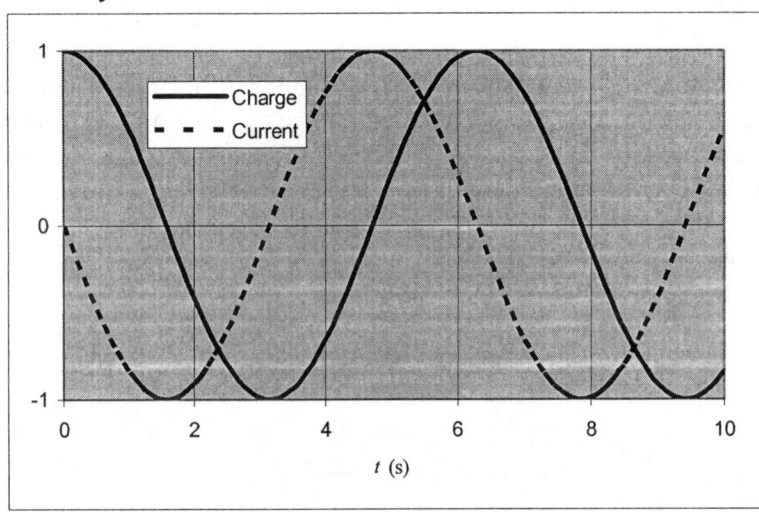

(b) The equation for the current is:

$$I = -\omega Q_0 \sin \omega t \qquad (1)$$

The sine and cosine functions are related through the identity:

$$-\sin \theta = \cos\left(\theta + \frac{\pi}{2}\right)$$

Use this identity to rewrite equation (1):

$$I = -\omega Q_0 \sin \omega t = \boxed{\omega Q_0 \cos\left(\omega t + \frac{\pi}{2}\right)}$$

showing that the current leads the charge by 90°.

RL Circuits with a Generator

*44 •• A resistor and an inductor are connected in parallel across an emf $\varepsilon = \varepsilon_{max}$ as shown in Figure 29-32. Show that (*a*) the current in the resistor is $I_R = (\varepsilon_{max}/R)\cos\omega t$, (*b*) the current in the inductor is $I_L = (\varepsilon_{max}/X_L)\cos(\omega t - 90°)$, and (*c*) $I = I_R + I_L = I_{max}\cos(\omega t - \delta)$, where $\tan\delta = R/X_L$ and $I_{max} = \varepsilon_{max}/Z$ with $Z^{-2} = R^{-2} + X_L^{-2}$.

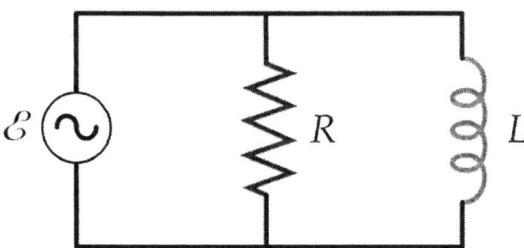

Figure 29-32 Problem 44

Picture the Problem We can apply Kirchhoff's loop rule to obtain expressions for I_R and I_L and then use trigonometric identities to show that $I = I_R + I_L = I_{max}\cos(\omega t - \delta)$, where $\tan\delta = R/X_L$ and $I_{max} = \varepsilon_{max}/Z$ with $Z^{-2} = R^{-2} + X_L^{-2}$.

(*a*) Apply Kirchhoff's loop rule to a clockwise loop that includes the source and the resistor:

$\varepsilon_{max}\cos\omega t - I_R R = 0$

Solve for I_R:

$$I_R = \boxed{\frac{\varepsilon_{max}}{R}\cos\omega t}$$

(*b*) Apply Kirchhoff's loop rule to a clockwise loop that includes the source and the inductor:

$\varepsilon_{max}\cos(\omega t - 90°) - I_L X_L = 0$

because the current lags the potential difference across the inductor by 90°.

Solve for I_L:

$$I_L = \boxed{\frac{\varepsilon_{max}}{X_L}\cos(\omega t - 90°)}$$

(*c*) Express the current drawn from the source in terms of I_{max} and the phase constant δ:

$I = I_R + I_L = I_{max}\cos(\omega t - \delta)$

Use a trigonometric identity to expand $\cos(\omega t - \delta)$:

$I = I_{max}(\cos\omega t\cos\delta + \sin\omega t\sin\delta)$
$= I_{max}\cos\omega t\cos\delta + I_{max}\sin\omega t\sin\delta$

228 Chapter 29

From our results in (a):

$$I = I_R + I_L = \frac{\mathcal{E}_{max}}{R}\cos\omega t$$
$$+ \frac{\mathcal{E}_{max}}{X_L}\cos(\omega t - 90°)$$
$$= \frac{\mathcal{E}_{max}}{R}\cos\omega t + \frac{\mathcal{E}_{max}}{X_L}\sin\omega t$$

A useful trigonometric identity is:

$$A\cos\omega t + B\sin\omega t = \sqrt{A^2 + B^2}\cos(\omega t - \delta)$$

where

$$\delta = \tan^{-1}\frac{B}{A}$$

Apply this identity to obtain:

$$I = \sqrt{\left(\frac{\mathcal{E}_{max}}{R}\right)^2 + \left(\frac{\mathcal{E}_{max}}{X_L}\right)^2}\cos(\omega t - \delta) \quad (1)$$

and

$$\delta = \tan^{-1}\left(\frac{\frac{\mathcal{E}_{max}}{X_L}}{\frac{\mathcal{E}_{max}}{R}}\right) = \tan^{-1}\left(\frac{R}{X_L}\right) \quad (2)$$

Simplify equation (1) and rewrite equation (2) to obtain:

$$I = \sqrt{\left(\frac{\mathcal{E}_{max}}{R}\right)^2 + \left(\frac{\mathcal{E}_{max}}{X_L}\right)^2}\cos(\omega t - \delta)$$
$$= \mathcal{E}_{max}\sqrt{\left(\frac{1}{R}\right)^2 + \left(\frac{1}{X_L}\right)^2}\cos(\omega t - \delta)$$
$$= \mathcal{E}_{max}\sqrt{\left(\frac{1}{Z}\right)^2}\cos(\omega t - \delta)$$
$$= \boxed{\frac{\mathcal{E}_{max}}{Z}\cos(\omega t - \delta)}$$

where

$$\tan\delta = \boxed{\frac{R}{X_L}} \text{ and } \frac{1}{Z^2} = \frac{1}{R^2} + \frac{1}{X_L^2}$$

Filters and Rectifiers

*48 •• The circuit shown in Figure 29-35 is called an *RC* high-pass filter because it transmits signals with a high-input frequency with greater amplitude than low-frequency signals. If the input voltage is $V_{in}(t) = V_{peak} \cos \omega t$, show that the output voltage is $V_{out}(t) = V_H \cos(\omega t - \delta)$ where

$$V_H = \frac{V_{peak}}{\sqrt{1 + \left(\frac{1}{\omega RC}\right)^2}}$$

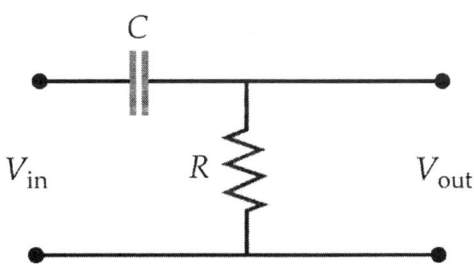

Figure 29-35 Problem 48

Picture the Problem We can use Kirchhoff's loop rule to obtain a differential equation relating the input, capacitor, and resistor voltages. We'll then assume a solution to this equation that is a linear combination of sine and cosine terms with coefficients that we can find by substitution in the differential equation. Repeating this process for the output side of the filter will yield the desired equation.

Apply Kirchhoff's loop rule to the input side of the filter to obtain:	$V_{in} - V - IR = 0$ where V is the potential difference across the capacitor.
Substitute for V_{in} and I to obtain:	$V_{peak} \cos \omega t - V - R\frac{dQ}{dt} = 0$
Because $Q = CV$:	$\frac{dQ}{dt} = \frac{d}{dt}[CV] = C\frac{dV}{dt}$
Substitute for dQ/dt to obtain:	$V_{peak} \cos \omega t - V - RC\frac{dV}{dt} = 0$ the differential equation describing the potential difference across the capacitor.
Assume a solution of the form:	$V = V_c \cos \omega t + V_s \sin \omega t$
Substitution of this assumed solution and its first derivative in the differential equations, followed by equating the coefficients of the sine and cosine terms, yields two coupled linear equations:	$V_c + \omega RC V_s = V_{peak}$ and $V_s - \omega RC V_c = 0$

230 Chapter 29

Solve these equations simultaneously to obtain:	$V_c = \dfrac{1}{1+(\omega RC)^2} V_{peak}$
	and
	$V_s = \dfrac{\omega RC}{1+(\omega RC)^2} V_{peak}$
Note that the output voltage is the voltage across the resistor and that it is phase shifted relative to the input voltage:	$V_{out} = V_H \cos(\omega t - \delta)$ where V_H is the amplitude of the signal.
Assume that V_H is of the form:	$V_H(t) = v_c \cos \omega t + v_s \sin \omega t$
The input, output, and capacitor voltages are related according to:	$V_H(t) = V_{in}(t) - V(t)$
Substitute for $V_H(t)$, $V_{peak}(t)$, and $V(t)$ and use the previously established values for V_c and V_s to obtain:	$v_c = V_{peak} - V_c$ and $v_s = -V_s$
Substitute for V_c and V_s to obtain:	$v_c = \dfrac{(\omega RC)^2}{1+(\omega RC)^2} V_{peak}$
	and
	$v_s = -\dfrac{\omega RC}{1+(\omega RC)^2} V_{peak}$
V_H, v_c, and v_s are related according to the Pythagorean relationship:	$V_H = \sqrt{v_c^2 + v_s^2}$
Substitute for v_c and v_s to obtain:	$V_H = \dfrac{\omega RC}{\sqrt{1+(\omega RC)^2}} V_{peak}$
	$= \boxed{\dfrac{V_{peak}}{\sqrt{1+\left(\dfrac{1}{\omega RC}\right)^2}}}$

*53 •• Show that the average power dissipated in the resistor of the high-pass filter of Problem 48 is given by

$$P_{ave} = \dfrac{V_{peak}^2}{2R}\left(\dfrac{(\omega RC)^2}{1+(\omega RC)^2}\right)$$

Picture the Problem We can express the instantaneous power dissipated in the resistor

Alternating-Current Circuits 231

and then use the fact that the average value of the square of the cosine function over one cycle is ½ to establish the given result.

The instantaneous power $P(t)$ dissipated in the resistor is:
$$P(t) = \frac{V_{out}^2}{R}$$

The output voltage V_{out} is:
$$V_{out} = V_H \cos(\omega t - \delta)$$

From Problem 48:
$$V_H = \frac{V_{peak}}{\sqrt{1 + \left(\frac{1}{\omega RC}\right)^2}}$$

Substitute in the expression for $P(t)$ to obtain:
$$P(t) = \frac{V_H^2}{R} \cos^2(\omega t - \delta)$$

$$= \frac{V_{peak}^2}{R\left[1 + \left(\frac{1}{\omega RC}\right)^2\right]} \cos^2(\omega t - \delta)$$

Because the average value of the square of the cosine function over one cycle is ½:
$$P_{ave} = \frac{V_{peak}^2}{2R\left[1 + \left(\frac{1}{\omega RC}\right)^2\right]}$$

Simplify this expression to obtain:
$$\boxed{P_{ave} = \frac{V_{peak}^2}{2R}\left(\frac{(\omega RC)^2}{1 + (\omega RC)^2}\right)}$$

*57 •• Using a spreadsheet program, make a graph of V_L versus $f = \omega/2\pi$ and δ versus f for the low-pass filter of Problem 55. Use $R = 10$ kΩ and $C = 5$ nF.

Picture the Problem We can use the expressions for V_L and δ derived in Problem 56 to plot the graphs of V_L versus f and δ versus f for the low-pass filter of Problem 55. We'll simplify the spreadsheet program by expressing both V_L and δ as functions of $f_{3 \text{ dB}}$.

From Problem 56 we have:
$$V_L = \frac{V_{peak}}{\sqrt{1 + (\omega RC)^2}}$$
and
$$\delta = \tan^{-1}(\omega RC)$$

Rewrite each of these expressions in terms of $f_{3 \text{ dB}}$ to obtain:
$$V_L = \frac{V_{peak}}{\sqrt{1 + (2\pi f RC)^2}} = \frac{V_{peak}}{\sqrt{1 + \left(\frac{f}{f_{3 \text{ dB}}}\right)^2}}$$

232 Chapter 29

and

$$\delta = \tan^{-1}(2\pi fRC) = \tan^{-1}\left(\frac{f}{f_{3\,dB}}\right)$$

A spreadsheet program to generate the data for graphs of V_L versus f and δ versus f for the low-pass filter is shown below. Note that V_{peak} has been arbitrarily set equal to 1 V. The formulas used to calculate the quantities in the columns are as follows:

Cell	Formula/Content	Algebraic Form
B1	2.00E+03	R
B2	5.00E−09	C
B3	1	V_{peak}
B4	(2*PI()*$B$1*$B$2)^−1	$f_{3\,dB}$
B8	$B$3/SQRT(1+((A8/$B$4)^2))	$\dfrac{V_{peak}}{\sqrt{1+\left(\dfrac{f}{f_{3\,dB}}\right)^2}}$
C8	ATAN(A8/$B$4)	$\tan^{-1}\left(\dfrac{f}{f_{3\,dB}}\right)$
D8	C8*180/PI()	δ in degrees

	A	B	C	D
1	R=	1.00E+04	ohms	
2	C=	5.00E−09	F	
3	V_peak=	1	V	
4	f_3 dB=	3.183	kHz	
5				
6	f(kHz)	V_out	delta(rad)	delta(deg)
7	0	1.000	0.000	0.0
8	1	0.954	0.304	17.4
9	2	0.847	0.561	32.1
10	3	0.728	0.756	43.3
54	47	0.068	1.503	86.1
55	48	0.066	1.505	86.2
56	49	0.065	1.506	86.3
57	50	0.064	1.507	86.4

A graph of V_{out} as a function of f follows:

Alternating-Current Circuits   233

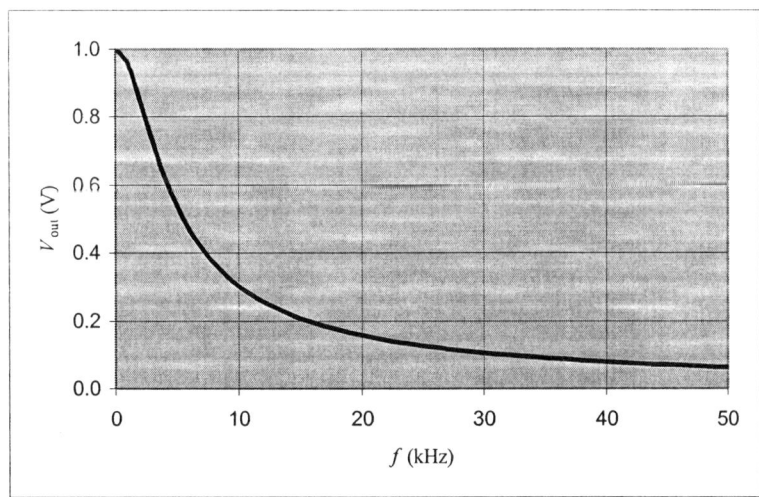

A graph of δ as a function of f follows:

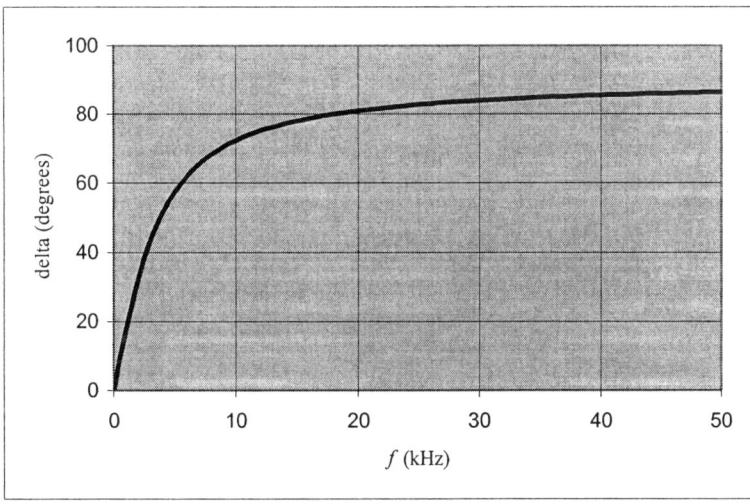

*59 ••• Show that the *trap* filter, shown in Figure 29-37, acts to reject signals at a frequency $\omega = 1/\sqrt{LC}$. How does the width of the frequency band rejected depend on the resistance R?

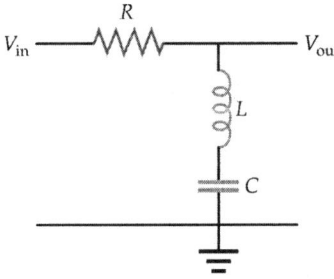

Figure 29-37 Problem 59

Picture the Problem We can apply Kirchhoff's loop rule to both the input side and output side of the trap filter to obtain an expression for the impedance of the trap. Requiring that the impedance of the trap be zero will yield the frequency at which the circuit rejects signals. Defining the bandwidth as $\Delta\omega = |\omega - \omega_{trap}|$ and requiring that

$|Z_{trap}| = R$ will yield an expression for the bandwidth and reveal its dependence on R.

Apply Kirchhoff's loop rule to the output of the trap circuit to obtain:
$$V_{out} - IX_L - IX_C = 0$$

Solve for V_{out}:
$$V_{out} = I(X_L + X_C) = IZ_{trap} \quad (1)$$
where $Z_{trap} = X_L + X_C$

Apply Kirchhoff's loop rule to the input of the trap circuit to obtain:
$$V_{in} - IR - IX_L - IX_C = 0$$

Solve for I:
$$I = \frac{V_{in}}{R + X_L + X_C} = \frac{V_{in}}{R + Z_{trap}}$$

Substitute for I in equation (1) to obtain:
$$V_{out} = V_{in} \frac{Z_{trap}}{R + Z_{trap}}$$

Because $X_L = i\omega L$ and $X_C = \frac{-i}{\omega C}$:
$$Z_{trap} = i\left(\omega L - \frac{1}{\omega C}\right)$$

Note that $Z_{trap} = 0$ and $V_{out} = 0$ provided:
$$\boxed{\omega = \frac{1}{\sqrt{LC}}}$$

Let the bandwidth $\Delta\omega$ be:
$$\Delta\omega = |\omega - \omega_{trap}| \quad (2)$$

Let the frequency bandwidth to be defined by the frequency at which $|Z_{trap}| = R$. Then:
$$\omega L - \frac{1}{\omega C} = R$$
or
$$\omega^2 LC - 1 = \omega RC$$

Because $\omega_{trap} = \frac{1}{\sqrt{LC}}$
$$\left(\frac{\omega}{\omega_{trap}}\right)^2 - 1 = \omega RC$$

For $\omega \approx \omega_{trap}$:
$$\left(\frac{\omega^2 - \omega_{trap}^2}{\omega_{trap}}\right) \approx \omega_{trap} RC$$

Solve for $\omega^2 - \omega_{trap}^2$:
$$\omega^2 - \omega_{trap}^2 = (\omega - \omega_{trap})(\omega + \omega_{trap})$$

Because $\omega \approx \omega_{trap}$, $\omega - \omega_{trap} \approx 2\omega_{trap}$:
$$\omega^2 - \omega_{trap}^2 \approx 2\omega_{trap}(\omega - \omega_{trap})$$

Substitute in equation (2) to obtain:
$$\Delta\omega = |\omega - \omega_{trap}| = \frac{RC\omega_{trap}^2}{2} = \boxed{\frac{R}{2L}}$$

LC Circuits with a Generator

*64 •••   One method for measuring the compressibility of a dielectric material uses an LC circuit with a parallel-plate capacitor. The dielectric is inserted between the plates and the change in resonance frequency is determined as the capacitor plates are subjected to a compressive stress. In such an arrangement, the resonance frequency is 120 MHz when a dielectric of thickness 0.1 cm and dielectric constant $\kappa = 6.8$ is placed between the capacitor plates. Under a compressive stress of 800 atm, the resonance frequency decreases to 116 MHz. Find Young's modulus of the dielectric material.

Picture the Problem We can use the definition of the capacitance of a dielectric-filled capacitor and the expression for the resonance frequency of an LC circuit to derive an expression for the fractional change in the thickness of the dielectric in terms of the resonance frequency and the frequency of the circuit when the dielectric is under compression. We can then use this expression for $\Delta t/t$ to calculate the value of Young's modulus for the dielectric material.

Use its definition to express Young's modulus of the dielectric material:
$$Y = \frac{\text{stress}}{\text{strain}} = \frac{\Delta P}{\Delta t/t} \qquad (1)$$

Letting t be the initial thickness of the dielectric, express the initial capacitance of the capacitor:
$$C_0 = \frac{\kappa \epsilon_0 A}{t}$$

Express the capacitance of the capacitor when it is under compression:
$$C_c = \frac{\kappa \epsilon_0 A}{t - \Delta t}$$

Express the resonance frequency of the capacitor before the dielectric is compressed:
$$\omega_0 = \frac{1}{\sqrt{C_0 L}} = \frac{1}{\sqrt{\frac{\kappa \epsilon_0 AL}{t}}}$$

When the dielectric is compressed:
$$\omega_c = \frac{1}{\sqrt{C_c L}} = \frac{1}{\sqrt{\frac{\kappa \epsilon_0 AL}{t - \Delta t}}}$$

Express the ratio of ω_c to ω_0 and simplify to obtain:

$$\frac{\omega_c}{\omega_0} = \frac{\sqrt{\frac{\kappa \epsilon_0 \, AL}{t}}}{\sqrt{\frac{\kappa \epsilon_0 \, AL}{t - \Delta t}}} = \sqrt{1 - \frac{\Delta t}{t}}$$

Expand the radical binomially to obtain:

$$\frac{\omega_c}{\omega_0} = \left(1 - \frac{\Delta t}{t}\right)^{1/2} \approx 1 - \frac{\Delta t}{2t}$$

provided $\Delta t \ll t$.

Solve for $\Delta t / t$:

$$\frac{\Delta t}{t} = 2\left(1 - \frac{\omega_c}{\omega_0}\right)$$

Substitute in equation (1) to obtain:

$$Y = \frac{\Delta P}{2\left(1 - \frac{\omega_c}{\omega_0}\right)}$$

Substitute numerical values and evaluate Y:

$$Y = \frac{(800\,\text{atm})(101.325\,\text{kPa/atm})}{2\left(1 - \frac{116\,\text{MHz}}{120\,\text{MHz}}\right)}$$

$$= \boxed{1.22 \times 10^9 \text{ N/m}^2}$$

RLC Circuits with a Generator

***69 ••** Show that the formula $P_{av} = R\mathcal{E}_{rms}^2 / Z^2$ gives the correct result for a circuit containing only a generator and (*a*) a resistor, (*b*) a capacitor, and (*c*) an inductor.

Picture the Problem The impedance of an ac circuit is given by $Z = \sqrt{R^2 + (X_L - X_C)^2}$. We can evaluate the given expression for P_{av} first for $X_L = X_C = 0$ and then for $R = 0$.

(*a*) For $X = 0$, $Z = R$ and:

$$P_{av} = \frac{R\mathcal{E}_{rms}^2}{Z^2} = \frac{R\mathcal{E}_{rms}^2}{R^2} = \boxed{\frac{\mathcal{E}_{rms}^2}{R}}$$

(*b*), (*c*) If $R = 0$, then:

$$P_{av} = \frac{R\mathcal{E}_{rms}^2}{Z^2} = \frac{(0)\mathcal{E}_{rms}^2}{(X_L - X_C)^2} = \boxed{0}$$

Remarks: Recall that there is no energy dissipation in an ideal inductor or capacitor.

Alternating-Current Circuits 237

*74 •• FM radio stations have carrier frequencies that are separated by 0.20 MHz. When the radio is tuned to a station, such as 100.1 MHz, the resonance width of the receiver circuit should be much smaller than 0.2 MHz, so that adjacent stations are not received. If f_0 = 100.1 MHz and Δf = 0.05 MHz, what is the Q factor for the circuit?

Picture the Problem We can use its definition, $Q = f_0/\Delta f$ to find the Q factor for the circuit.

Express the Q factor for the circuit:
$$Q = \frac{f_0}{\Delta f}$$

Substitute numerical values and evaluate Q:
$$Q = \frac{100.1\,\text{MHz}}{0.05\,\text{MHz}} = \boxed{2002}$$

*79 •• In the circuit shown in Figure 29-42, the ac generator produces an rms voltage of 115 V when operated at 60 Hz. What is the rms voltage across points (a) AB, (b) BC, (c) CD, (d) AC, and (e) BD?

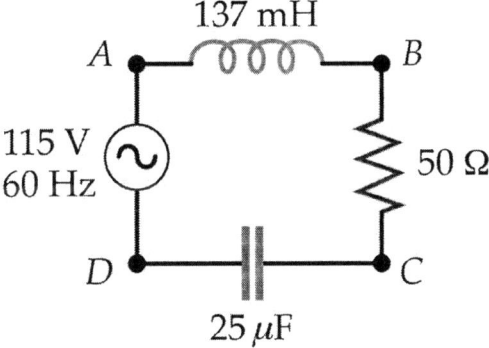

Figure 29-42 Problem 79

Picture the Problem We can find the rms current in the circuit and then use it to find the potential differences across each of the circuit elements. We can use phasor diagrams and our knowledge the phase shifts between the voltages across the three circuit elements to find the voltage differences across their combinations.

(a) Express the potential difference between points A and B in terms of I_{rms} and X_L:
$$V_{AB} = I_{rms} X_L \qquad (1)$$

Express I_{rms} in terms of ε and Z:
$$I_{rms} = \frac{\varepsilon}{Z} = \frac{\varepsilon}{\sqrt{R^2 + (X_L - X_C)^2}}$$

238   Chapter 29

Evaluate X_L and X_C to obtain:

$$X_L = 2\pi f L = 2\pi(60\,\text{s}^{-1})(137\,\text{mH})$$
$$= 51.6\,\Omega$$

and

$$X_C = \frac{1}{2\pi f C} = \frac{1}{2\pi(60\,\text{s}^{-1})(25\,\mu\text{F})}$$
$$= 106.1\,\Omega$$

Substitute numerical values and evaluate I_{rms}:

$$I_{rms} = \frac{115\,\text{V}}{\sqrt{(50\,\Omega)^2 + (51.6\,\Omega - 106.1\,\Omega)^2}}$$
$$= 1.55\,\text{A}$$

Substitute numerical values in equation (1) and evaluate V_{AB}:

$$V_{AB} = (1.55\,\text{A})(51.6\,\Omega) = \boxed{80.0\,\text{V}}$$

(b) Express the potential difference between points B and C in terms of I_{rms} and R:

$$V_{BC} = I_{rms}R = (1.55\,\text{A})(50\,\Omega)$$
$$= \boxed{77.5\,\text{V}}$$

(c) Express the potential difference between points C and D in terms of I_{rms} and X_C:

$$V_{CD} = I_{rms}X_C = (1.55\,\text{A})(106.1\,\Omega)$$
$$= \boxed{164\,\text{V}}$$

(d) The voltage across the inductor lags the voltage across the resistor as shown in the phasor diagram to the right:

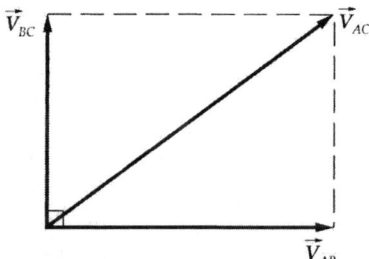

Use the Pythagorean theorem to find V_{AC}:

$$V_{AC} = \sqrt{V_{AB}^2 + V_{BC}^2}$$
$$= \sqrt{(80.0\,\text{V})^2 + (77.5\,\text{V})^2} = \boxed{111\,\text{V}}$$

(e) The voltage across the inductor lags the voltage across the resistor as shown in the phasor diagram to the right:

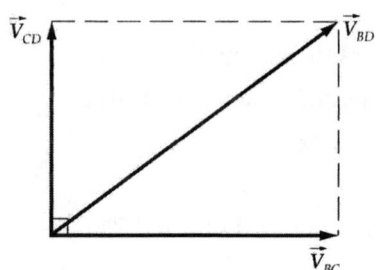

Alternating-Current Circuits   239

Use the Pythagorean theorem to find V_{BD}:

$$V_{BD} = \sqrt{V_{CD}^2 + V_{BC}^2}$$
$$= \sqrt{(164\,\text{V})^2 + (77.5\,\text{V})^2} = \boxed{181\,\text{V}}$$

*84 •• Show that Equation 29-48 can be written as

$$I_{max} = \frac{\omega \mathcal{E}_{max}}{\sqrt{L^2(\omega^2 - \omega_0^2)^2 + \omega^2 R^2}}$$

Picture the Problem We can substitute for X_L and X_C in Equation 29-48 and simplify the resulting equation to obtain the given equation for I_{max}.

Equation 29-48 is:

$$I_{max} = \frac{\mathcal{E}_{max}}{\sqrt{R^2 + (X_L - X_C)^2}}$$

Substitute for X_L and X_C to obtain:

$$I_{max} = \frac{\mathcal{E}_{max}}{\sqrt{R^2 + \left(\omega L - \dfrac{1}{\omega C}\right)^2}}$$

Simplify algebraically to obtain:

$$I_{max} = \frac{\mathcal{E}_{max}}{\sqrt{R^2 + \omega^2 L^2 \left(1 - \dfrac{1}{\omega^2 LC}\right)^2}} = \frac{\mathcal{E}_{max}}{\sqrt{R^2 + \omega^2 L^2 \left(1 - \dfrac{\omega_0^2}{\omega^2}\right)^2}} = \frac{\mathcal{E}_{max}}{\sqrt{R^2 + \dfrac{L^2}{\omega^2}(\omega^2 - \omega_0^2)^2}}$$

$$= \frac{\mathcal{E}_{max}}{\dfrac{1}{\omega}\sqrt{\omega^2 R^2 + L^2(\omega^2 - \omega_0^2)^2}} = \boxed{\frac{\omega \mathcal{E}_{max}}{\sqrt{\omega^2 R^2 + L^2(\omega^2 - \omega_0^2)^2}}}$$

*88 •• A method for measuring inductance is to connect the inductor in series with a known capacitance, a known resistance, an ac ammeter, and a variable-frequency signal generator. The frequency of the signal generator is varied and the emf is kept constant until the current is maximum. (a) If $C = 10\,\mu\text{F}$, $\mathcal{E}_{max} = 10\,\text{V}$, $R = 100\,\Omega$, and I is maximum at $\omega = 5000$ rad/s, what is L? (b) What is I_{max}?

Picture the Problem We can use the fact that when the current is a maximum, $X_L = X_C$, to find the inductance of the circuit. In (b), we can find Imax from $\mathcal{E}_{max}$ and the impedance of the circuit at resonance.

240   Chapter 29

(a) Relate X_L and X_C at resonance:

$$X_L = X_C \text{ or } \omega_0 L = \frac{1}{\omega_0 C}$$

Solve for L to obtain:

$$L = \frac{1}{\omega_0^2 C}$$

Substitute numerical values and evaluate L:

$$L = \frac{1}{(5000 \text{ s}^{-1})^2 (10 \mu F)} = \boxed{4.00 \text{ mH}}$$

(b) Noting that, at resonance, $X = 0$, express I_{max} in terms of the applied emf and the impedance of the circuit at resonance:

$$I_{max} = \frac{\mathcal{E}_{max}}{Z} = \frac{10 \text{ V}}{100 \Omega} = \boxed{0.100 \text{ A}}$$

*90 ••   In the circuit shown in Figure 29-44, $R = 10 \Omega$, $R_L = 30 \Omega$, $L = 150$ mH, and $C = 8 \mu F$; the frequency of the ac source is 10 Hz and its amplitude is 100 V. (a) Using phasor diagrams, determine the impedance of the circuit when switch S is closed. (b) Determine the impedance of the circuit when switch S is open. (c) What are the voltages across the load resistor R_L when switch S is closed and when it is open? (d) Repeat Parts (a), (b), and (c) with the frequency of the source changed to 1000 Hz. (e) Which arrangement is a better low-pass filter, S open or S closed?

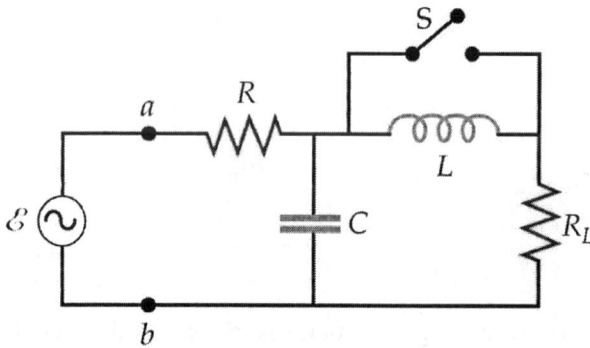

Figure 29-44 Problem 90

Picture the Problem Because we'll need to use it repeatedly in solving this problem, we'll begin by using complex numbers to derive an expression for the impedance Z_p of the parallel combination of C with L and R_L in series. The total impedance of the circuit is then $Z = R + Z_p$. We can apply Kirchhoff's loop rule to obtain expressions for the voltages across the load resistor with S either open or closed.

Use complex numbers to relate Z_p to R_L, X_L, and X_C:

$$\frac{1}{Z_p} = \frac{1}{-iX_C} + \frac{1}{R_L + iX_L}$$

$$= \frac{R_L + i(X_L - X_C)}{X_C X_L - iR_L X_C}$$

or

$$Z_p = \frac{X_C X_L - iR_L X_C}{R_L + i(X_L - X_C)}$$

Multiple the numerator and denominator of this fraction by the complex conjugate of $R_L + i(X_L - X_C)$:

$$Z_p = \frac{X_C X_L - iR_L X_C}{R_L + i(X_L - X_C)} \cdot \frac{R_L - i(X_L - X_C)}{R_L - i(X_L - X_C)}$$

Simplify to obtain:

$$Z_p = \frac{R_L X_C^2}{R_L^2 + (X_L - X_C)^2} - i\frac{X_C[R_L^2 + X_L(X_L - X_C)]}{R_L^2 + (X_L - X_C)^2} \quad (1)$$

(a) **S is closed**. Because L is shorted:

$$X_L = 0$$

Evaluate X_C:

$$X_C = \frac{1}{2\pi f C} = \frac{1}{2\pi(10\,\text{s}^{-1})(8\,\mu\text{F})}$$

$$= 1.99\,\text{k}\Omega$$

Substitute numerical values in equation (1) and evaluate Z_p, Z, $|Z|$, and δ:

$$Z_p = 30\,\Omega - i(0.452\,\Omega),$$
$$Z = 40\,\Omega - i(0.452\,\Omega),$$
and
$$|Z| = \sqrt{(40\,\Omega)^2 + (0.452\,\Omega)^2} = \boxed{40.0\,\Omega}$$

In Problem 29-77 we showed that for a parallel combination of a resistor and capacitor, the phase angle δ is given by:

$$\delta = \tan^{-1}\left(\frac{R}{X_C}\right)$$

Substitute numerical values and evaluate δ:

$$\delta = \tan^{-1}\left(\frac{40\,\Omega}{-0.452\,\Omega}\right) = \boxed{-89.4°}$$

No phasor diagram is shown because it is

242   Chapter 29

impossible to represent it to scale.

(b) S is open; i.e., the inductor is in the circuit. Find X_L:

$$X_L = \omega L = 2\pi f L = 2\pi(10\,\text{s}^{-1})(0.15\,\text{H})$$
$$= 9.42\,\Omega$$

Substitute numerical values in equation (1) and evaluate Z_p, Z, $|Z|$, and δ:

$$Z_p = 30.3\,\Omega + i(9.01\,\Omega),$$
$$Z = 40.3\,\Omega + i(9.01\,\Omega),$$
$$|Z| = \sqrt{(40.3\,\Omega)^2 + (9.01\,\Omega)^2} = \boxed{41.3\,\Omega}$$

and

$$\delta = \tan^{-1}\left(\frac{X}{R}\right) = \tan^{-1}\left(\frac{9.01\,\Omega}{40.3\,\Omega}\right)$$
$$= \boxed{12.6°}$$

The phasor diagram for this case is shown to the right.

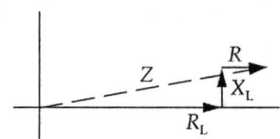

(c) S is closed. Apply Kirchhoff's loop rule to a loop including the source, R, and R_L:

$$\mathcal{E} - IR - V_{R_L} = 0$$

Solve for V_{R_L}:

$$V_{R_L} = \mathcal{E} - IR$$

Express the current I in the circuit:

$$I = \frac{\mathcal{E}}{Z}$$

Substitute and simplify to obtain:

$$V_{R_L} = \mathcal{E} - \frac{\mathcal{E}R}{Z} = \left(1 - \frac{R}{|Z|}\right)\mathcal{E}_{max}\cos(\omega t - \delta)$$

From (a) we have:

$$Z_p = 30\,\Omega - i(0.452\,\Omega),$$
$$Z = 40\,\Omega - i(0.452\,\Omega),$$
$$|Z| = 40.0\,\Omega, \text{ and}$$
$$\delta = \tan^{-1}\left(\frac{-0.452}{40}\right) = -0.647° \approx 0°$$

Alternating-Current Circuits 243

Substitute numerical values to obtain:	$V_{R_L} = \left(1 - \dfrac{10\,\Omega}{40\,\Omega}\right)(100\,\text{V})\cos\!\left[(20\,\text{s}^{-1})\pi t\right]$
	$= \boxed{(75\,\text{V})\cos\!\left[(20\,\text{s}^{-1})\pi t\right]}$

S is open. Apply Kirchhoff's loop rule to a loop including the source, R, L, and R_L when S is open:

$$\mathcal{E} - IR - IX_L - V_{R_L} = 0$$

Solve for V_{R_L}:

$$V_{R_L} = \mathcal{E} - IR - IX_L = \mathcal{E} - I(R + X_L)$$

Express the current I in the circuit:

$$I = \dfrac{\mathcal{E}}{Z}$$

Substitute to obtain:

$$V_{R_L} = \left(1 - \dfrac{R + X_L}{|Z|}\right)\mathcal{E}_{\max}\cos(\omega t - \delta)$$

Substitute numerical values and evaluate Z_p and Z:

$$Z_p = 30.3\,\Omega + i(9.01\,\Omega),$$
$$Z = 40.3\,\Omega + i(9.01\,\Omega),$$
$$|Z| = 41.3\,\Omega,$$

and

$$\delta = \tan^{-1}\!\left(\dfrac{X_L}{R + R_L}\right) = \tan^{-1}\!\left(\dfrac{9.42\,\Omega}{40.3\,\Omega}\right)$$
$$= 13.2°$$

Substitute numerical values and evaluate V_{R_L}:

$$V_{R_L} = \left(1 - \dfrac{10\,\Omega + 9.42\,\Omega}{41.3\,\Omega}\right)$$
$$\times (100\,\text{V})\cos\!\left[(20\,\text{s}^{-1})\pi t - 13.2°\right]$$
$$= \boxed{(53.0\,\text{V})\cos\!\left[(20\,\text{s}^{-1})\pi t - 13.2°\right]}$$

(d) Find X_L and X_C when $f = 1000$ Hz:

$$X_L = 2\pi(1000\,\text{s}^{-1})(0.15\,\text{H}) = 942\,\Omega$$

and

$$X_C = \dfrac{1}{2\pi(1000\,\text{s}^{-1})(8\,\mu\text{F})} = 19.9\,\Omega$$

S is closed. $X_L = 0$, and Z_p simplifies to:

$$Z_p = \dfrac{R_L X_C^2}{R_L^2 + X_C^2} - i\dfrac{R_L^2 X_C}{R_L^2 + X_C^2}$$

Substitute numerical values in equation (1) and evaluate Z_p, Z, $|Z|$, and δ:

$Z_p = 9.17\,\Omega - i(13.8\,\Omega)$,
$Z = 19.17\,\Omega - i(13.8\,\Omega)$,
$|Z| = \sqrt{(19.17\,\Omega)^2 + (13.8\,\Omega)^2} = \boxed{23.6\,\Omega}$

and

$\delta = \tan^{-1}\left(\dfrac{-13.8\,\Omega}{19.17\,\Omega}\right) = \boxed{-35.7°}$

A phasor diagram for this circuit is shown to the right.

S is open. Substitute numerical values in equation (1) and evaluate Z_p, Z, $|Z|$, and δ:

$Z_p = 0.0140\,\Omega - i(20.3\,\Omega)$,
$Z = 10.0\,\Omega - i(20.3\,\Omega)$,
$|Z| = \sqrt{(10.0\,\Omega)^2 + (20.3\,\Omega)^2} = \boxed{22.6\,\Omega}$

and

Find the total impedance, its magnitude, and phase angle for the circuit:

$Z = 10.0\,\Omega - i(20.4\,\Omega)$,
$Z = \sqrt{(10.0\,\Omega)^2 + (20.4\,\Omega)^2} = \boxed{22.7\,\Omega}$

and

$\delta = \tan^{-1}\left(\dfrac{-20.4\,\Omega}{10\,\Omega}\right) = \boxed{-63.9°}$

The phasor diagram is shown to the right.

(e)

> The load voltage at the higher frequency is much more attenuated with S open, while opening S does not reduce the low frequency load voltage significantly. Therefore, S open is the better arrangement for a low-pass filter.

*97 ••• (a) Show that Equation 29-47 can be written as

$$\tan\delta = \dfrac{Q(\omega^2 - \omega_0^2)}{\omega\omega_0}$$

(b) Show that near resonance

$$\tan\delta \approx \frac{2Q(\omega-\omega_0)}{\omega}$$

(c) Sketch a plot of δ versus x, where $x = \omega/\omega_0$, for a circuit with high Q and for one with low Q.

Picture the Problem We can manipulate Equation 29-47 into a form that has the ratio of L to R in it and then use the definition of Q to eliminate L and R. In (b) we can approximate $\omega^2 - \omega_0^2$, near resonance, as $2\omega_0\Delta\omega$ and substitute in the result from (a) to obtain the desired result.

(a) From Equation 29-47:
$$\tan\delta = \frac{\omega L - 1/\omega C}{R} = \frac{\omega^2 L - 1/C}{\omega R}$$
$$= \frac{L(\omega^2 - 1/LC)}{\omega R} = \frac{L(\omega^2 - \omega_0^2)}{\omega R}$$

Express Q in terms of ω_0, L and R:
$$Q = \frac{\omega_0 L}{R}$$

Solve for L/R to obtain:
$$\frac{L}{R} = \frac{Q}{\omega_0}$$

Substitute to obtain:
$$\boxed{\tan\delta = \frac{Q(\omega^2 - \omega_0^2)}{\omega\omega_0}} \quad (1)$$

(b) Near resonance:
$$\omega^2 - \omega_0^2 = (\omega + \omega_0)(\omega - \omega_0)$$
$$\approx 2\omega_0\Delta\omega$$

Substitute in equation (1) to obtain:
$$\tan\delta = \frac{Q(2\omega_0\Delta\omega)}{\omega\omega_0} = \boxed{\frac{2Q(\omega - \omega_0)}{\omega}}$$

(c) A following graph of δ as a function of $x = \omega/\omega_0$ was plotted using a spreadsheet program. The solid curve is for a high-Q circuit and the dashed curve is for a low-Q circuit.

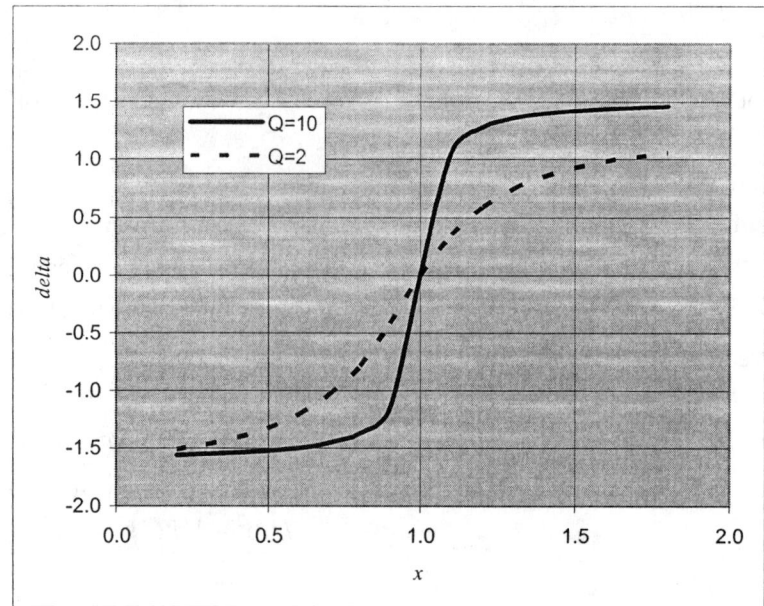

*102 ••• One method for measuring the magnetic susceptibility of a sample uses an LC circuit consisting of an air-core solenoid and a capacitor. The resonant frequency of the circuit without the sample is determined and then measured again with the sample inserted in the solenoid. Suppose the solenoid is 4 cm long, 0.3 cm in diameter, and has 400 turns of fine wire. Assume that the sample that is inserted in the solenoid is also 4 cm long and fills the air space. Neglect end effects. (In practice, a test sample of known susceptibility of the same shape as the unknown is used to calibrate the instrument.) (*a*) What is the inductance of the empty solenoid? (*b*) What should be the capacitance of the capacitor so that the resonance frequency of the circuit without a sample is 6.0000 MHz? (*c*) When a sample is inserted in the solenoid, the resonance frequency drops to 5.9989 MHz. Determine the sample's susceptibility.

Picture the Problem We can use $L = \mu_0 n^2 A \ell$ to determine the inductance of the empty solenoid and the resonance condition to find the capacitance of the sample-free circuit when the resonance frequency of the circuit is 6.0000 MHz. By expressing L as a function of f_0 and then evaluating df_0/dL and approximating the derivative with $\Delta f_0/\Delta L$, we can evaluate χ from its definition.

(*a*) Express the inductance of an air-core solenoid: $\qquad L = \mu_0 n^2 A \ell$

Substitute numerical values and evaluate L:

$$L = (4\pi \times 10^{-7}\, \text{N/A}^2)\left(\frac{400}{0.04\,\text{m}}\right)^2 \frac{\pi}{4}(0.003\,\text{m})^2 (0.04\,\text{m}) = \boxed{35.5\,\mu\text{H}}$$

Alternating-Current Circuits 247

(b) Express the condition for resonance in the LC circuit:	$X_L = X_C$ or $$2\pi f_0 L = \frac{1}{2\pi f_0 C}$$	(1)
Solve for C to obtain:	$$C = \frac{1}{4\pi^2 f_0^2 L}$$	
Substitute numerical values and evaluate C:	$$C = \frac{1}{4\pi^2 (6\,\text{MHz})(35.5\,\mu\text{H})} = \boxed{119\,\mu\text{F}}$$	
(c) Express the sample's susceptibility in terms of L and ΔL:	$$\chi = \frac{\Delta L}{L}$$	(2)
Solve equation (1) for f_0:	$$f_0 = \frac{1}{2\pi\sqrt{LC}}$$	
Differentiate f_0 with respect to L:	$$\frac{df_0}{dL} = \frac{1}{2\pi\sqrt{C}}\frac{d}{dL}L^{-1/2} = -\frac{1}{4\pi\sqrt{C}}L^{-3/2}$$ $$= -\frac{1}{4\pi L\sqrt{LC}} = -\frac{f_0}{2L}$$	
Approximate df_0/dL by $\Delta f_0/\Delta L$:	$$\frac{\Delta f_0}{\Delta L} = -\frac{f_0}{2L} \text{ or } \frac{\Delta f_0}{f_0} = -\frac{\Delta L}{2L}$$	
Substitute in equation (2) to obtain:	$$\chi = -2\frac{\Delta f_0}{f_0}$$	
Substitute numerical values and evaluate χ:	$$\chi = -2\left(\frac{5.9989\,\text{MHz} - 6.0000\,\text{MHz}}{6.0000\,\text{MHz}_0}\right)$$ $$= \boxed{3.67\times 10^{-4}}$$	

The Transformer

*105 • An ac voltage of 24 V is required for a device whose impedance is 12 Ω. (a) What should the turn ratio of a transformer be, so the device can be operated from a 120-V line? (b) Suppose the transformer is accidentally connected reversed (i.e., with the secondary winding across the 120-V line and the 12-Ω load across the primary). How much current will then flow in the primary winding?

248 Chapter 29

Picture the Problem Let the subscript 1 denote the primary and the subscript 2 the secondary. We can use $V_2 N_1 = V_1 N_2$ and $N_1 I_1 = N_2 I_2$ to find the turn ratio and the primary current when the transformer connections are reversed.

(*a*) Relate the number of primary and secondary turns to the primary and secondary voltages:

$$V_2 N_1 = V_1 N_2 \qquad (1)$$

Solve for and evaluate the ratio N_2/N_1:

$$\frac{N_2}{N_1} = \frac{V_2}{V_1} = \frac{24\,\text{V}}{120\,\text{V}} = \boxed{\frac{1}{5}}$$

(*b*) Relate the current in the primary to the current in the secondary and to the turns ratio:

$$I_1 = \frac{N_2}{N_1} I_2$$

Express the current in the primary winding in terms of the voltage across it and its impedance:

$$I_2 = \frac{V_2}{Z_2}$$

Substitute to obtain:

$$I_1 = \frac{N_2}{N_1} \frac{V_2}{Z_2}$$

Substitute numerical values and evaluate I_1:

$$I_1 = \left(\frac{5}{1}\right)\left(\frac{120\,\text{V}}{12\,\Omega}\right) = \boxed{50.0\,\text{A}}$$

---

*110 •• An audio oscillator (ac source) with an internal resistance of 2000 Ω and an open-circuit rms output voltage of 12 V is to be used to drive a loudspeaker with a resistance of 8 Ω. What should be the ratio of primary to secondary turns of a transformer, so that maximum power is transferred to the speaker? Suppose a second identical speaker is connected in parallel with the first speaker. How much power is then supplied to the two speakers combined?

Picture the Problem Note: In a simple circuit maximum power transfer from source to load requires that the load resistance equals the internal resistance of the source. We can use Ohm's law and the relationship between the primary and secondary currents and the primary and secondary voltages and the turns ratio of the transformer to derive an expression for the turns ratio as a function of the effective resistance of the circuit and the resistance of the speaker(s).

Alternating-Current Circuits 249

Express the effective loudspeaker resistance at the primary of the transformer:	$R_{eff} = \dfrac{V_1}{I_1}$
Relate V_1 to V_2, N_1, and N_2:	$V_1 = V_2 \dfrac{N_1}{N_2}$
Express I_1 in terms of I_2, N_1, and N_2:	$I_1 = I_2 \dfrac{N_2}{N_1}$
Substitute to obtain:	$R_{eff} = \dfrac{V_2 \dfrac{N_1}{N_2}}{I_2 \dfrac{N_2}{N_1}} = \left(\dfrac{V_2}{I_2}\right)\left(\dfrac{N_1}{N_2}\right)^2$
Solve for N_1/N_2:	$\dfrac{N_1}{N_2} = \sqrt{\dfrac{I_2 R_{eff}}{V_2}} = \sqrt{\dfrac{R_{eff}}{R_2}}$ (1)
Evaluate N_1/N_2 for $R_{eff} = R_{int}$:	$\dfrac{N_1}{N_2} = \sqrt{\dfrac{2000\,\Omega}{8\,\Omega}} = \boxed{15.8}$
Express the power delivered to the two speakers connected in parallel:	$P_{sp} = I_1^2 R_{eff}$ (2)
Find the equivalent resistance R_{sp} of the two 8-Ω speakers in parallel:	$\dfrac{1}{R_{sp}} = \dfrac{1}{8\,\Omega} + \dfrac{1}{8\,\Omega} = \dfrac{2}{8\,\Omega} = \dfrac{1}{4\,\Omega}$ and $R_{sp} = 4\,\Omega$
Solve equation (1) for R_{eff} to obtain:	$R_{eff} = R_2 \left(\dfrac{N_1}{N_2}\right)^2$
Substitute numerical values and evaluate R_{eff}:	$R_{eff} = (4\,\Omega)(15.8)^2 = 999\,\Omega$
Find the current drawn from the source:	$I_1 = \dfrac{V}{R_{tot}} = \dfrac{12\,\text{V}}{2000\,\Omega + 999\,\Omega} = 4.00\,\text{mA}$

250 Chapter 29

Substitute numerical values in equation (2) and evaluate the power delivered to the parallel speakers:

$$P_{sp} = (4\,\text{mA})^2(999\,\Omega) = \boxed{16.0\,\text{mW}}$$

General Problems

*115 •• Figure 29-47 shows the voltage V versus time t for a *square-wave* voltage. If $V_0 = 12$ V, (*a*) what is the rms voltage of this waveform? (*b*) If this alternating waveform is rectified by eliminating the negative voltages, so that only the positive voltages remain, what now is the rms voltage of the rectified waveform?

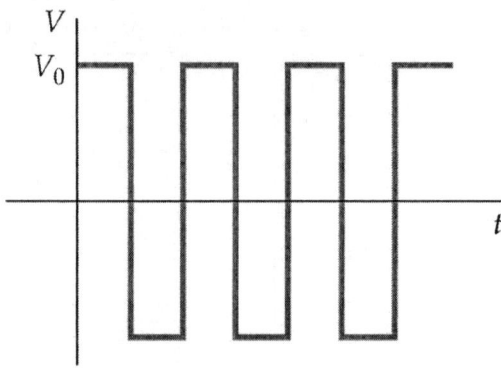

Figure 29-47 Problem 115

Picture the Problem The average of any quantity over a time interval ΔT is the integral of the quantity over the interval divided by ΔT. We can use this definition to find both the average of the voltage squared, $(V^2)_{av}$ and then use the definition of the rms voltage.

(*a*) From the definition of V_{rms} we have:

$$V_{rms} = \sqrt{(V_0^2)_{av}}$$

Noting that $-V_0^2 = V_0^2$, evaluate V_{rms}:

$$V_{rms} = \sqrt{V_0^2} = V_0 = \boxed{12.0\,\text{V}}$$

(*b*) Noting that the voltage during the second half of each cycle is now zero, express the voltage during the first half cycle of the time interval $\frac{1}{2}\Delta T$:

$$V = V_0$$

Express the square of the voltage during this half cycle:

$$V^2 = V_0^2$$

Alternating-Current Circuits 251

Calculate $(V^2)_{av}$ by integrating V^2 from $t = 0$ to $t = \frac{1}{2}\Delta T$ and dividing by ΔT:

$$(V^2)_{av} = \frac{V_0^2}{\Delta T}\int_0^{\frac{1}{2}\Delta T} dt = \frac{V_0^2}{\Delta T}[t]_0^{\frac{1}{2}\Delta T} = \frac{1}{2}V_0^2$$

Substitute to obtain:

$$V_{rms} = \sqrt{\frac{1}{2}V_0^2} = \frac{V_0}{\sqrt{2}} = \frac{12\,\text{V}}{\sqrt{2}} = \boxed{8.49\,\text{V}}$$

*120 •• Repeat Problem 119 if the resistor R is replaced by a 2-μF capacitor.

Picture the Problem We can apply Kirchhoff's loop rule to obtain an expression for charge on the capacitor as a function of time. Differentiating this expression with respect to time will give us the current in the circuit. We can then find I_{max} and I_{min} by considering the conditions under which the time-dependent factor in I will be a maximum or a minimum. We can use the maximum value of the current to find I_{rms}.

Apply Kirchhoff's loop rule to obtain:

$$\mathcal{E}_1 + \mathcal{E}_2 - \frac{q(t)}{C} = 0$$

Substitute numerical values and solve for $q(t)$:

$$q(t) = (2\,\mu\text{F})(20\,\text{V})\cos(1131\text{s}^{-1})t + (2\,\mu\text{F})(18\,\text{V})$$
$$= (40\,\mu\text{C})\cos(1131\text{s}^{-1})t + 36\,\mu\text{C}$$

Differentiate this expression with respect to t to obtain the current as a function of time:

$$I = \frac{dq}{dt} = \frac{d}{dt}[(40\,\mu\text{C})\cos(1131\text{s}^{-1})t + 36\,\mu\text{C}]$$
$$= \boxed{-(45.2\,\text{mA})\sin(1131\text{s}^{-1})t}$$

Express the condition that must be satisfied if the current is to be a minimum:

$$\sin(1131\text{s}^{-1})t = 1$$
and
$$I_{min} = \boxed{-45.2\,\text{mA}}$$

Express the condition that must be satisfied if the current is to be a maximum:

$$\sin(1131\text{s}^{-1})t = -1$$
and
$$I_{max} = \boxed{45.2\,\text{mA}}$$

Because the dc source sees the capacitor as an open circuit and the average value of the sine function over a period is zero:

$$I_{av} = \boxed{0}$$

Because the peak current is 45.2 mA:

$$I_{rms} = \frac{I_{max}}{\sqrt{2}} = \frac{45.2\,\text{mA}}{\sqrt{2}} = \boxed{32.0\,\text{mA}}$$

Chapter 30
Maxwell's Equations and Electromagnetic Waves

Conceptual Problems

*1 • True or false:

(a) Maxwell's equations apply only to fields that are constant over time.
(b) The wave equation can be derived from Maxwell's equations.
(c) Electromagnetic waves are transverse waves.
(d) In an electromagnetic wave in free space, the electric and magnetic fields are in phase.
(e) In an electromagnetic wave in free space, the electric and magnetic field vectors $\vec{E}$ and $\vec{B}$ are equal in magnitude.
(f) In an electromagnetic wave in free space, the electric and magnetic energy densities are equal.

(a) False. Maxwell's equations apply to both time-independent and time-dependent fields.

(b) True

(c) True

(d) True

(e) False. The magnitudes of the electric and magnetic field vectors are related according to $E = cB$.

(f) True

*4 • Are the frequencies of ultraviolet radiation greater or less than those of infrared radiation?

Determine the Concept The frequencies of ultraviolet radiation are greater than those of infrared radiation (see Table 30-1).

*8 • A helium-neon laser has a red beam. It is shone in turn on a red plastic filter (of the kind used for theater lighting) and a green plastic filter. (A red theater-lighting filter transmits only red light.) On which filter will the laser exert a larger force?

Determine the Concept A red plastic filter absorbs all the light incident on it except for the red light and a green plastic filter absorbs all the light incident on it except for the green light. If the red beam is incident on a red filter it will pass through, whereas, if it is

254 Chapter 30

incident on the green filter it will be absorbed. Because the green filter absorbs more energy than does the red filter, the laser beam will exert a greater force on the green filter.

Estimation and Approximation

*12 ••   Estimate the radiation pressure force exerted on the earth by the sun, and compare the radiation pressure force to the gravitational attraction of the sun. At the earth's orbit the intensity of sunlight is 1.37 kW/m².

Picture the Problem We can find the radiation pressure force from the definition of pressure and the relationship between the radiation pressure and the intensity of the radiation from the sun. We can use Newton's law of gravitation to find the gravitational force the sun exerts on the earth.

The radiation pressure exerted on the earth is given by:
$$P_r = \frac{F_r}{A} \Rightarrow F_r = P_r A$$
where A is the cross-sectional area of the earth.

Express the radiation pressure in terms of the intensity of the radiation I from the sun:
$$P_r = \frac{I}{c}$$

Substituting for P_r and A yields:
$$F_r = \frac{I \pi R^2}{c}$$

Substitute numerical values and evaluate F_r:
$$F_r = \frac{\pi (1370 \text{ W/m}^2)(6370 \text{ km})^2}{3 \times 10^8 \text{ m/s}}$$
$$= \boxed{5.82 \times 10^8 \text{ N}}$$

The gravitational force exerted on the earth by the sun is given by:
$$F = \frac{G m_{sun} m_{earth}}{r^2}$$
where r is the radius of the earth's orbit.

Substitute numerical values and evaluate F:

$$F = \frac{(6.67 \times 10^{-11} \text{ N} \cdot \text{m}^2 / \text{kg}^2)(1.99 \times 10^{30} \text{ kg})(5.98 \times 10^{24} \text{ kg})}{(1.5 \times 10^{11} \text{ m})^2} = 3.53 \times 10^{22} \text{ N}$$

Express the ratio of the force due radiation pressure F_r to the gravitational force F:
$$\frac{F_r}{F} = \frac{5.82 \times 10^8 \text{ N}}{3.53 \times 10^{22} \text{ N}} = 1.65 \times 10^{-14}$$

$$\boxed{\text{The gravitational force is greater by a factor of approximately } 10^{14}.}$$

Maxwell's Equations and Electromagnetic Waves 255

*13 ••   Repeat Problem 12 for the planet Mars. Which planet has the larger ratio of radiation pressure to gravitational attraction? Why?

Picture the Problem We can find the radiation pressure force from the definition of pressure and the relationship between the radiation pressure and the intensity of the radiation from the sun. We can use Newton's law of gravitation to find the gravitational force the sun exerts on Mars.

The radiation pressure exerted on Mars is given by:
$$P_r = \frac{F_r}{A} \Rightarrow F_r = P_r A$$
where A is the cross-sectional area of Mars.

Express the radiation pressure on Mars in terms of the intensity of the radiation I_{Mars} from the sun:
$$P_r = \frac{I_{Mars}}{c}$$

Substituting for P_r and A yields:
$$F_r = \frac{I_{Mars} \pi R_{Mars}^2}{c}$$

Express the ratio of the solar constant at the earth I_{earth} to the solar constant I_{Mars} at Mars:
$$\frac{I_{Mars}}{I_{earth}} = \left(\frac{r_{earth}}{r_{Mars}}\right)^2 \Rightarrow I_{Mars} = I_{earth}\left(\frac{r_{earth}}{r_{Mars}}\right)^2$$

Substitute for I_{Mars} to obtain:
$$F_r = \frac{I_{earth} \pi R_{Mars}^2}{c}\left(\frac{r_{earth}}{r_{Mars}}\right)^2$$

Substitute numerical values and evaluate F_r:

$$F_r = \frac{\pi(1370\text{ W/m}^2)(3395\text{ km})^2}{3\times 10^8\text{ m/s}}\left(\frac{1.50\times 10^{11}\text{ m}}{2.29\times 10^{11}\text{ m}}\right)^2 = \boxed{7.09\times 10^7\text{ N}}$$

The gravitational force exerted on Mars by the sun is given by:
$$F = \frac{Gm_{sun}m_{Mars}}{r^2} = \frac{Gm_{sun}(0.11m_{earth})}{r^2}$$
where r is the radius of Mars' orbit.

Substitute numerical values and evaluate F:

$$F = \frac{(6.67\times 10^{-11}\text{ N}\cdot\text{m}^2/\text{kg}^2)(1.99\times 10^{30}\text{ kg})(0.11)(5.98\times 10^{24}\text{ kg})}{(2.29\times 10^{11}\text{ m})^2} = 1.66\times 10^{21}\text{ N}$$

Express the ratio of the force due radiation pressure F_r to the gravitational force F:
$$\frac{F_r}{F} = \frac{7.09\times 10^7\text{ N}}{1.66\times 10^{21}\text{ N}} = 4.27\times 10^{-14}$$

256 Chapter 30

> Because the ratio of these forces is 1.65×10^{-14} for the earth and 4.27×10^{-14} for Mars, Mars has the larger ratio. The reason that the ratio is higher for Mars is that the dependence of the radiation pressure on the distance from the Sun is the same for both forces (r^{-2}), whereas the dependence on the radii of the planets is different. Radiation pressure varies as R^2, whereas the gravitational force varies as R^3 (assuming that the two planets have the same density, an assumption that is nearly true). Consequently, the ratio of the forces goes as $R^2 / R^3 = R^{-1}$. Because Mars is smaller than earth, the ratio is larger.

*14 •• In the new field of laser cooling and trapping, the forces associated with radiation pressure are used to slow down atoms from thermal speeds of hundreds of meters per second at room temperature to speeds of just a few meters per second or slower. An isolated atom will absorb radiation only at specific resonant frequencies. If the frequency of the laser-beam radiation is one of the resonant frequencies of the target atom, then the radiation is absorbed via a process called resonance absorption. The effective cross-sectional area of the atom for resonant absorption is approximately equal to λ^2, where λ is the wavelength of the laser beam. (a) Estimate the acceleration of a rubidium atom (atomic mass 85 g/mol) in a laser beam whose wavelength is 780 nm and intensity is 10 W/m². (b) About how long would it take such a light beam to slow a rubidium atom in a gas at room temperature (300 K) down to near-zero velocity?

Picture the Problem We can use Newton's 2nd law to express the acceleration of an atom in terms of the net force acting on the atom and the relationship between radiation pressure and the intensity of the beam to find the net force. Once we know the acceleration of an atom, we can use the definition of acceleration to find the stopping time for a rubidium atom at room temperature.

(a) Apply $\sum F = ma$ to the atom to obtain:

$F = ma$

where F is the force exerted by the laser beam.

The radiation pressure P_r and intensity of the beam I are related according to:

$$P_r = \frac{F}{A} = \frac{I}{c}$$

Solve for F to obtain:

$$F = \frac{IA}{c} = \frac{I\lambda^2}{c}$$

Substitute for F in the expression of Newton's 2nd law to obtain:

$$\frac{I\lambda^2}{c} = ma$$

Solve for a:

$$a = \frac{I\lambda^2}{mc}$$

Substitute numerical values and evaluate a:

$$a = \frac{(10 \text{ W/m}^2)(780 \text{ nm})^2}{\left(85 \frac{\text{g}}{\text{mol}} \times \frac{1 \text{ mol}}{6.02 \times 10^{23} \text{ particles}}\right)(3 \times 10^8 \text{ m/s})} = \boxed{1.44 \times 10^5 \text{ m/s}^2}$$

(b) Using the definition of acceleration, express the stopping time Δt of the atom:

$$\Delta t = \frac{v_{\text{final}} - v_{\text{initial}}}{a}$$

Because $v_{\text{final}} \approx 0$:

$$\Delta t \approx \frac{-v_{\text{initial}}}{a}$$

Using the rms speed as the initial speed of an atom, relate v_{initial} to the temperature of the gas:

$$v_{\text{initial}} = v_{\text{rms}} = \sqrt{\frac{3kT}{m}}$$

Substitute in the expression for the stopping time to obtain:

$$\Delta t = -\frac{1}{a}\sqrt{\frac{3kT}{m}}$$

Substitute numerical values and evaluate Δt:

$$\Delta t = -\frac{1}{-1.44 \times 10^5 \text{ m/s}^2}\sqrt{\frac{3(1.38 \times 10^{-23} \text{ J/K})(300 \text{ K})}{\left(85 \frac{\text{g}}{\text{mol}} \times \frac{1 \text{ mol}}{6.02 \times 10^{23} \text{ particles}}\right)}} = \boxed{2.06 \text{ ms}}$$

Maxwell's Displacement Current

*19 •• Current of 10 A flows into a capacitor having plates with areas of 0.5 m². (a) What is the displacement current between the plates? (b) What is dE/dt between the plates for this current? (c) What is the line integral of $\vec{B} \cdot d\vec{\ell}$ around a circle of radius 10 cm that lies within the plates and parallel to the plates?

Picture the Problem We can use the conservation of charge to find I_d, the definitions of the displacement current and electric flux to find dE/dt, and Ampere's law to evaluate $\vec{B} \cdot d\vec{\ell}$ around the given path.

(a) From conservation of charge we know that:

$$I_d = I = \boxed{10.0 \text{ A}}$$

(b) Express the displacement current I_d:

$$I_\text{d} = \epsilon_0 \frac{d\phi_\text{e}}{dt} = \epsilon_0 \frac{d}{dt}[EA] = \epsilon_0 A \frac{dE}{dt}$$

Substitute for dE/dt:

$$\frac{dE}{dt} = \frac{I_\text{d}}{\epsilon_0 A}$$

Substitute numerical values and evaluate dE/dt:

$$\frac{dE}{dt} = \frac{10\,\text{A}}{(8.85 \times 10^{-12}\,\text{C}^2/\text{N} \cdot \text{m}^2)(0.5\,\text{m}^2)}$$

$$= \boxed{2.26 \times 10^{12}\,\frac{\text{V}}{\text{m} \cdot \text{s}}}$$

(c) Apply Ampère's law to a circular path of radius r between the plates and parallel to their surfaces to obtain:

$$\oint_C \vec{B} \cdot d\vec{\ell} = \mu_0 I_\text{enclosed}$$

Assuming that the displacement current is uniformly distributed:

$$\frac{I_\text{enclosed}}{\pi r^2} = \frac{I_\text{d}}{A} \Rightarrow I_\text{enclosed} = \frac{\pi r^2}{A} I_\text{d}$$

where R is the radius of the circular plates.

Substitute for I_enclosed to obtain:

$$\oint_C \vec{B} \cdot d\vec{\ell} = \frac{\mu_0 \pi r^2}{A} I_\text{d}$$

Substitute numerical values and evaluate $\oint_C \vec{B} \cdot d\vec{\ell}$:

$$\oint_C \vec{B} \cdot d\vec{\ell} = \frac{(4\pi \times 10^{-7}\,\text{N/A}^2)\pi(0.1\,\text{m})^2(10\,\text{A})}{0.5\,\text{m}^2} = \boxed{7.90 \times 10^{-7}\,\text{T} \cdot \text{m}}$$

*23 ••• Show that the generalized form of Ampère's law (Equation 30-4) and the Biot–Savart law give the same result in a situation in which they both can be used. Figure 30-13 shows two charges $+Q$ and $-Q$ on the x axis at $x = -a$ and $x = +a$, with a current $I = -dQ/dt$ along the line between them. Point P is on the y axis at $y = R$. (a) Use the Biot–Savart law to show that the magnitude of B at point P is

$$B = \frac{\mu_0 I a}{2\pi R} \frac{1}{\sqrt{R^2 + a^2}}$$

(b) Consider a circular strip of radius r and width dr in the yz plane with its center at the origin. Show that the flux of the electric field through this strip is

$$E_x dA = \frac{Q}{\epsilon_0} a(r^2 + a^2)^{-3/2} r dr$$

(c) Use your result from Part (b) to find the total flux ϕ_e through a circular area of radius R. Show that

$$\epsilon_0 \phi_e = Q \left(1 - \frac{a}{\sqrt{a^2 + R^2}}\right)$$

(d) Find the displacement current I_d, and show that

$$I + I_d = I \frac{a}{\sqrt{a^2 + R^2}}$$

(e) Finally, show that Equation 30-4 gives the same result for B as the result found in Part (a).

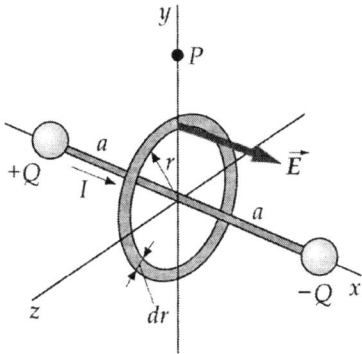

Figure 30-13 Problem 23

Picture the Problem We can follow the step-by-step instructions in the problem statement to show that Equation 30-4 gives the same result for B as that given in Part (a).

(a) Express the magnetic field at P using the expression for B due to a straight wire segment:

$$B_P = \frac{\mu_0}{4\pi} \frac{I}{R} (\sin\theta_1 + \sin\theta_2)$$

where

$$\sin\theta_1 = \sin\theta_2 = \frac{a}{\sqrt{R^2 + a^2}}$$

Substitute for $\sin\theta_1$ and $\sin\theta_2$ to obtain:

$$B_P = \frac{\mu_0}{4\pi} \frac{I}{R} \frac{2a}{\sqrt{R^2 + a^2}}$$

$$= \boxed{\frac{\mu_0 I a}{2\pi R} \frac{1}{\sqrt{R^2 + a^2}}}$$

(b) Express the electric flux through the circular strip of radius r and width dr in the yz plane:

$$d\phi_e = E_x dA = E_x(2\pi r\,dr)$$

The electric field due to the dipole is:

$$E_x = \frac{2kQ}{r^2+a^2}\cos\theta_1 = \frac{2kQa}{(r^2+a^2)^{3/2}}$$

Substitute for E_x to obtain:

$$d\phi_e = E_x dA = \frac{2kQa}{(r^2+a^2)^{3/2}}(2\pi r\,dr)$$

$$= \frac{2Qa}{4\pi\epsilon_0 (r^2+a^2)^{3/2}}(2\pi r\,dr)$$

$$= \boxed{\frac{Qa}{\epsilon_0(r^2+a^2)^{3/2}}r\,dr}$$

(c) Multiply both sides of the expression for ϕ_e by ϵ_0:

$$\epsilon_0\,d\phi_e = \frac{Qa}{(r^2+a^2)^{3/2}}r\,dr$$

Integrate r from 0 to R to obtain:

$$\epsilon_0\phi_e = Qa\int_0^R \frac{r\,dr}{(r^2+a^2)^{3/2}} = Qa\left(\frac{-1}{\sqrt{R^2+a^2}}+\frac{1}{a}\right) = \boxed{Q\left(1-\frac{a}{\sqrt{R^2+a^2}}\right)}$$

(d) The displacement current is defined to be:

$$I_d = \epsilon_0 \frac{d\phi_e}{dt} = \frac{d}{dt}\left[Q\left(1-\frac{a}{\sqrt{R^2+a^2}}\right)\right]$$

$$= \left(1-\frac{a}{\sqrt{R^2+a^2}}\right)\frac{dQ}{dt}$$

$$= -I\left(1-\frac{a}{\sqrt{R^2+a^2}}\right)$$

The total current is the sum of I and I_d:

$$I+I_d = I - I\left(1-\frac{a}{\sqrt{R^2+a^2}}\right)$$

$$= \boxed{I\frac{a}{\sqrt{R^2+a^2}}}$$

(e) Apply Equation 30-4 (the generalized form of Ampere's law) to obtain:

$$\oint_C \vec{B}\cdot d\vec{\ell} = 2\pi R B = \mu_0(I+I_d)$$

Solve for B:
$$B = \frac{\mu_0}{2\pi R}(I + I_d)$$

Substitute for $I + I_d$ from (d) to obtain:
$$B = \frac{\mu_0}{2\pi R}\left(I \frac{a}{\sqrt{R^2 + a^2}}\right)$$
$$= \boxed{\frac{\mu_0 I a}{2\pi R} \frac{1}{\sqrt{R^2 + a^2}}}$$

Maxwell's Equations and the Electromagnetic Spectrum

*25 • Find the wavelength for (a) a typical AM radio wave with a frequency of 1000 kHz and (b) a typical FM radio wave with a frequency of 100 MHz.

Picture the Problem We can use $c = f\lambda$ to find the wavelengths corresponding to the given frequencies.

Solve $c = f\lambda$ for λ:
$$\lambda = \frac{c}{f}$$

(a) For $f = 1000$ kHz:
$$\lambda = \frac{3 \times 10^8 \text{ m/s}}{1000 \times 10^3 \text{ s}^{-1}} = \boxed{300 \text{ m}}$$

(b) For $f = 100$ MHz:
$$\lambda = \frac{3 \times 10^8 \text{ m/s}}{100 \times 10^6 \text{ s}^{-1}} = \boxed{3.00 \text{ m}}$$

*26 • What is the frequency of a 3-cm microwave?

Picture the Problem We can use $c = f\lambda$ to find the frequency corresponding to the given wavelength.

Solve $c = f\lambda$ for f:
$$f = \frac{c}{\lambda}$$

Substitute numerical values and evaluate f:
$$f = \frac{3 \times 10^8 \text{ m/s}}{3 \times 10^{-2} \text{ m}} = 10^{10} \text{ Hz} = \boxed{10.0 \text{ GHz}}$$

Electric Dipole Radiation

*32 ••• At a distance of 30 km from a radio station broadcasting at a frequency of 0.8 MHz, the intensity of the electromagnetic wave is 2×10^{-13} W/m^2. The transmitting antenna is a vertical dipole. What is the total power radiated by the station?

Picture the Problem The intensity of radiation from an electric dipole is given by $I_0(\sin^2\theta)/r^2$, where θ is the angle between the electric dipole moment and the position vector $\vec{r}$. We can integrate the intensity to express the total power radiated by the antenna and use this result to evaluate I_0. Knowing I_0 we can find the total power radiated by the station.

From the definition of intensity we have:
$$dP = I dA$$
and
$$P_{tot} = \iint I(r,\theta) dA$$
where, in polar coordinates,
$$dA = r^2 \sin\theta \, d\theta \, d\phi$$

Substitute for dA to obtain:
$$P_{tot} = \int_0^{2\pi}\int_0^{\pi} I(r,\theta) r^2 \sin\theta \, d\theta \, d\phi$$

Express the intensity of the signal as a function of r and θ:
$$I(r,\theta) = I_0 \frac{\sin^2\theta}{r^2} \quad (1)$$

Substitute for $I(r,\theta)$:
$$P_{tot} = I_0 \int_0^{2\pi}\int_0^{\pi} \sin^3\theta \, d\theta \, d\phi$$

From integral tables we find that:
$$\int_0^{\pi} \sin^3\theta \, d\theta = -\tfrac{1}{3}\cos\theta(\sin^2\theta + 2)\Big|_0^{\pi} = \frac{4}{3}$$

Substitute and integrate with respect to ϕ to obtain:
$$P_{tot} = \frac{4}{3}I_0 \int_0^{2\pi} d\phi = \frac{4}{3}I_0 [\phi]_0^{2\pi} = \frac{8\pi}{3}I_0$$

From equation (1) we have:
$$I_0 = \frac{I(r,\theta) r^2}{\sin^2\theta}$$

Substitute to obtain:
$$P_{tot} = \frac{8\pi}{3}\frac{I(r,\theta) r^2}{\sin^2\theta}$$

or, because $\theta = 90°$,

Maxwell's Equations and Electromagnetic Waves 263

$$P_{tot} = \frac{8\pi}{3} I(r) r^2$$

Substitute numerical values and evaluate P_{tot}:

$$P_{tot} = \frac{8\pi}{3} (2 \times 10^{-13} \text{ W/m}^2)(30 \text{ km})^2$$

$$= \boxed{1.51 \text{ mW}}$$

Energy and Momentum in an Electromagnetic Wave

*38 •   The rms value of the magnitude of the magnetic field in an electromagnetic wave is $B_{rms} = 0.245\ \mu T$. Find (a) E_{rms}, (b) the average energy density, and (c) the intensity.

Picture the Problem Given B_{rms}, we can find E_{rms} using $E_{rms} = cB_{rms}$. The average energy density of the wave is given by $u_{av} = E_{rms}B_{rms}/\mu_0 c$ and the intensity of the wave by $I = u_{av}c$

(a) Express E_{rms} in terms of B_{rms}:

$$E_{rms} = cB_{rms}$$

Substitute numerical values and evaluate E_{rms}:

$$E_{rms} = (3 \times 10^8 \text{ m/s})(0.245\ \mu T)$$

$$= \boxed{73.5 \text{ V/m}}$$

(b) The average energy density u_{av} is given by:

$$u_{av} = \frac{E_{rms} B_{rms}}{\mu_0 c}$$

Substitute numerical values and evaluate u_{av}:

$$u_{av} = \frac{(73.5 \text{ V/m})(0.245\ \mu T)}{(4\pi \times 10^{-7} \text{ N/A}^2)(3 \times 10^8 \text{ m/s})}$$

$$= \boxed{47.8 \text{ nJ/m}^3}$$

(c) Express the intensity as the product of the average energy density and the speed of light in a vacuum:

$$I = u_{av} c$$

Substitute numerical values and evaluate I:

$$I = (47.8 \text{ nJ/m}^3)(3 \times 10^8 \text{ m/s})$$

$$= \boxed{14.3 \text{ W/m}^2}$$

264 Chapter 30

*41 •• An AM radio station radiates an isotropic sinusoidal wave with an average power of 50 kW. What are the amplitudes of E_{max} and B_{max} at a distance of (a) 500 m, (b) 5 km, and (c) 50 km?

Picture the Problem We can use $I = P_{av}/4\pi r^2$ and $I = E_{rms}B_{rms}/\mu_0$ to express E_{rms} in terms of P_{av} and the distance r from the station.

Express the intensity I of the radiation as a function of its average power and the distance r from the station:

$$I = \frac{P_{av}}{4\pi r^2}$$

The intensity is also given by:

$$I = \frac{E_{rms}B_{rms}}{\mu_0} = \frac{E_{rms}^2}{c\mu_0} = \frac{E_{max}^2}{2c\mu_0}$$

Equate these expressions to obtain:

$$\frac{P_{av}}{4\pi r^2} = \frac{E_{max}^2}{2c\mu_0}$$

Solve for E_{max}:

$$E_{max} = \sqrt{\frac{c\mu_0 P_{av}}{2\pi}}\left(\frac{1}{r}\right)$$

(a) Substitute numerical values and evaluate E_{max} for $r = 500$ m:

$$E_{max}(500\,\text{m}) = \sqrt{\frac{(3\times 10^8\,\text{m/s})(4\pi\times 10^{-7}\,\text{N/A}^2)(50\,\text{kW})}{2\pi}}\left(\frac{1}{500\,\text{m}}\right) = \boxed{3.46\,\text{V/m}}$$

Use $B_{max} = E_{max}/c$ to evaluate B_{max}:

$$B_{max} = \frac{3.46\,\text{V/m}}{3\times 10^8\,\text{m/s}} = \boxed{11.5\,\text{nT}}$$

(b) Substitute numerical values and evaluate E_{max} for $r = 5$ km:

$$E_{max}(5\,\text{km}) = \sqrt{\frac{(3\times 10^8\,\text{m/s})(4\pi\times 10^{-7}\,\text{N/A}^2)(50\,\text{kW})}{2\pi}}\left(\frac{1}{5\,\text{km}}\right) = \boxed{0.346\,\text{V/m}}$$

Use $B_{max} = E_{max}/c$ to evaluate B_{max}:

$$B_{max} = \frac{0.346\,\text{V/m}}{3\times 10^8\,\text{m/s}} = \boxed{1.15\,\text{nT}}$$

(c) Substitute numerical values and evaluate E_{max} for $r = 50$ km:

Maxwell's Equations and Electromagnetic Waves   265

$$E_{max}(500\,m) = \sqrt{\frac{(3\times10^8\,m/s)(4\pi\times10^{-7}\,N/A^2)(50\,kW)}{2\pi}\left(\frac{1}{50\,km}\right)} = \boxed{0.0346\,V/m}$$

Use $B_{max} = E_{max}/c$ to evaluate B_{max}:

$$B_{max} = \frac{0.0346\,V/m}{3\times10^8\,m/s} = \boxed{0.115\,nT}$$

*43 ••   Instead of sending power by a 750-kV, 1000-A transmission line, one desires to beam this energy via an electromagnetic wave. The beam has a uniform intensity within a cross-sectional area of 50 m². What are the rms values of the electric and the magnetic fields?

Picture the Problem We can use $I = E_{rms}B_{rms}/\mu_0$ and $B_{rms} = E_{rms}/c$ to express E_{rms} in terms of I. We can then use $B_{rms} = E_{rms}/c$ to find B_{rms}.

Express the intensity I of the radiation as a function of its average power and the distance r from the station:

$$I = \frac{E_{rms}B_{rms}}{\mu_0} = \frac{E_{rms}^2}{c\mu_0}$$

Solve for E_{rms}:

$$E_{rms} = \sqrt{c\mu_0 I}$$

Use the definition of intensity to relate the intensity of the electromagnetic wave to the power in the beam:

$$I = \frac{P}{A} = \frac{I_{trans.\,line}V}{A}$$

Substitute for I to obtain:

$$E_{rms} = \sqrt{\frac{c\mu_0 I_{trans.\,line}V}{A}}$$

Substitute numerical values and evaluate E_{rms}:

$$E_{rms} = \sqrt{\frac{(3\times10^8\,m/s)(4\pi\times10^{-7}\,N/A^2)(10^3\,A)(750\,kV)}{50\,m^2}} = \boxed{75.2\,kV/m}$$

Use $B_{rms} = E_{rms}/c$ to evaluate B_{rms}:

$$B_{rms} = \frac{75.2\,kV/m}{3\times10^8\,m/s} = \boxed{0.251\,mT}$$

266 Chapter 30

*45 ••   The electric field of an electromagnetic wave oscillates in the *y* direction and the Poynting vector is given by

$$\vec{S}(x,t) = (100 \text{ W/m}^2)\cos^2[10x - (3\times 10^9)t]\hat{i}$$

where *x* is in meters and *t* is in seconds. (*a*) What is the direction of propagation of the wave? (*b*) Find the wavelength and the frequency. (*c*) Find the electric and magnetic fields.

Picture the Problem We can determine the direction of propagation of the wave, its wavelength, and its frequency by examining the argument of the cosine function. We can find *E* from $|\vec{S}| = E^2/\mu_0 c$ and *B* from $B = E/c$. Finally, we can use the definition of the Poynting vector and the given expression for $\vec{S}$ to find $\vec{E}$ and $\vec{B}$.

(*a*) | Because the argument of the cosine function is of the form $kx - \omega t$, the wave propagates in the positive *x* direction. |

(*b*) Examining the argument of the cosine function, we note that the wave number *k* of the wave is:

$$k = \frac{2\pi}{\lambda} = 10 \text{ m}^{-1}$$

Solve for and evaluate λ:

$$\lambda = \frac{2\pi}{10 \text{ m}^{-1}} = \boxed{0.628 \text{ m}}$$

Examining the argument of the cosine function, we note that the angular frequency ω of the wave is:

$$\omega = 2\pi f = 3\times 10^9 \text{ s}^{-1}$$

Solve for and evaluate *f* to obtain:

$$f = \frac{3\times 10^9 \text{ s}^{-1}}{2\pi} = \boxed{477 \text{ MHz}}$$

(*c*) Express the magnitude of $\vec{S}$ in terms of *E*:

$$|\vec{S}| = \frac{E^2}{\mu_0 c}$$

Solve for *E*:

$$E = \sqrt{\mu_0 c |\vec{S}|}$$

Substitute numerical values and evaluate *E*:

$$E = \sqrt{(3\times 10^8 \text{ m/s})(4\pi\times 10^{-7} \text{ N/A}^2)(100 \text{ W/m}^2)} = 194 \text{ V/m}$$

Because $\vec{S}(x,t) = (100 \text{ W/m}^2) \cos^2[10x - (3 \times 10^9)t] \hat{i}$ and $\vec{S} = \dfrac{1}{\mu_0} \vec{E} \times \vec{B}$:

$$\boxed{\vec{E}(x,t) = (194 \text{ V/m}) \cos[10x - (3 \times 10^9)t] \hat{j}}$$

Use $B = E/c$ to evaluate B:

$$B = \dfrac{194 \text{ V/m}}{3 \times 10^8 \text{ m/s}} = 0.647 \text{ }\mu\text{T}$$

Because $\vec{S} = \dfrac{1}{\mu_0} \vec{E} \times \vec{B}$, the direction of $\vec{B}$ must be such that the cross product of $\vec{E}$ with $\vec{B}$ is in the positive x direction:

$$\boxed{\vec{B}(x,t) = (0.647 \text{ }\mu\text{T}) \cos[10x - (3 \times 10^9)t] \hat{k}}$$

The Wave Equation for Electromagnetic Waves

*52 ••• (a) Using arguments similar to those given in the text, show that for a plane wave, in which E and B are independent of y and z,

$$\dfrac{\partial E_z}{\partial x} = \dfrac{\partial B_y}{\partial t} \text{ and } \dfrac{\partial B_y}{\partial x} = \mu_0 \epsilon_0 \dfrac{\partial E_z}{\partial t}$$

(b) Show that E_z and B_y also satisfy the wave equation.

Picture the Problem We can use Figures 30-10 and 30-11 and a derivation similar to that in the text to obtain the given results.

In Figure 30-11, replace B_z by E_z. For Δx small:

$$E_z(x_2) = E_z(x_1) + \dfrac{\partial E_z}{\partial x} \Delta x$$

Evaluate the line integral of $\vec{E}$ around the rectangular area $\Delta x \Delta z$:

$$\oint \vec{E} \cdot d\vec{\ell} \approx -\dfrac{\partial E_z}{\partial x} \Delta x \Delta z \quad (1)$$

Express the magnetic flux through the same area:

$$\int B_n dA = B_y \Delta x \Delta z$$

268 Chapter 30

Apply Faraday's law to obtain:
$$\oint \vec{E} \cdot d\vec{\ell} \approx -\frac{\partial}{\partial t}\int_S B_n dA = -\frac{\partial}{\partial t}(B_y \Delta x \Delta z)$$
$$= -\frac{\partial B_y}{\partial t}\Delta x \Delta z$$

Substitute in equation (1) to obtain:
$$-\frac{\partial E_z}{\partial x}\Delta x \Delta z = -\frac{\partial B_y}{\partial t}\Delta x \Delta z$$

or
$$\boxed{\frac{\partial E_z}{\partial x} = \frac{\partial B_y}{\partial t}}$$

In Figure 30-10, replace E_y by B_y and evaluate the line integral of $\vec{B}$ around the rectangular area $\Delta x \Delta z$:
$$\oint \vec{B} \cdot d\vec{\ell} = \mu_0 \epsilon_0 \int_S E_n dA$$

provided there are no conduction currents.

Evaluate these integrals to obtain:
$$\boxed{\frac{\partial B_y}{\partial x} = \mu_0 \epsilon_0 \frac{\partial E_z}{\partial t}}$$

(b) Using the first result obtained in (a), find the second partial derivative of E_z with respect to x:
$$\frac{\partial}{\partial x}\left(\frac{\partial E_z}{\partial x}\right) = \frac{\partial}{\partial x}\left(\frac{\partial B_y}{\partial t}\right)$$

or
$$\frac{\partial^2 E_z}{\partial x^2} = \frac{\partial}{\partial t}\left(\frac{\partial B_y}{\partial x}\right)$$

Use the second result obtained in (a) to obtain:
$$\frac{\partial^2 E_z}{\partial x^2} = \frac{\partial}{\partial t}\left(\mu_0 \epsilon_0 \frac{\partial E_z}{\partial t}\right) = \mu_0 \epsilon_0 \frac{\partial^2 E_z}{\partial t^2}$$

or, because $\mu_0 \epsilon_0 = 1/c^2$,
$$\boxed{\frac{\partial^2 E_z}{\partial x^2} = \frac{1}{c^2}\frac{\partial^2 E_z}{\partial t^2}}.$$

Using the second result obtained in (a), find the second partial derivative of B_y with respect to x:
$$\frac{\partial}{\partial x}\left(\frac{\partial B_y}{\partial x}\right) = \mu_0 \epsilon_0 \frac{\partial}{\partial x}\left(\frac{\partial E_z}{\partial t}\right)$$

or
$$\frac{\partial^2 B_y}{\partial x^2} = \mu_0 \epsilon_0 \frac{\partial}{\partial t}\left(\frac{\partial E_z}{\partial x}\right)$$

Use the second result obtained in (a) to obtain:
$$\frac{\partial^2 B_y}{\partial x^2} = \mu_0 \epsilon_0 \frac{\partial}{\partial t}\left(\frac{\partial B_y}{\partial t}\right) = \mu_0 \epsilon_0 \frac{\partial^2 B_y}{\partial t^2}$$

or, because $\mu_0 \epsilon_0 = 1/c^2$,

$$\boxed{\frac{\partial^2 B_y}{\partial x^2} = \frac{1}{c^2} \frac{\partial^2 B_y}{\partial t^2}}.$$

General Problems

*57 •• A circular loop of wire can be used to detect electromagnetic waves. Suppose a 100-MHz FM radio station radiates 50 kW uniformly in all directions. What is the maximum rms voltage induced in a loop of radius 30 cm at a distance of 10^5 m from the station?

Picture the Problem The maximum rms voltage induced in the loop is given by $\mathcal{E}_{rms} = A\omega B_0 / \sqrt{2}$, where A is the area of the loop, B_0 is the amplitude of the magnetic field, and ω is the angular frequency of the wave. We can use the definition of density and the expression for the intensity of an electromagnetic wave to derive an expression for B_0.

The maximum induced rms emf occurs when the plane of the loop is perpendicular to $\vec{B}$:

$$\mathcal{E}_{rms} = \frac{A\omega B_0}{\sqrt{2}} = \frac{\pi R^2 \omega B_0}{\sqrt{2}} \quad (1)$$

where R is the radius of loop of wire.

From the definition of intensity we have:

$$I = \frac{P}{4\pi r^2}$$

where r is the distance from the transmitter.

The intensity is also given by:

$$I = \frac{E_0 B_0}{2\mu_0} = \frac{B_0^2 c}{2\mu_0}$$

Substitute to obtain:

$$\frac{B_0^2 c}{2\mu_0} = \frac{P}{4\pi r^2}$$

Solve for B_0:

$$B_0 = \frac{1}{r}\sqrt{\frac{\mu_0 P}{2\pi c}}$$

Substitute in equation (1) to obtain:

$$\mathcal{E}_{rms} = \frac{\pi R^2 (2\pi f)}{\sqrt{2}\, r} \sqrt{\frac{\mu_0 P}{2\pi c}}$$

$$= \frac{R^2 f}{\sqrt{2}\, r} \sqrt{\frac{2\pi^3 \mu_0 P}{c}}$$

270   Chapter 30

Substitute numerical values and evaluate ε_{rms}:

$$\varepsilon_{rms} = \frac{(0.3\,\text{m})^2(100\,\text{MHz})}{\sqrt{2}(10^5\,\text{m})}\sqrt{\frac{2\pi^3(4\pi \times 10^{-7}\,\text{N/A}^2)(50\,\text{kW})}{3 \times 10^8\,\text{m/s}}} = \boxed{7.25\,\text{mV}}$$

*61 •• The electric fields of two harmonic waves of angular frequency ω_1 and ω_2 are given by $\vec{E}_1 = E_{1,0}\cos(k_1 x - \omega_1 t)\hat{j}$ and $\vec{E}_2 = E_{2,0}\cos(k_2 x - \omega_2 t + \delta)\hat{j}$. Find (a) the instantaneous Poynting vector for the resultant wave motion and (b) the time-average Poynting vector. If the direction of propagation of the second wave is reversed so $\vec{E}_2 = E_{2,0}\cos(k_2 x + \omega_2 t + \delta)\hat{j}$, find (c) the instantaneous Poynting vector for the resultant wave motion and (d) the time-average Poynting vector.

Picture the Problem We can use the definition of the Poynting vector and the relationship between $\vec{B}$ and $\vec{E}$ to find the instantaneous Poynting vectors for each of the resultant wave motions and the fact that the time average of the cross product term is zero for $\omega_1 \neq \omega_2$, and ½ for the square of cosine function to find the time-averaged Poynting vectors.

(a) Because $\vec{E}_1$ and $\vec{E}_2$ propagate in the x direction:

$$\vec{E} \times \vec{B} = \mu_0 S\hat{i} \Rightarrow \vec{B} = B\hat{k}$$

Express B in terms of E_1 and E_2:

$$B = \frac{1}{c}(E_1 + E_2)$$

Substitute for E_1 and E_2 to obtain:

$$\vec{B} = \frac{1}{c}\left[E_{1,0}\cos(k_1 x - \omega_1 t) + E_{2,0}\cos(k_2 x - \omega_2 t + \delta)\right]\hat{k}$$

Express the instantaneous Poynting vector for the resultant wave motion:

$$\vec{S} = \frac{1}{\mu_0}\left(E_{1,0}\cos(k_1 x - \omega_1 t) + E_{2,0}\cos(k_2 x - \omega_2 t + \delta)\right)\hat{j}$$

$$\times \frac{1}{c}\left(E_{1,0}\cos(k_1 x - \omega_1 t) + E_{2,0}\cos(k_2 x - \omega_2 t + \delta)\right)\hat{k}$$

$$= \frac{1}{\mu_0 c}\left(E_{1,0}\cos(k_1 x - \omega_1 t) + E_{2,0}\cos(k_2 x - \omega_2 t + \delta)\right)^2(\hat{j} \times \hat{k})$$

$$= \boxed{\frac{1}{\mu_0 c}\left[E_{1,0}^2\cos^2(k_1 x - \omega_1 t) + 2E_{1,0}E_{2,0}\cos(k_1 x - \omega_1 t)\right.\\ \left.\times \cos(k_2 x - \omega_2 t + \delta) + E_{2,0}^2\cos^2(k_2 x - \omega_2 t + \delta)\right]\hat{i}}$$

(b) The time average of the cross product term is zero for $\omega_1 \neq \omega_2$, and the time average of the square of the cosine terms is ½:

$$\vec{S}_{av} = \boxed{\frac{1}{2\mu_0 c}\left[E_{1,0}^2 + E_{2,0}^2\right]\hat{i}}$$

(c) In this case $\vec{B}_2 = -B\hat{k}$ because the wave with $k = k_2$ propagates in the $-\hat{i}$ direction. The magnetic field is then:

$$\vec{B} = \frac{1}{c}\left[E_{1,0}\cos(k_1 x - \omega_1 t) - E_{2,0}\cos(k_2 x + \omega_2 t + \delta)\right]\hat{k}$$

Express the instantaneous Poynting vector for the resultant wave motion:

$$\vec{S} = \frac{1}{\mu_0}\left(E_{1,0}\cos(k_1 x - \omega_1 t) + E_{2,0}\cos(k_2 x - \omega_2 t + \delta)\right)\hat{j}$$

$$\times \frac{1}{c}\left(E_{1,0}\cos(k_1 x - \omega_1 t) - E_{2,0}\cos(k_2 x + \omega_2 t + \delta)\right)\hat{k}$$

$$= \boxed{\frac{1}{\mu_0 c}\left[E_{1,0}^2 \cos^2(k_1 x - \omega_1 t) - E_{2,0}^2 \cos^2(k_2 x + \omega_2 t + \delta)\right]\hat{i}}$$

(d) The time average of the square of the cosine terms is ½:

$$\vec{S}_{av} = \boxed{\frac{1}{2\mu_0 c}\left[E_{1,0}^2 - E_{2,0}^2\right]\hat{i}}$$

*62 •• At the surface of the earth, there is an approximate average solar flux of 0.75 kW/m². A family wishes to construct a solar energy conversion system to power their home. If the conversion system is 30 percent efficient and the family needs a maximum of 25 kW, what effective surface area is needed for perfectly absorbing collectors?

Picture the Problem We can use the definitions of power and intensity to express the area of the surface as a function of P, I, and the efficiency ε.

Use the definition of power to relate the required surface area to the intensity of the solar radiation:

$$P = \frac{E}{t}\varepsilon = IA\varepsilon$$

where ε is the efficiency of the system.

Solve for A to obtain:

$$A = \frac{P}{I\varepsilon}$$

Substitute numerical values and evaluate A:

$$A = \frac{25\,\text{kW}}{0.3(0.75\,\text{kW/m}^2)} = \boxed{111\,\text{m}^2}$$

*65 ••• A long cylindrical conductor of length L, radius a, and resistivity ρ carries a steady current I that is uniformly distributed over its cross-sectional area. (a) Use Ohm's law to relate the electric field E in the conductor to I, ρ, and a. (b) Find the magnetic field B just outside the conductor. (c) Use the results for Part (a) and Part (b) to compute the Poynting vector $\vec{S} = (\vec{E} \times \vec{B})/\mu_0$ at r = a (the edge of the conductor). In what direction is $\vec{S}$? (d) Find the flux $\oint S_n dA$ through the surface of the conductor into the conductor, and show that the rate of energy flow into the conductor equals $I^2 R$, where R is the resistance of the cylinder. (Here S_n is the *inward* component of $\vec{S}$ perpendicular to the surface of the conductor.)

Picture the Problem We can use Ohm's law to relate the electric field E in the conductor to I, ρ, and a and Ampere's law to find the magnetic field B just outside the conductor. Knowing $\vec{E}$ and $\vec{B}$ we can find $\vec{S}$ and, using its normal component, show that the rate of energy flow into the conductor equals $I^2 R$, where R is the resistance.

(a) Apply Ohm's law to the cylindrical conductor to obtain:

$$V = IR = \frac{I \rho L}{A} = \frac{I \rho L}{\pi a^2} = EL$$

Solve for E:

$$E = \boxed{\frac{I \rho}{\pi a^2}}$$

(b) Apply Ampere's law to a circular path of radius a at the surface of the cylindrical conductor:

$$\oint_C \vec{B} \cdot d\vec{\ell} = B(2\pi a) = \mu_0 I_{enclosed} = \mu_0 I$$

Solve for B to obtain:

$$B = \boxed{\frac{\mu_0 I}{2\pi a}}$$

(c) The electric field at the surface of the conductor is in the direction of the current and the magnetic field at the surface is tangent to the surface. Use the results of (a) and (b) and the right-hand rule to evaluate $\vec{S}$:

$$\vec{S} = \frac{1}{\mu_0} \vec{E} \times \vec{B}$$

$$= \frac{1}{\mu_0} \left(\frac{I\rho}{\pi a^2}\right) \hat{u}_{parallel} \times \left(\frac{\mu_0 I}{2\pi a}\right) \hat{u}_{tangent}$$

$$= \boxed{-\frac{I^2 \rho}{2\pi^2 a^3} \hat{r}}$$

where $\hat{r}$ is a unit vector directed radially outward from the cylindrical conductor.

(d) The flux through the surface of the conductor into the conductor is:

$$\oint S_n dA = S(2\pi a L)$$

Substitute for S_n, the *inward* component of $\vec{S}$, and simplify to obtain:

$$\oint S_n \, dA = \frac{I^2 \rho}{2\pi^2 a^3}(2\pi a L) = \frac{I^2 \rho L}{\pi a^2}$$

Since $R = \dfrac{\rho L}{A} = \dfrac{\rho L}{\pi a^2}$:

$$\oint S_n \, dA = \boxed{I^2 R}$$

*67 ••• Small particles might be blown out of solar systems by the radiation pressure of sunlight. Assume that the particles are spherical with a radius r and a density of 1 g/cm^3 and that they absorb all the radiation in a cross-sectional area of πr^2. The particles are a distance R from the sun, which has a power output of 3.83×10^{26} W. What is the radius r for which the radiation force of repulsion just balances the gravitational force of attraction to the sun?

Picture the Problem We can use a condition for translational equilibrium to obtain an expression relating the forces due to gravity and radiation pressure that act on the particles. We can express the force due to radiation pressure in terms of the radiation pressure and the effective cross sectional area of the particles and the radiation pressure in terms of the intensity of the solar radiation. We can solve the resulting equation for r.

Apply the condition for translational equilibrium to the particle:

$$F_r - F_g = 0$$

or, since $F_r = P_r A$ and $F_g = mg$,

$$P_r A - \frac{GM_s m}{R^2} = 0 \quad (1)$$

The radiation pressure P_r depends on the intensity of the radiation I:

$$P_r = \frac{I}{c}$$

The intensity of the solar radiation at a distance R is:

$$I = \frac{P}{4\pi R^2}$$

Substitute to obtain:

$$P_r = \frac{P}{4\pi R^2 c}$$

Substitute for P_r, A, and m in equation (1):

$$\frac{P}{4\pi R^2 c}(\pi r^2) - \frac{\frac{4}{3}\pi r^3 \rho G M_s}{R^2} = 0$$

Solve for R to obtain:

$$r = \frac{3P}{16\pi \rho c G M_s}$$

Substitute numerical values and evaluate r:

274 Chapter 30

$$r = \frac{3(3.83 \times 10^{26} \text{ W})}{16\pi(1 \text{ g/cm}^3)(3 \times 10^8 \text{ m/s})(6.67 \times 10^{-11} \text{ N} \cdot \text{m}^2/\text{kg}^2)(1.99 \times 10^{30} \text{ kg})}$$

$$= \boxed{0.574 \text{ μm}}$$

*69 ••• An intense point source of light radiates 1 MW isotropically. The source is located 1 m above an infinite, perfectly reflecting plane. Determine the force that acts on the plane.

Picture the Problem Let the point source be a distance a above the plane. Consider a ring of radius r and thickness dr in the plane and centered at the point directly below the light source. Express the force of force on this ring and integrate the resulting expression to obtain F.

The intensity anywhere along this infinitesimal ring is $P/4\pi(r^2 + a^2)$ and the element of force dF on this ring of area $2\pi r dr$ is given by:

$$dF = \frac{P r dr}{c(r^2 + a^2)} \frac{a}{\sqrt{r^2 + a^2}}$$

$$= \frac{P a r dr}{c(r^2 + a^2)^{3/2}}$$

where we have taken into account that only the normal component of the incident radiation contributes to the force on the plane, and that the plane is a perfectly reflecting plane.

Integrate dF from $r = 0$ to $r = \infty$:

$$F = \frac{Pa}{c} \int_0^\infty \frac{r dr}{(r^2 + a^2)^{3/2}}$$

From integral tables:

$$\int_0^\infty \frac{r dr}{(r^2 + a^2)^{3/2}} = \left[\frac{-1}{\sqrt{r^2 + a^2}} \right]_0^\infty = \frac{1}{a}$$

Substitute to obtain:

$$F = \frac{Pa}{c}\left(\frac{1}{a}\right) = \frac{P}{c}$$

Substitute numerical values and evaluate F:

$$F = \frac{1 \text{ MW}}{3 \times 10^8 \text{ m/s}} = \boxed{3.33 \text{ mN}}$$

Chapter 31
Properties of Light

Conceptual Problems

*5 •• The density of the atmosphere decreases with height, as does the index of refraction. Explain how one can see the sun after it has set. Why does the setting sun appear flattened?

Determine the Concept The change in atmospheric density results in refraction of the light from the sun, bending it toward the earth. Consequently, the sun can be seen even after it is just below the horizon. Also, the light from the lower portion of the sun is refracted more than that from the upper portion, so the lower part appears to be slightly higher in the sky. The effect is an apparent flattening of the disk into an ellipse.

*10 •• We learned in Chapter 30, Section 30-3, that an oscillating electric dipole produces electromagnetic radiation (see Figure 30-8). Assuming that the light reflected off and refracted into the surface of a piece of transparent material is caused by such dipoles, show that the condition for Brewster's angle (Equation 31-21) is exactly the same as saying that the refracted ray is perpendicular to the axis of the radiating dipoles for light polarized in the plane of incidence.

Determine the Concept The diagram shows that the radiated intensity for a dipole is zero in the direction of the dipole moment. Because the dipole axis is in the same direction as the polarization, for light polarized parallel to plane of incidence, the dipole axis will point in the same direction as the reflected wave, i.e., in the direction described by Brewster's law. As the diagram indicates, there is zero field in the direction of the refracted ray. On the other hand, if the incoming wave is polarized perpendicular to the plane of incidence, the dipole axis will never point along the direction of propagation for the reflected or refracted wave.

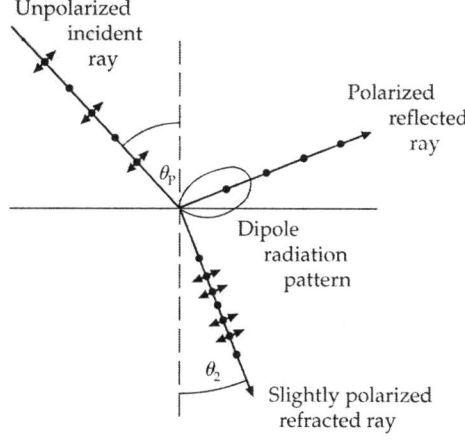

275

*14 •• It is a common experience that on a calm, sunny day one can hear voices of persons in a boat over great distances. Explain this phenomenon, keeping in mind that sound is reflected from the surface of the water and that the temperature of the air just above the water's surface is usually less than that at a height of 10 m or 20 m above the water.

Picture the Problem The sound is reflected specularly from the surface of the water (we assume it is calm). It is then refracted back toward the water in the region above the water because the speed of sound depends on the temperature of the air and is greater at the higher temperature. The pattern of the sound wave is shown schematically below.

Sources of Light

*24 •• Singly ionized helium is a hydrogen-like atom with a nuclear charge of $2e$. Its energy levels are given by $E_n = -4E_0/n^2$, where $E_0 = 13.6$ eV. If a beam of visible white light is sent through a gas of singly ionized helium, at what wavelengths will dark lines be found in the spectrum of the transmitted radiation?

Determine the Concept The energy difference between the ground state and the first excited state is $3E_0 = 40.8$ eV, corresponding to a wavelength of 30.4 nm. This is in the far ultraviolet, well outside the visible range of wavelengths. There will be no dark lines in the transmitted radiation.

The Speed of Light

*29 •• In Galileo's attempt to determine the speed of light, he and his assistant were located on hilltops about 3 km apart. Galileo flashed a light and received a return flash from his assistant. (*a*) If his assistant had an instant reaction, what time difference would Galileo need to be able to measure for this method to be successful? (*b*) How does this time compare with human reaction time, which is about 0.2 s?

Picture the Problem We can use the distance, rate, and time relationship to find the time difference Galileo would need to be able to measure the speed of light successfully.

(*a*) Relate the distance separating Galileo and his assistant to the speed of light and the time required for it travel to the assistant and back to Galileo:
$\qquad D = c\Delta t$

Solve for Δt:
$$\Delta t = \frac{D}{c}$$

Substitute numerical values and evaluate Δt:
$$\Delta t = \frac{2(3\,\text{km})}{3\times 10^8\,\text{m/s}} = \boxed{20.0\,\mu\text{s}}$$

(b) Express the ratio of the human reaction time to the transit time for the light:
$$\frac{\Delta t_{\text{reaction}}}{\Delta t} = \frac{0.2\,\text{s}}{20\,\mu\text{s}} = 10^4$$

or
$$\Delta t_{\text{reaction}} = \boxed{10^4 \Delta t}$$

Reflection and Refraction

*31 •• A ray of light is incident on one of a pair of mirrors set at right angles to each other. The plane of incidence is perpendicular to both mirrors. Show that after reflecting off each mirror the ray will emerge in the opposite direction, regardless of the angle of incidence.

Picture the Problem The diagram shows ray 1 incident on the vertical surface at an angle θ_1, reflected as ray 2, and incident on the horizontal surface at an angle of incidence θ_3. We'll prove that rays 1 and 3 are parallel by showing that $\theta_1 = \theta_4$, i.e., by showing that they make equal angles with the horizontal. Note that the law of reflection has been used in identifying equal angles of incidence and reflection.

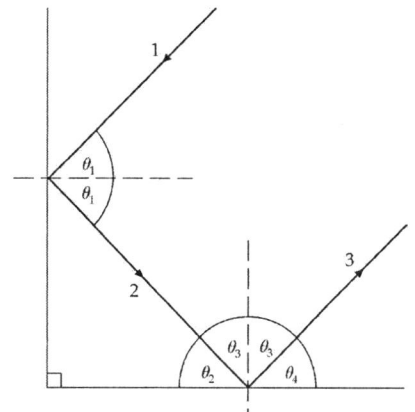

We know that the angles of the right triangle formed by ray 2 and the two mirror surfaces add up to 180°:
$$\theta_2 + 90° + 90° - \theta_1 = 180°$$
or
$$\theta_1 = \theta_2$$

The sum of θ_2 and θ_3 is 90°:
$$\theta_3 = 90° - \theta_2$$

Because $\theta_1 = \theta_2$:
$$\theta_3 = 90° - \theta_1$$

The sum of θ_4 and θ_3 is 90°:
$$\theta_3 + \theta_4 = 90°$$

Substitute for θ_3 to obtain:
$$90° - \theta_1 + \theta_4 = 90° \Rightarrow \theta_1 = \boxed{\theta_4}$$

278   Chapter 31

*37 ••   Light is incident normally on a slab of glass with an index of refraction $n = 1.5$. Reflection occurs at both surfaces of the slab. Approximately what percentage of the incident light energy is transmitted by the slab?

Picture the Problem Let the subscript 1 refer to the medium to the left (air) of the first interface, the subscript 2 to glass, and the subscript 3 to the medium (air) to the right of the second interface. Apply the equation relating the intensity of reflected light at normal incidence to the intensity of the incident light and the indices of refraction of the media on either side of the interface to both interfaces. We'll neglect multiple reflections at glass-air interfaces.

Express the intensity of the transmitted light in the second medium:

$$I_2 = I_1 - I_{r,1} = I_1 - \left(\frac{n_1 - n_2}{n_1 + n_2}\right)^2 I_1$$

$$= I_1\left[1 - \left(\frac{n_1 - n_2}{n_1 + n_2}\right)^2\right]$$

Express the intensity of the transmitted light in the third medium:

$$I_3 = I_2 - I_{r,2} = I_2 - \left(\frac{n_2 - n_3}{n_2 + n_3}\right)^2 I_2$$

$$= I_2\left[1 - \left(\frac{n_2 - n_3}{n_2 + n_3}\right)^2\right]$$

Substitute for I_2 to obtain:

$$I_3 = I_1\left[1 - \left(\frac{n_1 - n_2}{n_1 + n_2}\right)^2\right]\left[1 - \left(\frac{n_2 - n_3}{n_2 + n_3}\right)^2\right]$$

Solve for the ratio I_3/I_1:

$$\frac{I_3}{I_1} = \left[1 - \left(\frac{n_1 - n_2}{n_1 + n_2}\right)^2\right]\left[1 - \left(\frac{n_2 - n_3}{n_2 + n_3}\right)^2\right]$$

Substitute numerical values and evaluate I_3/I_1:

$$\frac{I_3}{I_1} = \left[1 - \left(\frac{1 - 1.5}{1 + 1.5}\right)^2\right]\left[1 - \left(\frac{1.5 - 1}{1.5 + 1}\right)^2\right]$$

$$= 0.922 = \boxed{92.2\%}$$

Properties of Light 279

*40 ••• Figure 31-56 shows a beam of light incident on a glass plate of thickness d and index of refraction n. (a) Find the angle of incidence so that the perpendicular separation between the ray reflected from the top surface and the ray reflected from the bottom surface and exiting the top surface is a maximum. (b) What is this angle of incidence if the index of refraction of the glass is 1.60? What is the separation of the two beams if the thickness of the glass plate is 4.0 cm?

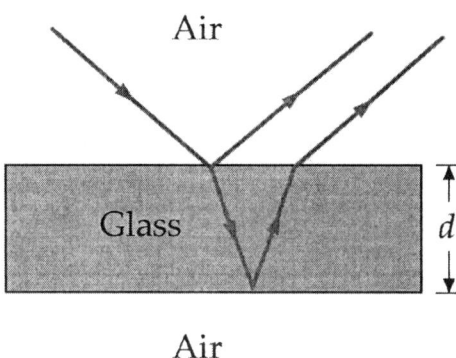

Figure 31-56 Problem 40

Picture the Problem Let x be the perpendicular separation between the two rays and let ℓ be the separation between the points of emergence of the two rays on the glass surface. We can use the geometry of the refracted and reflected rays to express x as a function of ℓ, d, θ_r, and θ_i. Setting the derivative of the resulting equation equal to zero will yield the value of θ_i that maximizes x.

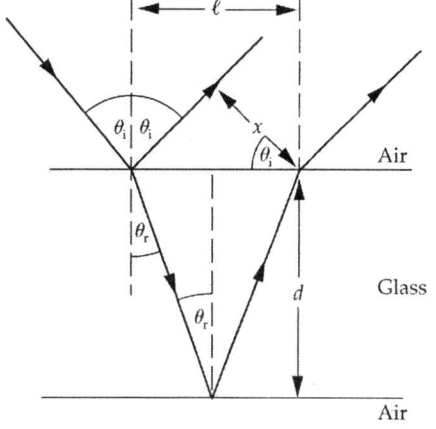

(a) Express ℓ in terms of d and the angle of refraction θ_r:

$\ell = 2d \tan \theta_r$

Express x as a function of ℓ, d, θ_r, and θ_i:

$x = 2d \tan \theta_r \cos \theta_i$

Differentiate x with respect to θ_i:

Chapter 31

$$\frac{dx}{d\theta_i} = 2d \frac{d}{d\theta_i}(\tan\theta_r \cos\theta_i) = 2d\left(-\tan\theta_r \sin\theta_i + \sec^2\theta_r \cos\theta_i \frac{d\theta_r}{d\theta_i}\right) \quad (1)$$

Apply Snell's law to the air-glass interface:

$$n_1 \sin\theta_i = n_2 \sin\theta_r \quad (2)$$

or, since $n_1 = 1$ and $n_2 = n$,

$$\sin\theta_i = n\sin\theta_r$$

Differentiate implicitly with respect to θ_i to obtain:

$$\cos\theta_i\, d\theta_i = n\cos\theta_r\, d\theta_r$$

or

$$\frac{d\theta_r}{d\theta_i} = \frac{1}{n}\frac{\cos\theta_i}{\cos\theta_r}$$

Substitute in equation (1) to obtain:

$$\frac{dx}{d\theta_i} = 2d\left(-\frac{\sin\theta_r}{\cos\theta_r}\sin\theta_i + \frac{1}{n}\frac{\cos\theta_i}{\cos^2\theta_r}\frac{\cos\theta_i}{\cos\theta_r}\right) = 2d\left(\frac{1}{n}\frac{\cos^2\theta_i}{\cos^3\theta_r} - \frac{\sin\theta_r \sin\theta_i}{\cos\theta_r}\right)$$

Substitute $1 - \sin^2\theta_i$ for $\cos^2\theta_i$ and $\frac{1}{n}\sin\theta_i$ for $\sin\theta_r$ to obtain:

$$\frac{dx}{d\theta_i} = 2d\left(\frac{1-\sin^2\theta_i}{n\cos^3\theta_r} - \frac{\sin^2\theta_i}{n\cos\theta_r}\right)$$

Multiply the second term in parentheses by $\cos^2\theta_r/\cos^2\theta_r$ and simplify to obtain:

$$\frac{dx}{d\theta_i} = 2d\left(\frac{1-\sin^2\theta_i}{n\cos^3\theta_r} - \frac{\sin^2\theta_i \cos^2\theta_r}{n\cos^3\theta_r}\right) = \frac{2d}{n\cos^3\theta_r}\left(1 - \sin^2\theta_i - \sin^2\theta_i \cos^2\theta_r\right)$$

Substitute $1 - \sin^2\theta_r$ for $\cos^2\theta_r$:

$$\frac{dx}{d\theta_i} = \frac{2d}{n\cos^3\theta_r}\left[1 - \sin^2\theta_i - \sin^2\theta_i(1 - \sin^2\theta_r)\right]$$

Substitute $\frac{1}{n}\sin\theta_i$ for $\sin\theta_r$ to obtain:

$$\frac{dx}{d\theta_i} = \frac{2d}{n\cos^3\theta_r}\left[1 - \sin^2\theta_i - \sin^2\theta_i\left(1 - \frac{1}{n^2}\sin^2\theta_i\right)\right]$$

Factor out $1/n^2$, simplify, and set equal to zero to obtain:

Properties of Light 281

$$\frac{dx}{d\theta_i} = \frac{2d}{n^3 \cos^3 \theta_r}\left[\sin^4 \theta_i - 2n^2 \sin^2 \theta_i + n^2\right] = 0 \text{ for extrema}$$

If $dx/d\theta_i = 0$, then it must be true that:

$$\sin^4 \theta_i - 2n^2 \sin^2 \theta_i + n^2 = 0$$

Solve this quartic equation for θ_i to obtain:

$$\boxed{\theta_i = \sin^{-1}\left(n\sqrt{1 - \sqrt{1 - \frac{1}{n^2}}}\right)}$$

(b) Evaluate θ_i for $n = 1.60$:

$$\theta_i = \sin^{-1}\left(1.6\sqrt{1 - \sqrt{1 - \frac{1}{(1.6)^2}}}\right)$$

$$= \boxed{48.5°}$$

In (a) we showed that:

$$x = 2d \tan\theta_r \cos\theta_i$$

Solve equation (2) for θ_r:

$$\theta_r = \sin^{-1}\left(\frac{n_1}{n_2}\sin\theta_i\right)$$

Substitute numerical values and evaluate θ_r:

$$\theta_r = \sin^{-1}\left(\frac{1}{1.6}\sin 48.5°\right) = 27.9°$$

Substitute numerical values and evaluate x:

$$x = 2(4\,\text{cm})\tan 27.9° \cos 48.5°$$

$$= \boxed{2.81\,\text{cm}}$$

Total Internal Reflection

*46 •• An optical fiber allows rays of light to propagate long distances through total internal reflection. As shown in Figure 31-57, the fiber consists of a core material with index of refraction n_2 and radius b, surrounded by a cladding material of index $n_3 < n_2$. The numerical aperture of the fiber is defined as $\sin\theta_1$, where θ_1 is the angle of incidence of a ray of light impinging the end of the fiber that reflects off the core-cladding interface at the critical angle. Using the figure as a guide, show that the numerical aperture is given by $\sqrt{n_2^2 - n_3^2}$ assuming the ray is incident from air. (*Hint: Use of the Pythagorean theorem may be required.*)

282   Chapter 31

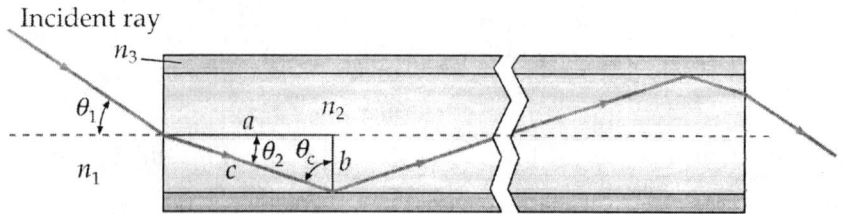

Figure 31-57 Problems 46, 47, and 48.

Picture the Problem We can use the geometry of the figure, the law of refraction at the air-n_1 interface, and the condition for total internal reflection at the n_1-n_2 interface to show that the numerical aperture is given by $\sqrt{n_2^2 - n_3^2}$.

Referring to the figure, note that:
$$\sin\theta_c = \frac{n_3}{n_2} = \frac{a}{c}$$

and
$$\sin\theta_2 = \frac{b}{c}$$

Apply the Pythagorean theorem to the right triangle to obtain:
$$a^2 + b^2 = c^2$$
or
$$\frac{a^2}{c^2} + \frac{b^2}{c^2} = 1$$

Solve for $\frac{b}{c}$:
$$\frac{b}{c} = \sqrt{1 - \frac{a^2}{c^2}}$$

Substitute for $\frac{a}{c}$ and $\frac{b}{c}$ to obtain:
$$\sin\theta_2 = \sqrt{1 - \frac{n_3^2}{n_2^2}}$$

Use the law of refraction to relate θ_1 and θ_2:
$$n_1 \sin\theta_1 = n_2 \sin\theta_2$$

Substitute for $\sin\theta_2$ and let $n_1 = 1$ (air) to obtain:
$$\sin\theta_1 = n_2\sqrt{1 - \frac{n_3^2}{n_2^2}} = \boxed{\sqrt{n_2^2 - n_3^2}}$$

Dispersion

*51 ••   A beam of light strikes the plane surface of silicate flint glass at an angle of incidence of 45°. The index of refraction of the glass varies with wavelength, as shown in the graph in Figure 31-26. How much smaller is the angle of refraction for violet light of wavelength 400 nm than that for red light of wavelength 700 nm?

Properties of Light   283

Picture the Problem We can apply Snell's law of refraction to express the angles of refraction for red and violet light in silicate flint glass.

Express the difference between the angle of refraction for violet light and for red light:
$$\Delta\theta = \theta_{r,red} - \theta_{r,violet} \quad (1)$$

Apply Snell's law of refraction to the interface to obtain:
$$\sin 45° = n\sin\theta_r$$

Solve for θ_r:
$$\theta_r = \sin^{-1}\left(\frac{1}{\sqrt{2}n}\right)$$

Substitute in equation (1):
$$\Delta\theta = \sin^{-1}\left(\frac{1}{\sqrt{2}n_{red}}\right) - \sin^{-1}\left(\frac{1}{\sqrt{2}n_{violet}}\right)$$

Substitute numerical values and evaluate $\Delta\theta$:
$$\Delta\theta = \sin^{-1}\left(\frac{1}{\sqrt{2}(1.60)}\right) - \sin^{-1}\left(\frac{1}{\sqrt{2}(1.66)}\right)$$
$$= 26.23° - 25.21° = \boxed{1.02°}$$

Polarization

*60 •• A stack of $N+1$ ideal polarizing sheets is arranged with each sheet rotated by an angle of $\pi/(2N)$ rad with respect to the preceding sheet. A plane, linearly polarized light wave of intensity I_0 is incident normally on the stack. The incident light is polarized along the transmission axis of the first sheet and is therefore perpendicular to the transmission axis of the last sheet in the stack. (*a*) Show that the transmitted intensity through the stack is given by the expression $I_0 \cos^{2N}\left(\frac{\pi}{2N}\right)$. (*b*) Using a spreadsheet or graphing program, plot the transmitted intensity as a function of N for values of N from 2 to 100. (*c*) What is the direction of polarization of the transmitted beam in each case?

Picture the Problem Let I_n be the intensity after the *n*th polarizing sheet and use $I = I_0 \cos^2\theta$ to find the ratio of I_{n+1} to I_n.

(*a*) Find the ratio of I_{n+1} to I_n:
$$\frac{I_{n+1}}{I_n} = \cos^2\frac{\pi}{2N}$$

284   Chapter 31

Because there are N such reductions of intensity:

$$\frac{I_{N+1}}{I_1} = \frac{I_{N+1}}{I_0} = \cos^{2N}\left(\frac{\pi}{2N}\right)$$

and

$$I_{N+1} = \boxed{I_0 \cos^{2N}\left(\frac{\pi}{2N}\right)}$$

(b) A spreadsheet program to graph I_{N+1}/I_0 as a function of N is shown below. The formulas used to calculate the quantities in the columns are as follows:

Cell	Content/Formula	Algebraic Form
A2	2	N
A3	A2 + 1	$N+1$
B2	(cos(PI()/(2*A2))^(2*A2)	$\cos^{2N}\left(\dfrac{\pi}{2N}\right)$

	A	B
1	N	I/I_0
2	2	0.250
3	3	0.422
4	4	0.531
5	5	0.605
95	95	0.974
96	96	0.975
97	97	0.975
98	98	0.975
99	99	0.975
100	100	0.976

A graph of I/I_0 as a function of N follows.

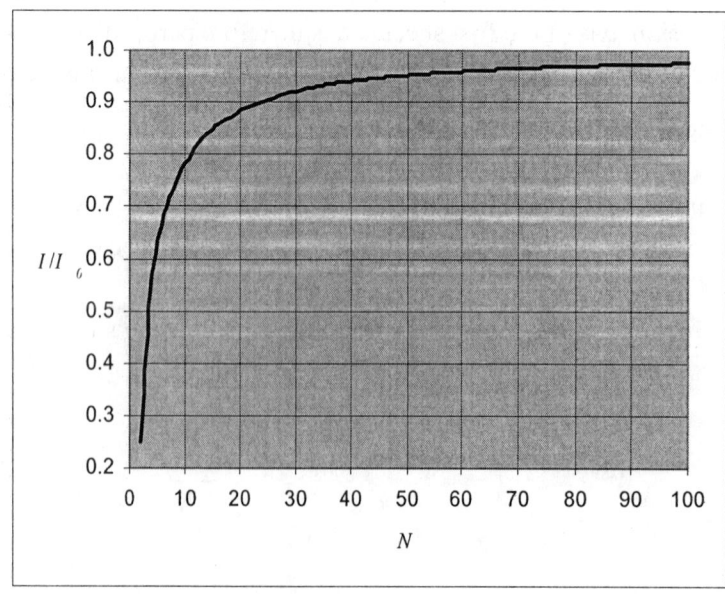

Properties of Light   285

(c) | In each case, the polarization of the transmitted beam is perpendicular to that of the incident beam. |

*62 ••   Show that a linearly polarized wave can be thought of as a superposition of a right and a left circularly polarized wave.

Picture the Problem A circularly polarized wave is said to be *right circularly polarized* if the electric and magnetic fields rotate clockwise when viewed along the direction of propagation and *left circularly polarized* if the fields rotate counterclockwise.

For a circularly polarized wave, the x and y components of the electric field are given by:

$E_x = E_0 \cos \omega t$

and

$E_y = E_0 \sin \omega t$ or $E_y = -E_0 \sin \omega t$

for left and right circular polarization, respectively.

For a wave polarized along the x axis:

$\vec{E}_{\text{right}} + \vec{E}_{\text{left}} = E_0 \cos \omega t\, \hat{i} + E_0 \cos \omega t\, \hat{i}$

$= \boxed{2E_0 \cos \omega t\, \hat{i}}$

*64 ••   Show that the electric field of a circularly polarized wave propagating in the x direction can be expressed by

$$\vec{E} = E_0 \sin(kx - \omega t)\hat{j} + E_0 \cos(kx - \omega t)\hat{k}$$

Picture the Problem We can use the components of $\vec{E}$ to show that $\vec{E}$ is constant in time and rotates with angular frequency ω.

Express the magnitude of $\vec{E}$ in terms of its components:

$E = \sqrt{E_x^2 + E_y^2}$

Substitute for E_x and E_y to obtain:

$E = \sqrt{[E_0 \sin(kx - \omega t)]^2 + [E_0 \cos(kx - \omega t)]^2} = \sqrt{E_0^2 [\sin^2(kx - \omega t) + \cos^2(kx - \omega t)]}$

$= E_0$

and the $\vec{E}$ vector rotates in the yz plane with angular frequency ω.

286 Chapter 31

General Problems

*68 •• Figure 31-58 shows two plane mirrors that make an angle θ with each other. Show that the angle between the incident and reflected rays is 2θ.

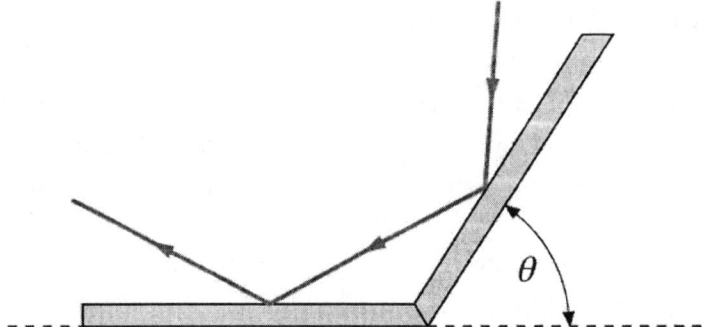

Figure 31-58 Problem 68

Picture the Problem Angle *ADE* is the angle between the direction of the incoming ray and that reflected by the two mirror surfaces. Note that triangle *ABC* is isosceles and that angles *CAB* and *ABC* are equal and their sum equals θ. Also from the law of reflection, angles *CAD* and *CBD* equal angle *ABC*. Because angle *BAD* is twice *BAC* and angle *DBA* is twice *CBA*, angle *ADE* is twice the angle θ.

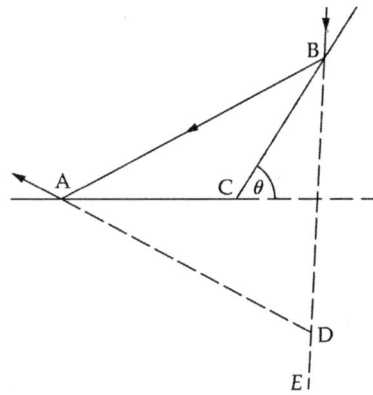

*71 •• A swimmer at the bottom of a pool 3 m deep looks up and sees a circle of light. If the index of refraction of the water in the pool is 1.33, find the radius of the circle.

Picture the Problem We can apply Snell's law to the water-air interface to express the critical angle θ_c in terms of the indices of refraction of water (n_1) and air (n_2) and then relate the radius of the circle to the depth d of the swimmer and θ_c.

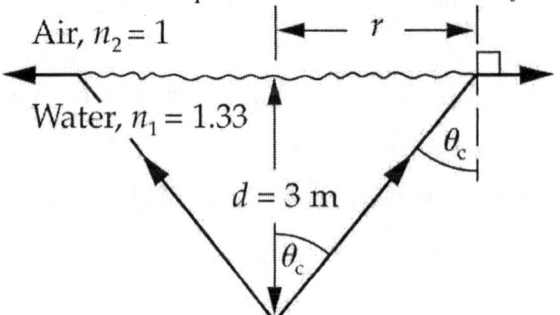

Relate the radius of the circle to the depth d of the point source and the critical angle θ_c:

$$r = d \tan \theta_c$$

Apply Snell's law to the water-air interface to obtain:

$$n_1 \sin \theta_c = n_2 \sin 90° = n_2$$

Solve for θ_c:

$$\theta_c = \sin^{-1}\left(\frac{n_2}{n_1}\right)$$

Substitute for θ_c to obtain:

$$r = d \tan\left[\sin^{-1}\left(\frac{n_2}{n_1}\right)\right]$$

Substitute numerical values and evaluate r:

$$r = (3\,\text{m})\tan\left[\sin^{-1}\left(\frac{1}{1.33}\right)\right] = \boxed{3.42\,\text{m}}$$

*76 •• A Brewster window is used in lasers to preferentially transmit light of one polarization, as shown in Figure 31-59. Show that if θ_{P1} is the polarizing angle for the n_1/n_2 interface, then θ_{P2} is the polarizing angle for the n_2/n_1 interface.

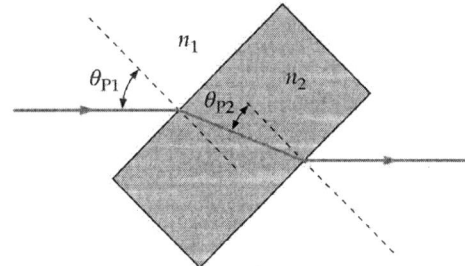

Figure 31-59 Problem 76

Picture the Problem Let the angle of refraction at the first interface by θ_1 and the angle of refraction at the second interface be θ_2. We can apply Snell's law at each interface and eliminate θ_1 and n_2 to show that $\theta_2 = \theta_{P2}$.

Apply Snell's Brewster's law at the n_1-n_2 interface:

$$\tan \theta_{P1} = \frac{n_2}{n_1}$$

Draw a reference triangle consistent with Brewster's law:

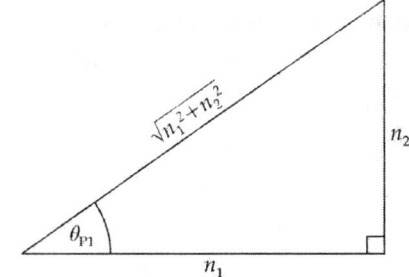

Apply Snell's law at the n_1-n_2 interface:

$$n_1 \sin\theta_{P1} = n_2 \sin\theta_1$$

Solve for θ_1 to obtain:

$$\theta_1 = \sin^{-1}\left(\frac{n_1}{n_2}\sin\theta_{P1}\right)$$

Referring to the reference triangle we note that:

$$\theta_1 = \sin^{-1}\left(\frac{n_1}{n_2}\frac{n_2}{\sqrt{n_1^2+n_2^2}}\right)$$

$$= \sin^{-1}\left(\frac{n_1}{\sqrt{n_1^2+n_2^2}}\right)$$

i.e., θ_1 is the complement of θ_{p1}.

Apply Snell's law at the n_2-n_1 interface:

$$n_2 \sin\theta_1 = n_1 \sin\theta_2$$

Solve for θ_2 to obtain:

$$\theta_2 = \sin^{-1}\left(\frac{n_2}{n_1}\sin\theta_1\right)$$

Refer to the reference triangle again to obtain:

$$\theta_2 = \sin^{-1}\left(\frac{n_2}{n_1}\frac{n_1}{\sqrt{n_1^2+n_2^2}}\right)$$

$$= \sin^{-1}\left(\frac{n_2}{\sqrt{n_1^2+n_2^2}}\right) = \boxed{\theta_{P2}}$$

Equate these expressions for $n_2 \sin\theta_1$ to obtain:

$$n_1 \sin\theta_P = n_1 \sin\theta_2 \Rightarrow \theta_2 = \boxed{\theta_P}$$

*79 •• (a) For a light ray inside a transparent medium that has a planar interface with a vacuum, show that the polarizing angle and the critical angle for internal reflection satisfy $\tan\theta_p = \sin\theta_c$. (b) Which angle is larger?

Properties of Light 289

Picture the Problem We can apply Snell's law at the critical angle and the polarizing angle to show that $\tan\theta_p = \sin\theta_c$.

(a) Apply Snell's law at the medium-vacuum interface.

$$n_1 \sin\theta_1 = n_2 \sin\theta_r$$

For $\theta_1 = \theta_c$, $n_1 = n$, and $n_2 = 1$:

$$n\sin\theta_c = \sin 90° = 1$$

For $\theta_1 = \theta_p$, $n_1 = n$, and $n_2 = 1$:

$$\tan\theta_p = \frac{n_2}{n_1} = \frac{1}{n} \Rightarrow n\tan\theta_p = 1$$

Because both expressions equal one:

$$\boxed{\tan\theta_p = \sin\theta_c}$$

(b) For any value of θ:

$$\tan\theta > \sin\theta \Rightarrow \boxed{\theta_p > \theta_c}$$

*85 •• Suppose rain falls vertically from a stationary cloud 10,000 m above a confused marathoner running in a circle with constant speed of 4 m/s. The rain has a terminal speed of 9 m/s. (a) What is the angle that the rain appears to make with the vertical to the marathoner? (b) What is the apparent motion of the cloud as observed by the marathoner? (c) A star on the axis of the earth's orbit appears to have a circular orbit of angular diameter of 41.2 seconds of arc. How is this angle related to the earth's speed in its orbit and the velocity of photons received from this distant star? (d) What is the speed of light as determined from the data in Part (c)?

Picture the Problem The angle that the rain appears to make with the vertical, according to the marathoner, is the angle whose tangent is the ratio of v_{runner} to v_{rain}. The circular motion of the star is analogous to the circular motion of the cloud with $v_{runner} = v_{earth}$ and $v_{rain} = c$.

(a) The angle that the rain appears to make with the vertical to the marathoner is given by:

$$\theta = \tan^{-1}\left(\frac{v_{runner}}{v_{rain}}\right)$$

Substitute numerical values and evaluate θ:

$$\theta = \tan^{-1}\left(\frac{4\,\text{m/s}}{9\,\text{m/s}}\right) = \boxed{24.0°}$$

(b) The cloud moves in a circle whose radius is given by:

$$R = H\tan\theta$$

Substitute numerical values and evaluate R:

$$R = (10\,\text{km})\tan 24° = \boxed{4.45\,\text{km}}$$

(c) Here $v_{\text{runner}} = v_{\text{earth}}$ and $v_{\text{rain}} = c$:

$$\theta = \boxed{\tan^{-1}\left(\frac{v_{\text{earth}}}{c}\right)} \qquad (1)$$

where $\theta = \frac{1}{2}(\text{angular diameter})$

(d) From equation (1):

$$c = \frac{v_{\text{earth}}}{\tan\theta} = \frac{2\pi R_{\text{earth-sun}}}{T_{\text{earth}}\tan\theta}$$

Convert 20.6″ to degrees:

$$20.6'' = 20.6'' \times \frac{1'}{60''} \times \frac{1°}{60'} = 5.722\times 10^{-3}\,°$$

Substitute numerical values and evaluate c:

$$c = \frac{2\pi(1.5\times 10^{11}\,\text{m})}{(1\,\text{y})(3.156\times 10^7\,\text{s/y})\tan(20.6'')} = \boxed{2.99\times 10^8\,\text{m/s}}$$

Substitute numerical values and evaluate c:

$$c = \frac{2\pi(1.5\times 10^{11}\,\text{m})}{(1\,\text{y})(3.156\times 10^7\,\text{s/y})\tan(5.722\times 10^{-3}\,°)} = \boxed{2.99\times 10^8\,\text{m/s}}$$

Chapter 32
Optical Images

Conceptual Problems

*4 •• Under what condition will a concave mirror produce (*a*) an upright image, (*b*) a virtual image, (*c*) an image smaller than the object, and (*d*) an image larger than the object?

Determine the Concept Let *s* be the object distance and *f* the focal length of the mirror.

(*a*) If $s < f$, the image is virtual, upright, and larger than the object.

(*b*) If $s < f$, the image is virtual, upright, and larger than the object.

(*c*) If $s > 2f$, the image is real, inverted, and smaller than the object.

(*d*) If $f < s < 2f$, the image is real, inverted, and larger than the object.

*9 • Under what conditions will the focal length of a thin lens be (*a*) positive and (*b*) negative? Consider both the case where the index of refraction of the lens is greater than and less than the surrounding medium.

Determine the Concept

(*a*) The lens will be positive if its index of refraction is greater than that of the surrounding medium and the lens is thicker in the middle than at the edges. Conversely, if the index of refraction of the lens is less than that of the surrounding medium, the lens will be positive if it is thinner at its center than at the edges.

(*b*) The lens will be negative if its index of refraction is greater than that of the surrounding medium and the lens is thinner at the center than at the edges. Conversely, if the index of refraction of the lens is less than that of the surrounding medium, the lens will be negative if it is thicker at the center than at the edges.

*14 • If an object is placed 25 cm from the eye of a farsighted person who does not wear corrective lenses, a sharp image is formed (*a*) behind the retina, and the corrective lens should be convex. (*b*) behind the retina, and the corrective lens should be concave. (*c*) in front of the retina, and the corrective lens should be convex. (*d*) in front of the retina, and the corrective lens should be concave.

292 Chapter 32

Determine the Concept The eye muscles of a farsighted person lack the ability to shorten the focal length of the lens in the eye sufficiently to form an image on the retina of the eye. A convex lens (a lens that is thicker in the middle than at the circumference) will bring the image forward onto the retina. (a) is correct.

*17 •   The image of a real object formed by a convex mirror (a) is always real and inverted. (b) is always virtual and enlarged. (c) may be real. (d) is always virtual and diminished.

Determine the Concept Referring to the ray diagram show below we note that the image is always virtual and diminished. (d) is correct.

*21 •   Explain the following statement: A microscope is an object magnifier, but a telescope is an angle magnifier.

Determine the Concept Microscopes ordinarily produce images (either the intermediate one produced by the objective or the one viewed through the eyepiece) that are larger than the object being viewed. A telescope, on the other hand, ordinarily produces images that are much reduced compared to the object. The object is normally viewed from a great distance and the telescope magnifies the angle subtended by the object.

Estimation and Approximation

*24 ••   Estimate the maximum value that could be usefully obtained for the magnification of a simple magnifier, using Equation 32-20. (*Hint: Think about the smallest focal length lens that could be made from glass and still be used as a magnifier.*)

Picture the Problem Because the focal length of a spherical lens depends on its radii of curvature and the magnification depends on the focal length, there is a practical upper limit to the magnification.

Use equation 32-20 to relate the magnification M of a simple

$$M = \frac{x_{np}}{f}$$

magnifier to its focal length f:

Use the lens-maker's equation to relate the focal length of a lens to its radii of curvature and the index of refraction of the material from which it is constructed:

$$\frac{1}{f} = (n-1)\left(\frac{1}{r_1} - \frac{1}{r_2}\right)$$

For a plano-convex lens, $r_2 = \infty$. Hence:

$$\frac{1}{f} = \frac{n-1}{r_1} \Rightarrow f = \frac{r_1}{n-1}$$

Substitute in the expression for M and simplify to obtain:

$$M = \frac{(n-1)x_{np}}{r_1}$$

Note that the smallest reasonable value for r_1 will maximize M.

A reasonable smallest value for the radius of a magnifier is 1 cm. Use this value and $n = 1.5$ to estimate M_{max}:

$$M_{max} = \frac{(1.5-1)(25\,\text{cm})}{1\,\text{cm}} = \boxed{12.5}$$

Plane Mirrors

*27 • Two plane mirrors make an angle of 90°. The light from an object point that is arbitrarily positioned in front of the mirrors produces images at three locations. For each image location draw two rays from the object that, upon one or two reflections, appear to come from the image location.

Determine the Concept Draw rays of light from the object that satisfy the law of reflection at the two mirror surfaces. Three virtual images are formed, as shown in the adjacent figure. The eye should be to the right and above the mirrors in order to see these images.

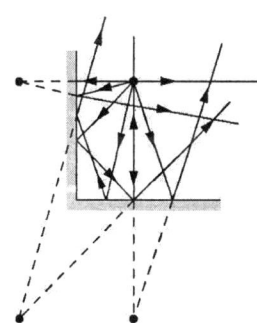

Spherical Mirrors

*30 •• A concave spherical mirror has a radius of curvature of 24 cm. Draw ray diagrams to locate the image (if one is formed) for an object at a distance of (*a*) 55 cm, (*b*) 24 cm, (*c*) 12 cm, and (*d*) 8 cm from the mirror. For each case, state whether the

294   Chapter 32

image is real or virtual; upright or inverted; and enlarged, reduced, or the same size as the object.

Picture the Problem The easiest rays to use in locating the image are 1) the ray parallel to the principal axis and passes through the focal point of the mirror, the ray that passes through the center of curvature of the spherical mirror and is reflected back on itself, and 2) the ray that passes through the focal point of the spherical mirror and is reflected parallel to the principal axis. We can use any two of these rays emanating from the top of the object to locate the image of the object.

(a) The ray diagram is shown to the right. The image is real, inverted, and reduced.

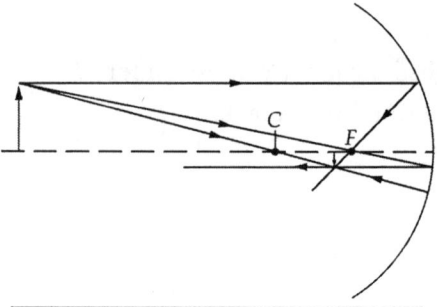

The image is real, inverted, and reduced.

(b) The ray diagram is shown to the right.

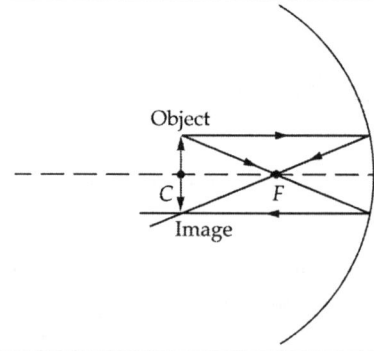

The image is real, inverted, and the same size as the object.

(c) The ray diagram is shown to the right. The object is at the focal point of the mirror.

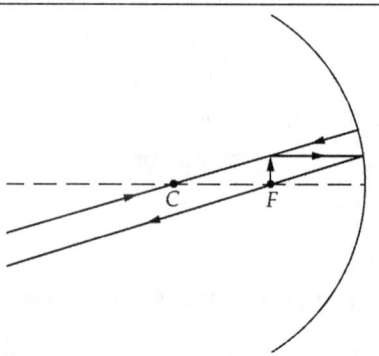

The emerging rays are parallel and do not form an image.

(*d*) The ray diagram is shown to the right.

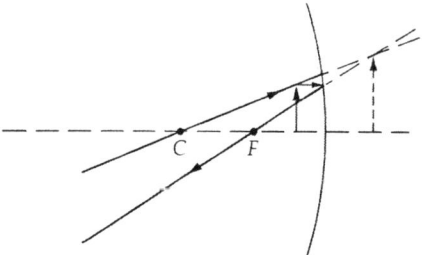

The image is virtual, erect, and enlarged.

*35 • A dentist wants a small mirror that will produce an upright image with a magnification of 5.5 when the mirror is located 2.1 cm from a tooth. (*a*) What should the radius of curvature of the mirror be? (*b*) Should the mirror be concave or convex?

Picture the Problem We can use the mirror equation and the definition of the lateral magnification to find the radius of curvature of the mirror.

(*a*) Express the mirror equation:

$$\frac{1}{s} + \frac{1}{s'} = \frac{1}{f} = \frac{2}{r}$$

Solve for *r*:

$$r = \frac{2ss'}{s'+s} \quad (1)$$

The lateral magnification of the mirror is given by:

$$m = -\frac{s'}{s}$$

Solve for *s'*:

$$s' = -ms$$

Substitute for *s'* in equation (1) to obtain:

$$r = \frac{-2ms}{1-m}$$

Substitute numerical values and evaluate *r*:

$$r = \frac{-2(5.5)(2.1\,\text{cm})}{1-5.5} = \boxed{5.13\,\text{cm}}$$

(*b*) The mirror must be concave. A convex mirror always produces a diminished virtual image.

*39 •• A concave mirror has a radius of curvature 6 cm. Draw rays parallel to the axis at 0.5 cm, 1 cm, 2 cm, and 4 cm above the axis, and find the points at which the reflected rays cross the axis. (Use a compass to draw the mirror and a protractor to find the angle of reflection for each ray.) (*a*) What is the spread *δx* of the points where these

296 Chapter 32

rays cross the axis? (*b*) By what percentage could this spread be reduced if the edge of the mirror were blocked off so that parallel rays more than 2 cm from the axis could not strike the mirror?

Picture the Problem

(*a*) The figure to the right shows the mirror and the four rays drawn to scale. Using a calibrated ruler, the spread of the crossing points is $\delta x \approx 1.0$ cm. Note that the triangles formed by the center of curvature, the point of reflection on the mirror, and the point of intersection of the reflected ray and the mirror axis are isosceles triangles.

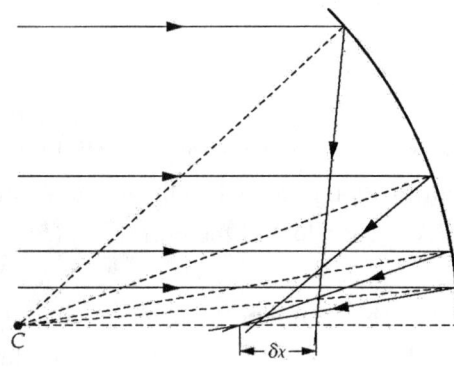

Express the equal angles of the isosceles triangles:

$$\theta_r = \sin^{-1}\left(\frac{y}{R}\right)$$

where *y* is the distance of the incoming ray from the mirror axis and *R* is the radius of curvature of the mirror.

Using the law of cosines, the distance between the point of intersection and the mirror is given by:

$$d = R\left\{1 - \left[2\cos\left(\sin^{-1}\left(\frac{y}{R}\right)\right)\right]^{-1}\right\}$$

Evaluate *d* for *y*/*R* = 2/3:

$$d = (6\,\text{cm})\left\{1 - \left[2\cos\left(\sin^{-1}\left(\frac{2}{3}\right)\right)\right]^{-1}\right\}$$

$$= 1.975\,\text{cm}$$

Evaluate *d* for *y*/*R* = 1/12:

$$d = (6\,\text{cm})\left\{1 - \left[2\cos\left(\sin^{-1}\left(\frac{1}{12}\right)\right)\right]^{-1}\right\}$$

$$= 2.990\,\text{cm}$$

Express the spread δx:

$$\delta x = 2.990\,\text{cm} - 1.975\,\text{cm} = \boxed{1.01\,\text{cm}}$$

in good agreement with the result obtained above.

(b) Evaluate d for $y/R = 1/3$:

$$d = (6\,\text{cm})\left\{1 - \left[2\cos\left(\sin^{-1}\left(\frac{1}{3}\right)\right)\right]^{-1}\right\}$$

$$= 2.818\,\text{cm}$$

Express the new spread $\delta x'$:

$\delta x' = 2.990\,\text{cm} - 2.818\,\text{cm} = 0.172\,\text{cm}$

Express the ratio of $\delta x'$ to δx:

$$\frac{\delta x'}{\delta x} = \frac{0.172\,\text{cm}}{1.01\,\text{cm}} = 17.0\%$$

By blocking off the edges of the mirror so that only paraxial rays within 2 cm of the mirror axis are reflected, the spread is reduced by 83.0%.

Images Formed by Refraction

*44 •• A very long glass rod of 3.5-cm diameter has one end ground to a convex spherical surface of radius 7.2 cm. Its index of refraction is 1.5. (a) A point object in air is on the axis of the rod 35 cm from the surface. Find the image and state whether the image is real or virtual. Repeat (b) for an object 6.5 cm from the surface and (c) an object very far from the surface. Draw a ray diagram for each case.

Picture the Problem We can use the equation for refraction at a single surface to find the images corresponding to these three object positions. The signs of the image distances will tell us whether the images are real or virtual and the ray diagrams will confirm the correctness of our analytical solutions.

Use the equation for refraction at a single surface to relate the image and object distances:

$$\frac{n_1}{s} + \frac{n_2}{s'} = \frac{n_2 - n_1}{r} \quad (1)$$

Here we have $n_1 = 1$ and $n_2 = n = 1.5$. Therefore:

$$\frac{1}{s} + \frac{n}{s'} = \frac{n-1}{r}$$

Solve for s':

$$s' = \frac{nrs}{s(n-1) - r}$$

(a) Substitute numerical values ($s = 35$ cm and $r = 7.2$ cm) and evaluate s':

$$s' = \frac{(1.5)(7.2\,\text{cm})(35\,\text{cm})}{(35\,\text{cm})(1.5 - 1) - (7.2\,\text{cm})}$$

$$= \boxed{36.7\,\text{cm}}$$

where the positive distance tells us that the

298 Chapter 32

image is 36.7 cm in back of the surface and is real.

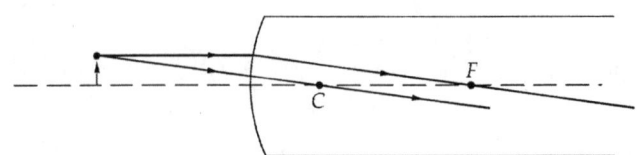

(b) Substitute numerical values ($s = 6.5$ cm and $r = 7.2$ cm) and evaluate s':

$$s' = \frac{(1.5)(7.2\,\text{cm})(6.5\,\text{cm})}{(6.5\,\text{cm})(1.5-1)-(7.2\,\text{cm})}$$

$$= \boxed{-17.8\,\text{cm}}$$

where the minus sign tells us that the image is 17.8 cm in front of the surface and is virtual.

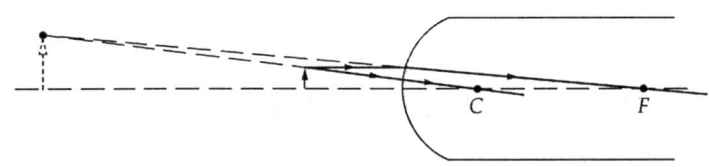

(c) When $s = \infty$, equation (1) becomes:

$$\frac{n}{s'} = \frac{n-1}{r}$$

Solve for s':

$$s' = \frac{nr}{n-1}$$

Substitute numerical values and evaluate s':

$$s' = \frac{(1.5)(7.2\,\text{cm})}{1.5-1} = \boxed{21.6\,\text{cm}}$$

i.e., the image is at the focal point, is real, and of zero size.

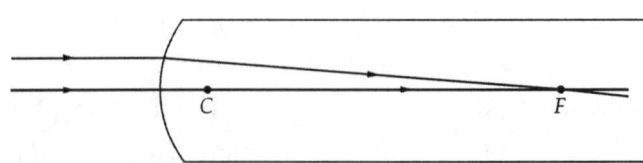

*49 ••   A glass rod 96 cm long with an index of refraction of 1.6 has its ends ground to convex spherical surfaces of radii 8 cm and 16 cm. A point object is in air on the axis of the rod 20 cm from the end with the 8-cm radius. (*a*) Find the image distance due to refraction at the first surface. (*b*) Find the final image due to refraction at both surfaces. (*c*) Is the final image real or virtual?

Picture the Problem We can use the equation for refraction at a single surface to find the images due to refraction at the ends of the glass rod. The image formed by the refraction at the first surface will serve as the object for the second surface. The sign of the final image distance will tell us whether the image is real or virtual.

(*a*) Use the equation for refraction at a single surface to relate the image and object distances at the first surface:

$$\frac{n_1}{s} + \frac{n_2}{s'} = \frac{n_2 - n_1}{r} \quad (1)$$

Solve for s':

$$s' = \frac{n_2 r s}{s(n_2 - n_1) - n_1 r}$$

Substitute numerical values and evaluate s':

$$s' = \frac{(1.6)(8\,\text{cm})(20\,\text{cm})}{(20\,\text{cm})(1.6-1) - (8\,\text{cm})}$$

$$= \boxed{64.0\,\text{cm}}$$

(*b*) The object for the second surface is 96 cm − 64 cm = 32 cm from the surface whose radius is 16 cm. Substitute numerical values and evaluate s':

$$s' = \frac{(1)(-16\,\text{cm})(32\,\text{cm})}{(32\,\text{cm})(1-1.6) - (1.6)(-16\,\text{cm})}$$

$$= \boxed{-80.0\,\text{cm}}$$

(*c*) $\boxed{\text{The final image is 96 cm} - 80\,\text{cm} = 16\,\text{cm from the surface whose radius is 8 cm and is virtual.}}$

Thin Lenses

*53 •   A double concave lens of index of refraction 1.45 has radii of magnitudes 30 cm and 25 cm. An object is located 80 cm to the left of the lens. Find (*a*) the focal length of the lens, (*b*) the location of the image, and (*c*) the magnification of the image. (*d*) Is the image real or virtual? Is the image upright or inverted?

Picture the Problem We can use the lens-maker's equation to find the focal length of the

300 Chapter 32

lens and the thin-lens equation to locate the image. We can use $m = -\dfrac{s'}{s}$ to find the lateral magnification of the image.

(a) The lens-maker's equation is:
$$\frac{1}{f} = (n-1)\left(\frac{1}{r_1} - \frac{1}{r_2}\right)$$

where the numerals 1 and 2 denote the first and second surfaces, respectively.

Substitute numerical values to obtain:
$$\frac{1}{f} = (1.45 - 1)\left(\frac{1}{-30\,\text{cm}} - \frac{1}{25\,\text{cm}}\right)$$

Solve for f:
$$f = \boxed{-30.3\,\text{cm}}$$

(b) Use the thin-lens equation to relate the image and object distances:
$$\frac{1}{s} + \frac{1}{s'} = \frac{1}{f}$$

Solve for s':
$$s' = \frac{fs}{s - f}$$

Substitute numerical values and evaluate s':
$$s' = \frac{(-30.3\,\text{cm})(80\,\text{cm})}{80\,\text{cm} - (-30.3\,\text{cm})} = \boxed{-22.0\,\text{cm}}$$

(c) The lateral magnification of the image is given by:
$$m = -\frac{s'}{s}$$

Substitute numerical values and evaluate m:
$$m = -\frac{-22\,\text{cm}}{80\,\text{cm}} = \boxed{0.275}$$

(d) Because $s' < 0$ and $m > 0$, the image is $\boxed{\text{virtual and upright.}}$

*55 • An object 3 cm high is placed 25 cm in front of a thin lens of power 10 D. Draw a precise ray diagram to find the position and the size of the image, and check your results using the thin-lens equation.

Picture the Problem The parallel and central rays were used to locate the image in the diagram shown below. The power P of the lens, in diopters, can be found from $P = 1/f$ and the size of the image from $m = \dfrac{y'}{y} = -\dfrac{s'}{s}$.

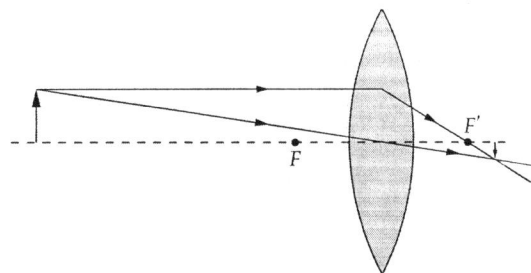

The image is real, inverted, and diminished.

The thin-lens equation is:
$$\frac{1}{s}+\frac{1}{s'}=\frac{1}{f}$$

Solve for s':
$$s'=\frac{fs}{s-f}$$

Use the definition of the power of the lens to find its focal length:
$$f=\frac{1}{P}=\frac{1}{10\,\text{m}^{-1}}=0.1\,\text{m}=10\,\text{cm}$$

Substitute numerical values and evaluate s':
$$s'=\frac{(10\,\text{cm})(25\,\text{cm})}{25\,\text{cm}-10\,\text{cm}}=\boxed{16.7\,\text{cm}}$$

Use the lateral magnification equation to relate the height of the image y' to the height y of the object and the image and object distances:
$$m=\frac{y'}{y}=-\frac{s'}{s}$$

Solve for y':
$$y'=-\frac{s'}{s}y$$

Substitute numerical values and evaluate y':
$$y'=-\frac{16.7\,\text{cm}}{25\,\text{cm}}(3\,\text{cm})=\boxed{-2.00\,\text{cm}}$$

Because $s'>0$ and $y'=-2.00\,\text{cm}$, the image is real, inverted, and diminished in agreement with the ray diagram.

*59 •• Two converging lenses, each of focal length 10 cm, are separated by 35 cm. An object is 20 cm to the left of the first lens. (*a*) Find the position of the final image using both a ray diagram and the thin-lens equation. (*b*) Is the image real or virtual? Is the image upright or inverted? (*c*) What is the overall lateral magnification of the image?

302 Chapter 32

Picture the Problem We can apply the thin-lens equation to find the image formed in the first lens and then use this image as the object for the second lens.

(*a*) The parallel, central, and focal rays were used to locate the image formed by the first lens and the parallel and central rays to locate the image formed by the second lens.

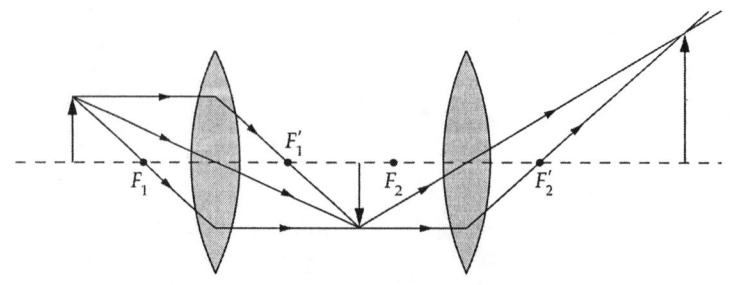

Apply the thin-lens equation to express the location of the image formed by the first lens:

$$s_1' = \frac{f_1 s_1}{s_1 - f_1} \qquad (1)$$

Substitute numerical values and evaluate s_1':

$$s_1' = \frac{(10\,\text{cm})(20\,\text{cm})}{20\,\text{cm} - 10\,\text{cm}} = 20\,\text{cm}$$

Find the lateral magnification of the first image:

$$m_1 = -\frac{s_1'}{s} = -\frac{20\,\text{cm}}{20\,\text{cm}} = -1$$

Because the lenses are separated by 35 cm, the object distance for the second lens is 35 cm − 20 cm = 15 cm. Equation (1) applied to the second lens is:

$$s_2' = \frac{f_2 s_2}{s_2 - f_2}$$

Substitute numerical values and evaluate s_2':

$$s_2' = \frac{(10\,\text{cm})(15\,\text{cm})}{15\,\text{cm} - 10\,\text{cm}} = 30\,\text{cm}$$

and the final image is $\boxed{85.0\,\text{cm}}$ from the object.

Find the lateral magnification of the second image:

$$m_2 = -\frac{s_2'}{s} = -\frac{30\,\text{cm}}{15\,\text{cm}} = -2$$

Because $s_2' > 0$ and $m = m_1 m_2 = 2$, the image is real, erect, and twice the size of the object.

The overall lateral magnification of the image is the product of the magnifications of each image:

$$m = m_1 m_2 = (-1)(-2) = \boxed{2.00}$$

*64 •• An object is 15 cm in front of a positive lens of focal length 15 cm. A second positive lens of focal length 15 cm is 20 cm from the first lens. Find the final image and draw a ray diagram.

Picture the Problem We can apply the thin-lens equation to find the image formed in the first lens and then use this image as the object for the second lens.

Apply the thin-lens equation to express the location of the image formed by the first lens:

$$s_1' = \frac{f_1 s_1}{s_1 - f_1} \qquad (1)$$

Substitute numerical values and evaluate s_1':

$$s_1' = \frac{(15\,\text{cm})(15\,\text{cm})}{15\,\text{cm} - 15\,\text{cm}} = \infty$$

With $s_1' = \infty$, the thin-lens equation applied to the second lens becomes:

$$\frac{1}{s_2'} = \frac{1}{f_2} \Rightarrow s_2' = f_2 = \boxed{15.0\,\text{cm}}$$

A ray diagram is shown below:

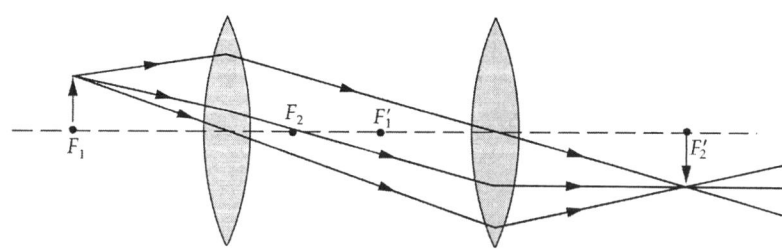

The final image is 50 cm from the object, real, inverted, and the same size as the object.

Aberrations

*70 • Chromatic aberration is a common defect of (*a*) concave and convex lenses. (*b*) concave lenses only. (*c*) concave and convex mirrors. (*d*) all lenses and mirrors.

Determine the Concept Chromatic aberrations are a consequence of the differential refraction of light of differing wavelengths by lenses. $\boxed{(a)\text{ is correct.}}$

304 Chapter 32

The Eye

*73 •• The Model Eye I: A simple model for the eye is a lens with variable power P located a fixed distance d in front of a screen, with the space between the lens and the screen filled by air. Refer to Figure 32-60. The "eye" can focus for all values of s such that $x_{np} \leq s \leq x_{fp}$. This "eye" is said to be normal if it can focus on very distant objects. (a) Show that for a normal "eye," the minimum value of P is

$$P_{min} = \frac{1}{d}$$

(b) Show that the maximum value of P is

$$P_{max} = \frac{1}{x_{np}} + \frac{1}{d}$$

(c) The difference $A = P_{max} - P_{min}$ is called the accommodation. Find the minimum power and accommodation for a model eye with $d = 2.5$ cm and $x_{np} = 25$ cm.

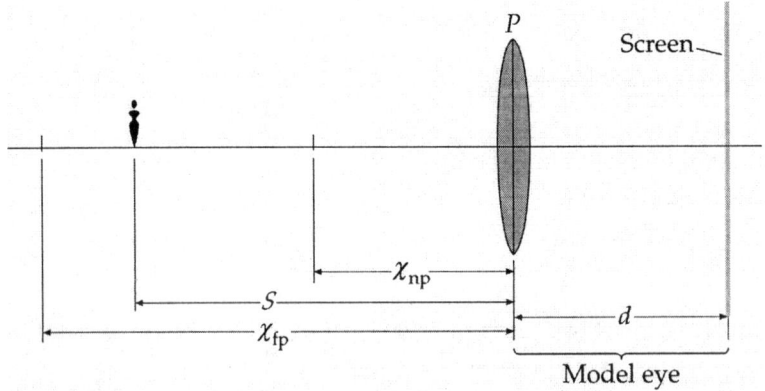

Figure 32-60 Problems 73, 74, and 75

Picture the Problem The thin-lens equation relates the image and object distances to the power of a lens.

(a) Use the thin-lens equation to relate the image and object distances to the power of the lens:

$$\frac{1}{s} + \frac{1}{s'} = \frac{1}{f} = P$$

Because $s' = d$ and, for a distance object, $s = \infty$:

$$P_{min} = \frac{1}{s'} = \boxed{\frac{1}{d}}$$

(b) If x_{np} is the closest distance an object could be and still remain in clear focus on the screen, equation (1) becomes:

$$P_{max} = \boxed{\dfrac{1}{x_{np}} + \dfrac{1}{d}}$$

(c) Use our result in (a) to obtain:

$$P_{min} = \dfrac{1}{2.5\,\text{cm}} = \boxed{40.0\,\text{D}}$$

Use the results of (a) and (b) to express the accommodation of the model eye:

$$A = P_{max} - P_{min} = \dfrac{1}{x_{np}} + \dfrac{1}{d} - \dfrac{1}{d} = \dfrac{1}{x_{np}}$$

Substitute numerical values and evaluate A:

$$A = \dfrac{1}{25\,\text{cm}} = \boxed{4.00\,\text{D}}$$

*79 • If two point objects close together are to be seen as two distinct objects, the images must fall on the retina on two different cones that are not adjacent. That is, there must be an unactivated cone between them. The separation of the cones is about 1 μm. Model the eye as a uniform 2.5-cm-diameter sphere with a refractive index of 1.34. (a) What is the smallest angle the two points can subtend? (See Figure 32-61.) (b) How close together can two points be if they are 20 m from the eye?

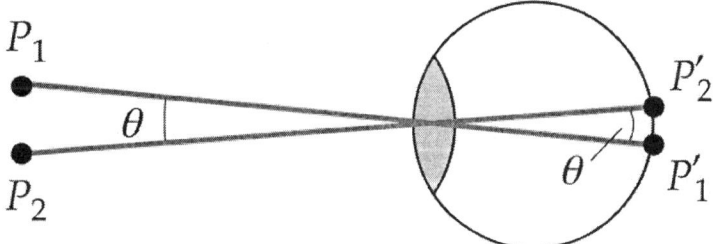

Figure 32-61 Problem 79

Picture the Problem We can use the relationship between a distance measured along the arc of a circle and the angle subtended at its center to approximate the smallest angle the two points can subtend and the separation of the two points 20 m from the eye.

(a) Relate θ_{min} to the diameter of the eye and the distance between the activated cones:

$$d_{eye}\theta_{min} \approx 2\,\mu\text{m}$$

Solve for θ_{min}:

$$\theta_{min} = \dfrac{2\,\mu\text{m}}{d_{eye}}$$

306 Chapter 32

Substitute numerical values and evaluate θ_{min}:

$$\theta_{min} = \frac{2\,\mu m}{2.5\,cm} = \boxed{80.0\,\mu rad}$$

(b) Let D represent the separation of the points $R = 20$ m from the eye to obtain:

$$D = R\theta_{min} = (20\,m)(80\,\mu rad)$$
$$= \boxed{1.60\,mm}$$

The Simple Magnifier

*85 •   A person with a near-point distance of 30 cm uses a simple magnifier of power 20 D. What is the magnification obtained if the final image is at infinity?

Picture the Problem We can use the definitions of the magnifying power of a lens ($M = x_{np}/f$) and of the power of a lens ($P = 1/f$) to find the magnifying power of the given lens.

The magnifying power of the lens is given by:

$$M = \frac{x_{np}}{f} = Px_{np}$$

where P is the power of the lens.

Substitute numerical values and evaluate M:

$$M = (20\,m^{-1})(0.3\,m) = \boxed{6.00}$$

*90 ••   (a) Show that if the final image of a simple magnifier is to be at the near point of the eye rather than at infinity, the angular magnification is given by

$$M = \frac{x_{np}}{f} + 1$$

(b) Find the magnification of a 20-D lens for a person with a near point of 30 cm if the final image is at the near point. Draw a ray diagram for this situation.

Picture the Problem We can use the definition of the angular magnification of a lens and the thin-lens equation to show that $M = \frac{x_{np}}{f} + 1$.

(a) Express the angular magnification of the simple magnifier in terms of the angles subtended by the object and the image:

$$M = \frac{\theta}{\theta_0} \quad (1)$$

Solve the thin-lens equation for s:

$$s = \frac{fs'}{s' - f}$$

Because the image is virtual:

$$s' = -x_{np}$$

Substitute to obtain:

$$s = \frac{f(-x_{np})}{-x_{np} - f} = \frac{fx_{np}}{x_{np} + f}$$

Express the angle subtended by the object:

$$\theta_0 = \frac{y}{x_{np}}$$

where y is the height of the object.

Express the angle subtended by the image:

$$\theta = \frac{y}{s}$$

Substitute for s to obtain:

$$\theta = \frac{y}{\frac{fx_{np}}{x_{np} + f}} = \frac{y(x_{np} + f)}{fx_{np}}$$

Substitute in equation (1) and simplify:

$$M = \frac{\frac{y(x_{np} + f)}{fx_{np}}}{\frac{y}{x_{np}}} = \frac{x_{np} + f}{f} = \boxed{\frac{x_{np}}{f} + 1}$$

(b) In terms of the power of the magnifying lens:

$$M = x_{np}P + 1$$

The magnification of a 20-D lens for a person with a near point of 30 cm and the final image at the near point is:

$$M = (0.3\,\text{m})(20\,\text{m}^{-1}) + 1 = \boxed{7.00}$$

A ray diagram for this situation is shown to the right:

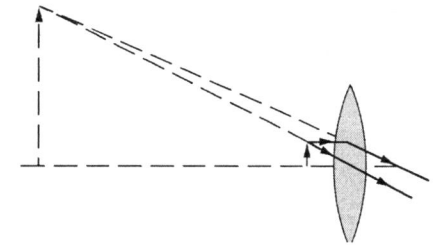

The Microscope

*93 •• A microscope has an objective of focal length 8.5 mm and an eyepiece that gives an angular magnification of 10 for a person whose near point is 25 cm. The tube length is 16 cm. (*a*) What is the lateral magnification of the objective? (*b*) What is the magnifying power of the microscope?

Picture the Problem The lateral magnification of the objective is $m_o = -L/f_o$ and the magnifying power of the microscope is $M = m_o M_e$.

(*a*) The lateral magnification of the objective is given by:
$$m_o = -\frac{L}{f_o}$$

Substitute numerical values and evaluate m_o:
$$m_o = -\frac{16\,\text{cm}}{8.5\,\text{mm}} = \boxed{-1.88}$$

(*b*) The magnifying power of the microscope is given by:
$$M = m_o M_e$$
where M_e is the angular magnification of the lens.

Substitute numerical values and evaluate M:
$$M = (-1.88)(10) = \boxed{-18.8}$$

*95 •• A compound microscope has an objective lens with a power of 45 D and an eyepiece with a power of 80 D. The lenses are separated by 28 cm. Assuming that the final image is formed 25 cm from the eye, what is the magnifying power?

Picture the Problem The magnifying power of a compound microscope is the product of the magnifying powers of the objective and the eyepiece.

Express the magnifying power of the microscope in terms of the magnifying powers of the objective and eyepiece:
$$M = m_o m_e \qquad (1)$$

From Problem 82, the magnification of the eyepiece is given by:
$$m_e = \frac{x_{np}}{f_e} + 1 = P_e x_{np} + 1$$

The magnification of the objective is given by:
$$m_o = -\frac{L}{f_o}$$

Optical Images 309

Substitute to obtain:
$$m_o = -\frac{D - f_o - f_e}{f_o}$$

Substitute for m_e and m_o in equation (1) to obtain:
$$M = (P_e x_{np} + 1)\left(-\frac{D - f_o - f_e}{f_o}\right)$$

Substitute numerical values and evaluate M:

$$M = [(80\,D)(0.25\,m) + 1]\left(-\frac{28\,cm - 2.22\,cm - 1.25\,cm}{2.22\,cm}\right) = \boxed{-232}$$

The Telescope

*99 •• The 200-in (5.1-m) mirror of the reflecting telescope at Mt. Palomar has a focal length of 1.68 m. (a) By what factor is the light-gathering power increased over the 40-in (1.016-m) diameter refracting lens of the Yerkes Observatory telescope? (b) If the focal length of the eyepiece is 1.25 cm, what is the magnifying power of this telescope?

Picture the Problem Because the light-gathering power of a mirror is proportional to its area, we can compare the light-gathering powers of these mirrors by finding the ratio of their areas. We can use the ratio of the focal lengths of the objective and eyepiece lenses to find the magnifying power of the Palomar telescope.

(a) Express the ratio of the light-gathering powers of the Palomar and Yerkes mirrors:

$$\frac{P_{Palomar}}{P_{Yerkes}} = \frac{A_{Palomar\ mirror}}{A_{Yerkes\ mirror}} = \frac{\frac{\pi}{4}d^2_{Palomar\ mirror}}{\frac{\pi}{4}d^2_{Yerkes\ mirror}}$$

$$= \frac{d^2_{Palomar\ mirror}}{d^2_{Yerkes\ mirror}}$$

Substitute numerical values and evaluate $P_{Palomar}/P_{Yerkes}$:

$$\frac{P_{Palomar}}{P_{Yerkes}} = \frac{(200\,in)^2}{(40\,in)^2} = 25.0$$

or

$$P_{Palomar} = \boxed{(25.0)P_{Yerkes}}$$

(b) Express the magnifying power of the Palomar telescope:

$$M = -\frac{f_o}{f_e}$$

310 Chapter 32

Substitute numerical values and evaluate M:
$$M = -\frac{1.68\,\text{m}}{1.25\,\text{cm}} = \boxed{-134}$$

General Problems

*105 •   A camera uses a positive lens to focus light from an object onto film. Unlike the eye, the camera lens has a fixed focal length, but the lens itself can be moved slightly to vary the image distance to the image on the film. A telephoto lens has a focal length of 200 mm. By how much must it move to change from focusing on an object at infinity to an object at a distance of 30 m?

Picture the Problem We can express the distance Δs that the lens must move as the difference between the image distances when the object is at 30 m and when it is at infinity and then express these image distances using the thin-lens equation.

Express the distance Δs that the lens must move to change from focusing on an object at infinity to one at a distance of 30 m:
$$\Delta s = s'_{30} - s'_{\infty}$$

Solve the thin-lens equation for s':
$$s' = \frac{fs}{s-f}$$

Substitute and simplify to obtain:
$$\Delta s = \frac{fs_{30}}{s_{30}-f} - \frac{fs_{\infty}}{s_{\infty}-f}$$
$$= \frac{fs_{30}}{s_{30}-f} - \frac{f}{1-f/s_{\infty}}$$
$$= f\left[\frac{s_{30}}{s_{30}-f} - 1\right]$$

Substitute numerical values and evaluate Δs:
$$\Delta s = (200\,\text{mm})\left[\frac{30\,\text{m}}{30\,\text{m}-0.2\,\text{m}} - 1\right]$$
$$= \boxed{1.34\,\text{mm}}$$

*110 ••   A scuba diver wears a diving mask with a face plate that bulges outward with a radius of curvature of 0.5 m. There is thus a convex spherical surface between the water and the air in the mask. A fish is 2.5 m in front of the diving mask. (*a*) Where does the fish appear to be? (*b*) What is the magnification of the image of the fish?

Picture the Problem We can use the equation for refraction at a single surface to locate the image of the fish and the expression for the magnification due to refraction at a spherical surface to find the magnification of the image.

(a) Use the equation describing refraction at a single surface to relate the image and object distances:
$$\frac{n_1}{s} + \frac{n_2}{s'} = \frac{n_2 - n_1}{r}$$

Solve for s':
$$s' = \frac{n_2 rs}{(n_2 - n_1)s - n_1 r}$$

Substitute numerical values and evaluate s':
$$s' = \frac{(1)(0.5\,\text{m})(2.5\,\text{m})}{(1 - 1.33)(2.5\,\text{m}) - (1.33)(0.5\,\text{m})}$$
$$= \boxed{-0.839\,\text{m}}$$

Note that the fish appears to be much closer to the diver than it actually is.

(b) Express the magnification due to refraction at a spherical surface:
$$m = -\frac{n_1 s'}{n_2 s}$$

Substitute numerical values and evaluate m:
$$m = -\frac{(1.33)(-0.839\,\text{m})}{(1)(2.5\,\text{m})} = \boxed{0.446}$$

Note that the fish appears to be smaller than it actually is.

*115 •• (a) Find the focal length of a *thick*, double convex lens with an index of refraction of 1.5, a thickness of 4 cm, and radii of +20 cm and −20 cm. (b) Find the focal length of this lens in water.

Picture the Problem Here we must consider refraction at each surface separately. To find the focal length we imagine the object at $s = \infty$, and find the image from the first refracting surface at s'_1. That image serves as the object for the second refracting surface. We'll find that this is a virtual image for the second refracting surface, i.e., s_2 is negative. Using the equation for refraction at a single surface a second time, we can locate the image formed by the second refracting surface by the virtual object at s_2. The location of that image is then the focal point of the thick lens. We'll let the numeral 1 denote the first surface and the numeral 2 the second surface. In part (b) we can proceed as in part (a) (except that now $n_1 = 1.33$ for the first refraction and $n_2 = 1.33$ for the second refraction) to determine the focal length in water, which we denote by f_w.

312 Chapter 32

(a) Use the equation for refraction at a single surface to relate s_1 and s_1':

$$\frac{n_1}{s_1} + \frac{n_2}{s_1'} = \frac{n_2 - n_1}{r_1}$$

For $s_1 = \infty$:

$$\frac{n_2}{s_1'} = \frac{n_2 - n_1}{r_1}$$

Solve for s_1':

$$s_1' = \frac{n_2 r_1}{n_2 - n_1} \quad (1)$$

Substitute numerical values and evaluate s_1':

$$s_1' = \frac{(1.5)(20\,\text{cm})}{1.5 - 1} = 60.0\,\text{cm}$$

The object distance s_2 for the second lens is:

$$s_2 = -(s_1' - 4\,\text{cm}) = -(60\,\text{cm} - 4\,\text{cm})$$
$$= -56\,\text{cm}$$

Solve the equation for refraction at a single surface for s_2':

$$s_2' = \frac{n_2 r_2 s_2}{(n_2 - n_1)s_2 - n_1 r_2} \quad (2)$$

Substitute numerical values and evaluate s_2':

$$s_2' = \frac{(1)(-20\,\text{cm})(-56\,\text{cm})}{(1 - 1.5)(-56\,\text{cm}) - (1.5)(-20\,\text{cm})}$$
$$= 19.3\,\text{cm}$$

Because f is measured from the center of the lens:

$$f = s_2' + 2\,\text{cm} = 19.3\,\text{cm} + 2\,\text{cm}$$
$$= \boxed{21.3\,\text{cm}}$$

(b) Substitute numerical values in equation (1) and evaluate s_1':

$$s_1' = \frac{(1.5)(20\,\text{cm})}{1.5 - 1.33} = 176\,\text{cm}$$

The object distance s_2 for the second lens is:

$$s_2 = -(s_1' - 4\,\text{cm}) = -(176\,\text{cm} - 4\,\text{cm})$$
$$= -172\,\text{cm}$$

Substitute numerical values in equation (2) and evaluate s'_2:

$$s_2' = \frac{(1.33)(-20\,\text{cm})(-172\,\text{cm})}{(1.33 - 1.5)(-172\,\text{cm}) - (1.5)(-20\,\text{cm})} = 77.2\,\text{cm}$$

Because f_w is measured from the center of the lens:

$$f_w = s_2' + 2\,\text{cm} = 77.2\,\text{cm} + 2\,\text{cm}$$
$$= \boxed{79.2\,\text{cm}}$$

Remarks: Note that if we use the expression given in Problem 114 we obtain $f_w = 83.3$ cm, in only moderate agreement with the exact result given above.

*120 ••• When a bright light source is placed 30 cm in front of a lens, there is an upright image 7.5 cm from the lens. There is also a faint inverted image 6 cm in front of the lens due to reflection from the front surface of the lens. When the lens is turned around, this weaker, inverted image is 10 cm in front of the lens. Find the index of refraction of the lens.

Picture the Problem The mirror surfaces must be concave to create inverted images on reflection. Therefore, the lens is a diverging lens. Let the numeral 1 denote the lens in its initial orientation and the numeral 2 the lens in its second orientation. We can use the mirror equation to find the magnitudes of the radii of the lens' surfaces, the thin-lens equation to find its focal length, and the lens maker's equation to find its index of refraction.

Solve the mirror equation for $|r_1|$:

$$|r_1| = \frac{2 s_1 s_1'}{s_1' + s_1}$$

Substitute numerical values and evaluate $|r_1|$:

$$|r_1| = \frac{2(30\,\text{cm})(6\,\text{cm})}{6\,\text{cm} + 30\,\text{cm}} = 10.0\,\text{cm}$$

Solve the mirror equation for $|r_2|$:

$$|r_2| = \frac{2 s_2 s_2'}{s_2' + s_2}$$

Substitute numerical values and evaluate $|r_2|$:

$$|r_2| = \frac{2(30\,\text{cm})(10\,\text{cm})}{10\,\text{cm} + 30\,\text{cm}} = 15.0\,\text{cm}$$

Solve the thin-lens equation for f:

$$f = \frac{ss'}{s' + s}$$

Substitute numerical values and evaluate f:

$$f = \frac{(30\,\text{cm})(-7.5\,\text{cm})}{-7.5\,\text{cm} + 30\,\text{cm}} = -10.0\,\text{cm}$$

Solve the lens-maker's equation for n to obtain:

$$n = \frac{1}{f\left(\frac{1}{r_1} - \frac{1}{r_2}\right)} + 1$$

314   Chapter 32

Because the lens is a diverging lens, $r_1 = -10$ cm and $r_2 = 15$ cm. Substitute numerical values and evaluate n:

$$n = \cfrac{1}{(-10\,\text{cm})\left(\cfrac{1}{-10\,\text{cm}} - \cfrac{1}{15\,\text{cm}}\right)} + 1$$

$$= \boxed{1.60}$$

*125 ••• The lateral magnification of a spherical mirror or a thin lens is given by $m = -s'/s$. Show that for objects of small horizontal extent, the longitudinal magnification is approximately $-m^2$. (*Hint:* Show that $ds'/ds = -s'^2/s^2$.)

Picture the Problem We examine the amount by which the image distance s' changes due to a change in s.

Solve the thin-lens equation for s':

$$s' = \left(\frac{1}{f} - \frac{1}{s}\right)^{-1}$$

Differentiate s' with respect to s:

$$\frac{ds'}{ds} = \frac{d}{ds}\left[\left(\frac{1}{f} - \frac{1}{s}\right)^{-1}\right] = -\frac{1}{\left(\frac{1}{f} - \frac{1}{s}\right)^2}\frac{1}{s^2} = -\frac{s'^2}{s^2} = -m^2$$

$\boxed{\text{The image of an object of length } \Delta s \text{ will have a length } -m^2 \Delta s.}$

Chapter 33
Interference and Diffraction

Conceptual Problems

*1 • When destructive interference occurs, what happens to the energy in the light waves?

Determine the Concept The energy is distributed nonuniformly in space; in some regions the energy is below average (destructive interference), in others it is higher than average (constructive interference).

*6 • A loop of wire is dipped in soapy water and held so that the soap film is vertical. (*a*) Viewed by reflection with white light, the top of the film appears black. Explain why. (*b*) Below the black region are colored bands. Is the first band red or violet? (*c*) Describe the appearance of the film when it is viewed by *transmitted* light.

(*a*) The phase change on reflection from the front surface of the film is 180°; the phase change on reflection from the back surface of the film is 0°. As the film thins toward the top, the phase change associated with the film's thickness becomes negligible and the two reflected waves interfere destructively.

(*b*) The first constructive interference will arise when $t = \lambda/4$. Therefore, the first band will be violet (shortest visible wavelength).

(*c*) When viewed in transmitted light, the top of the film is white, since no light is reflected. The colors of the bands are those complementary to the colors seen in reflected light; i.e., the top band will be red.

*10 • A double-slit interference experiment is set up in a chamber that can be evacuated. Using monochromatic light, an interference pattern is observed when the chamber is open to air. As the chamber is evacuated, one will note that (*a*) the interference fringes remain fixed. (*b*) the interference fringes move closer together. (*c*) the interference fringes move farther apart. (*d*) the interference fringes disappear completely.

Determine the Concept The distance on the screen to *m*th bright fringe is given by $y_m = m\dfrac{\lambda L}{d}$, where L is the distance from the slits to the screen and d is the separation of the slits. Because the index of refraction of air is slightly larger than the index of refraction of a vacuum, the introduction of air reduces λ to λ/n and decreases y_m. Because the separation of the fringes is $y_m - y_{m-1}$, the separation of the fringes decreases and (b) is correct.

Estimation and Approximation

*12 • It is claimed that the Great Wall of China is the only human object that can be seen from space with the naked eye. Make an argument in support of this claim based on the resolving power of the human eye. Evaluate the validity of your argument for observers both in low-earth orbit (~400 km altitude) and on the moon.

Picture the Problem We'll assume that the diameter of the pupil of the eye is 5 mm and that the wavelength of light is 600 nm. Then we can use the expression for the minimum angular separation of two objects than can be resolved by the eye and the relationship between this angle and the width of an object and the distance from which it is viewed to support the claim.

Relate the width w of an object that can be seen at a height h to the critical angular separation α_c:

$$\tan\alpha_c = \frac{w}{h}$$

Solve for w:

$$w = h\tan\alpha_c$$

The minimum angular separation α_c of two point objects that can just be resolved by an eye depends on the diameter D of the eye and the wavelength λ of light:

$$\alpha_c = 1.22\frac{\lambda}{D}$$

Substitute for α_c in the expression for w to obtain:

$$w = h\tan\left(1.22\frac{\lambda}{D}\right)$$

In low-earth orbit:

$$w = (400\,\text{km})\tan\left(1.22\frac{600\,\text{nm}}{5\,\text{mm}}\right) = 58.6\,\text{m}$$

> Because the width of the Great Wall is about 5 m, a naked eye would not be able to see it from the moon.

At a distance equal to that of the distance of the moon from earth:

$$w = (3.84\times10^8\,\text{m})\tan\left(1.22\frac{600\,\text{nm}}{5\,\text{mm}}\right) = 56.2\,\text{km}$$

> Because the width of the Great Wall is about 5 m, a naked eye would not be able to see it from the moon.

Interference and Diffraction 317

*16 •• Human hair has a diameter of approximately 70 μm. If we illuminate a hair using a helium-neon laser with wavelength λ = 632.8 nm and intercept the light scattered from the hair on a screen 10 m away, what will be the separation of the first diffraction peak from the center? (The diffraction pattern of a hair with diameter d is the same as the diffraction pattern of a single slit with width $a = d$.)

Picture the Problem The diagram shows the hair whose diameter $d = a$, the screen a distance L from the hair, and the separation Δy of the first diffraction peak from the center. We can use the geometry of the experiment to relate Δy to L and a and the condition for diffraction maxima to express θ in terms of the diameter of the hair and the wavelength of the light illuminating the hair.

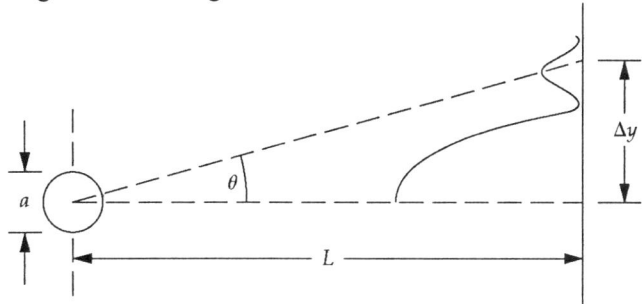

Relate θ to Δy:

$$\tan\theta = \frac{\Delta y}{L}$$

Solve for Δy:

$$\Delta y = L\tan\theta$$

Diffraction maxima occur where:

$$a\sin\theta = (m+\tfrac{1}{2})\lambda$$
where $m = 1, 2, 3, \ldots$

Solve for θ to obtain:

$$\theta = \sin^{-1}\left[\frac{(m+\tfrac{1}{2})\lambda}{a}\right]$$

Substitute for θ in the expression for Δy to obtain:

$$\Delta y = L\tan\left\{\sin^{-1}\left[\frac{(m+\tfrac{1}{2})\lambda}{a}\right]\right\}$$

For the first peak, $m = 1$. Substitute numerical values and evaluate Δy:

$$\Delta y = (10\,\text{m})\tan\left\{\sin^{-1}\left[\frac{(1+\tfrac{1}{2})(632.8\,\text{nm})}{70\,\mu\text{m}}\right]\right\} = \boxed{13.6\,\text{cm}}$$

318 Chapter 33

Phase Difference and Coherence

*19 •• Two coherent microwave sources that produce waves of wavelength 1.5 cm are in the *xy* plane, one on the *y* axis at *y* = 15 cm and the other at *x* = 3 cm, *y* = 14 cm. If the sources are in phase, find the difference in phase between the two waves from these sources at the origin.

Picture the Problem The difference in phase depends on the path difference according to $\delta = \dfrac{\Delta r}{\lambda} 360°$. The path difference is the difference in the distances of (0, 15 cm) and (3 cm, 14 cm) from the origin.

Relate a path difference Δr to a phase shift δ:
$$\delta = \dfrac{\Delta r}{\lambda} 360°$$

The path difference Δr is:
$$\Delta r = 15\,\text{cm} - \sqrt{(3\,\text{cm})^2 + (14\,\text{cm})^2}$$
$$= 0.682\,\text{cm}$$

Substitute numerical values and evaluate δ:
$$\delta = \dfrac{0.682\,\text{cm}}{1.5\,\text{cm}} 360° = \boxed{164°}$$

Interference in Thin Films

*21 •• The diameters of fine fibers can be accurately measured using interference patterns. Two optically flat pieces of glass of length *L* are arranged with the wire between them, as shown in Figure 33-40. The setup is illuminated by monochromatic light, and the resulting interference fringes are detected. Suppose that *L* = 20 cm and that yellow sodium light ($\lambda \approx 590$ nm) is used for illumination. If 19 bright fringes are seen along this 20-cm distance, what are the limits on the diameter of the wire? (*Hint:* The nineteenth fringe might not be right at the end, but you do not see a twentieth fringe at all.)

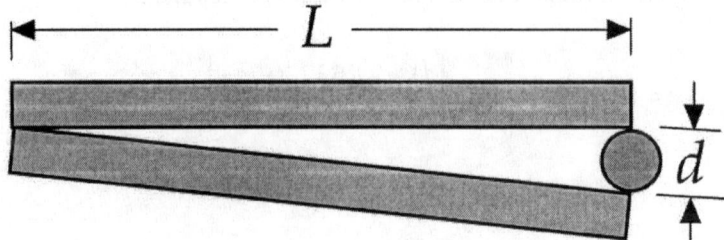

Figure 33-40 Problem 21

Picture the Problem The condition that one sees *m* fringes requires that the path difference between light reflected from the bottom surface of the top slide and the top

Interference and Diffraction 319

surface of the bottom slide is an integer multiple of a wavelength of the light.

The mth fringe occurs when the path difference $2d$ equals m wavelengths:

$$2d = m\lambda \Rightarrow d = \frac{m\lambda}{2}$$

Because the nineteenth (but not the twentieth) bright fringe can be seen, the limits on d must be:

$$\left(m - \tfrac{1}{2}\right)\frac{\lambda}{2} < d < \left(m + \tfrac{1}{2}\right)\frac{\lambda}{2}$$

where $m = 19$

Substitute numerical values to obtain:

$$\left(19 - \tfrac{1}{2}\right)\frac{590\,\text{nm}}{2} < d < \left(19 + \tfrac{1}{2}\right)\frac{590\,\text{nm}}{2}$$

or

$$\boxed{5.46\,\mu\text{m} < d < 5.75\,\mu\text{m}}$$

*26 •• A film of oil of index of refraction $n = 1.45$ floats on water ($n = 1.33$). When illuminated with white light at normal incidence, light of wavelengths 700 nm and 500 nm is predominant in the reflected light. Determine the thickness of the oil film.

Picture the Problem Because the index of refraction of air is less than that of the oil, there is a phase shift of π rad ($\tfrac{1}{2}\lambda$) in the light reflected at the air-oil interface. Because the index of refraction of the oil is greater than that of the glass, there is no phase shift in the light reflected from the oil-glass interface. We can use the condition for constructive interference to determine m for $\lambda = 700$ nm and then use this value in our equation describing constructive interference to find the thickness t of the oil film.

Express the condition for constructive interference between the waves reflected from the air-oil interface and the oil-glass interface:

$$2t + \tfrac{1}{2}\lambda' = \lambda', 2\lambda', 3\lambda', \ldots$$

or

$$2t = \tfrac{1}{2}\lambda', \tfrac{3}{2}\lambda', \tfrac{5}{2}\lambda', \ldots = \left(m + \tfrac{1}{2}\right)\lambda' \quad (1)$$

where λ' is the wavelength of light in the oil and $m = 0, 1, 2, \ldots$

Substitute for λ' and solve for λ to obtain:

$$\lambda = \frac{2nt}{m + \tfrac{1}{2}}$$

Substitute the predominant wavelengths to obtain:

$$700\,\text{nm} = \frac{2nt}{m + \tfrac{1}{2}} \quad \text{and} \quad 500\,\text{nm} = \frac{2nt}{m + \tfrac{3}{2}}$$

Divide the first of these equations by the second to obtain:

$$\frac{700\,\text{nm}}{500\,\text{nm}} = \frac{\dfrac{2nt}{m+\frac{1}{2}}}{\dfrac{2nt}{m+\frac{3}{2}}} = \frac{m+\frac{3}{2}}{m+\frac{1}{2}}$$

Solve for *m*:

$m = 2$ for $\lambda = 700$ nm

Solve equation (1) for *t*:

$$t = \left(m + \tfrac{1}{2}\right)\frac{\lambda}{2n}$$

Substitute numerical values and evaluate *t*:

$$t = \left(2 + \tfrac{1}{2}\right)\frac{700\,\text{nm}}{2(1.45)} = \boxed{603\,\text{nm}}$$

Newton's Rings

*27 •• A Newton's ring apparatus consists of a plano-convex glass lens with radius of curvature R that rests on a flat glass plate, as shown in Figure 33-42. The thin film is air of variable thickness. The pattern is viewed by reflected light.
(*a*) Show that for a thickness *t* the condition for a bright (constructive) interference ring is

$$t = \left(m + \tfrac{1}{2}\right)\frac{\lambda}{2},\, m = 0,1,2,\ldots$$

(*b*) Apply the Pythagorean Theorem to the triangle of sides r, $R - t$, and hypotenuse R to show that for $t \ll R$, the radius of a fringe is related to *t* by

$$r = \sqrt{2tR}$$

(*c*) How would the transmitted pattern look in comparison with the reflected one? (*d*) Use $R = 10$ m and a diameter of 4 cm. How many bright fringes would you see if the apparatus were illuminated by yellow sodium light ($\lambda \approx 590$ nm) and viewed by reflection? (*e*) What would be the diameter of the sixth bright fringe? (*f*) If the glass used in the apparatus has an index of refraction $n = 1.5$ and water ($n_W = 1.33$) is placed between the two pieces of glass, what change will take place in the bright-fringe pattern?

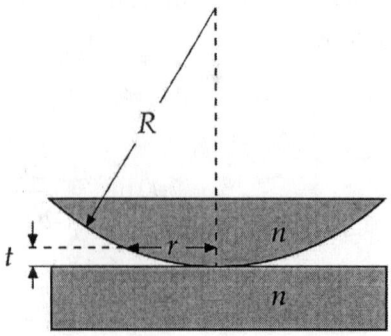

Figure 33-42 Problem 27

Picture the Problem This arrangement is essentially identical to a "thin film" configuration, except that the "film" is air. A phase change of 180° ($\frac{1}{2}\lambda$) occurs at the top of the flat glass plate. We can use the condition for constructive interference to derive the result given in (*a*) and use the geometry of the lens on the plate to obtain the result given in (*b*). We can then use these results in the remaining parts of the problem.

(*a*) The condition for constructive interference is:

$$2t + \tfrac{1}{2}\lambda = \lambda, 2\lambda, 3\lambda, \ldots$$

or

$$2t = \tfrac{1}{2}\lambda, \tfrac{3}{2}\lambda, \tfrac{5}{2}\lambda, \ldots = (m + \tfrac{1}{2})\lambda$$

where λ is the wavelength of light in air and $m = 0, 1, 2, \ldots$

Solve for *t*:

$$\boxed{t = (m + \tfrac{1}{2})\frac{\lambda}{2}, m = 0, 1, 2, \ldots} \qquad (1)$$

(*b*) From Figure 33-39 we have:

$$r^2 + (R - t)^2 = R^2$$

or

$$R^2 = r^2 + R^2 - 2Rt + t^2$$

For $t \ll R$ we can neglect the last term to obtain:

$$R^2 \approx r^2 + R^2 - 2Rt$$

Solve for *r*:

$$r = \boxed{\sqrt{2Rt}} \qquad (2)$$

(*c*) $\boxed{\text{The transmitted pattern is complementary to the reflected pattern.}}$

(*d*) Square equation (2) and substitute for *t* from equation (1) to obtain:

$$r^2 = (m + \tfrac{1}{2})R\lambda$$

Solve for *m*:

$$m = \frac{r^2}{R\lambda} - \frac{1}{2}$$

Substitute numerical values and evaluate *m*:

$$m = \frac{(2\,\text{cm})^2}{(10\,\text{m})(590\,\text{nm})} - \frac{1}{2} = 67$$

and so there will be $\boxed{68}$ bright fringes.

(*e*) The diameter of the mth fringe is:

$$D = 2r = 2\sqrt{(m + \tfrac{1}{2})R\lambda}$$

322   Chapter 33

Noting that $m = 5$ for the sixth fringe, substitute numerical values and evaluate D:

$$D = 2\sqrt{\left(5 + \tfrac{1}{2}\right)(10\,\text{m})(590\,\text{nm})}$$
$$= \boxed{1.14\,\text{cm}}$$

(f) The wavelength of the light in the film becomes $\lambda_{\text{air}}/n = 444$ nm. The separation between fringes is reduced and the number of fringes that will be seen is increased by the factor $n = 1.33$.

Two-Slit Interference Pattern

*30 • Two narrow slits separated by 1 mm are illuminated by light of wavelength 600 nm, and the interference pattern is viewed on a screen 2 m away. Calculate the number of bright fringes per centimeter on the screen.

Picture the Problem The number of bright fringes per unit distance is the reciprocal of the separation of the fringes. We can use the expression for the distance on the screen to the mth fringe to find the separation of the fringes.

Express the number N of bright fringes per centimeter in terms of the separation of the fringes:

$$N = \frac{1}{\Delta y} \quad (1)$$

Express the distance on the screen to the mth and $(m + 1)$st bright fringe:

$$y_m = m\frac{\lambda L}{d} \text{ and } y_{m+1} = (m+1)\frac{\lambda L}{d}$$

Subtract the second of these equations from the first to obtain:

$$\Delta y = \frac{\lambda L}{d}$$

Substitute in equation (1) to obtain:

$$N = \frac{d}{\lambda L}$$

Substitute numerical values and evaluate N:

$$N = \frac{1\,\text{mm}}{(600\,\text{nm})(2\,\text{m})} = \boxed{8.33\,\text{cm}^{-1}}$$

*35 •• White light falls at an angle of 30° to the normal of a plane containing a pair of slits separated by 2.5 μm. What visible wavelengths give a bright interference maximum in the transmitted light in the direction normal to the plane? (See Problem 34.)

Picture the Problem Let the separation of the slits be d. We can find the total path difference when the light is incident at an angle ϕ and set this result equal to an integer multiple of the wavelength of the light to relate the angle of incidence on the slits to the direction of the transmitted light and its wavelength.

Express the total path difference:	$\Delta \ell = d \sin \phi + d \sin \theta$
The condition for constructive interference is:	$\Delta \ell = m\lambda$ where m is an integer.
Substitute to obtain:	$d \sin \phi + d \sin \theta = m\lambda$
Divide both sides of the equation by d to obtain:	$\sin \phi + \sin \theta = \dfrac{m\lambda}{d}$
Set $\theta = 0$ and solve for λ:	$\lambda = \dfrac{d \sin \phi}{m}$
Substitute numerical values and simplify to obtain:	$\lambda = \dfrac{(2.5\,\mu m)\sin 30°}{m} = \dfrac{1.25\,\mu m}{m}$

Evaluate λ for positive integral values of m:

m	λ (nm)
1	1250
2	625
3	417
4	313

From the table we can see that 625 nm and 417 nm are in the visible portion of the electromagnetic spectrum.

Diffraction Pattern of a Single Slit

*39 •• Measuring the distance to the moon (lunar ranging) is routinely done by firing short-pulse lasers and measuring the time it takes for the pulses to reflect back from the moon. A pulse is fired from the earth; to send it out, the pulse is expanded so that it fills the aperture of a 6-in-diameter telescope. (*a*) Assuming the only thing spreading the beam out to be diffraction, how large will the beam be when it reaches the moon, 382,000 km away? (*b*) The pulse is reflected off a retroreflecting mirror left by the Apollo 11 astronauts. If the diameter of the mirror is 20 in, how large will the beam be when it gets back to the earth? (*c*) What fraction of the power of the beam is reflected back to the earth? (*d*) If the beam is refocused on return by the same 6-in telescope, what fraction of the original beam energy is recaptured? Ignore any atmospheric losses.

Picture the Problem The diagram shows the beam expanding as it travels to the moon and that portion of it that is reflected from the mirror on the moon expanding as it returns to earth. We can express the diameter of the beam at the moon as the product of the beam

324 Chapter 33

divergence angle and the distance to the moon and use the equation describing diffraction at a circular aperture to find the beam divergence angle. We can follow this same procedure to find the diameter of the beam when it gets back to the earth. In Parts (*c*) and (*d*) we can use the dependence of the power in a beam on its cross-sectional area to find the fraction of the power of the beam that is reflected back to earth and the fraction of the original beam energy that is recaptured upon return to earth.

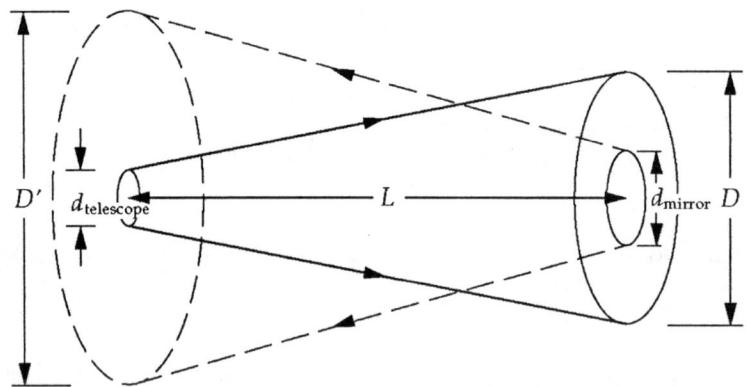

(*a*) Relate the diameter D of the beam at the moon to the distance to the moon L and the beam divergence angle θ:

$$D \approx \theta L$$

The angle θ subtended by the first diffraction minimum is related to the wavelength λ of the light and the diameter of the telescope opening $d_{telescope}$ by:

$$\sin\theta = 1.22 \frac{\lambda}{d_{telescope}}$$

Because $\theta \ll 1$, $\sin\theta \approx \theta$ and:

$$\theta \approx 1.22 \frac{\lambda}{d_{telescope}}$$

Substitute for θ in equation (1) to obtain:

$$D = \frac{1.22 L \lambda}{d_{telescope}}$$

Substitute numerical values and evaluate D:

$$D = (3.82 \times 10^8 \text{ m}) \left[\frac{1.22(500 \text{ nm})}{6 \text{ in} \times \frac{2.54 \text{ cm}}{\text{in}} \times \frac{1 \text{ m}}{10^2 \text{ cm}}} \right] = \boxed{1.53 \text{ km}}$$

(*b*) The portion of the beam reflected back to the earth will be that portion incident on the mirror, so the diffraction angle is:

$$\theta \approx 1.22 \frac{\lambda}{d_{mirror}}$$

The beam will expand back to:

$$D' = L\left[1.22\frac{\lambda}{d_{mirror}}\right]$$

Substitute numerical values and evaluate D':

$$D' = (3.82\times10^8\text{ m})\left[\frac{1.22(500\text{ nm})}{20\text{ in}\times\frac{2.54\text{ cm}}{\text{in}}\times\frac{1\text{ m}}{10^2\text{ cm}}}\right] = \boxed{459\text{ m}}$$

(c) Because the power of the beam is proportional to its cross-sectional area, the fraction of the power that is reflected back to the earth is the ratio of the area of the mirror to the area of the expanded beam at the moon:

$$\frac{P'}{P} = \frac{A_{mirror}}{A_{beam}} = \frac{\frac{\pi}{4}d_{mirror}^2}{\frac{\pi}{4}D^2} = \left(\frac{d_{mirror}}{D}\right)^2$$

Substitute for D to obtain:

$$\frac{P'}{P} = \left(\frac{d_{mirror}}{\frac{1.22L\lambda}{d_{telescope}}}\right)^2 = \left(\frac{d_{mirror}d_{telescope}}{1.22L\lambda}\right)^2 \quad (1)$$

Substitute numerical values and evaluate P'/P:

$$\frac{P'}{P} = \left[\frac{(20\text{ in})(6\text{ in})\left(\frac{2.54\text{ cm}}{\text{in}}\right)^2}{1.22(3.82\times10^8\text{ m})(500\text{ nm})}\right]^2$$

$$= \boxed{1.10\times10^{-7}}$$

(d) The angular spread of the beam from reflection from the 20-in mirror is given by:

$$\theta \approx 1.22\frac{\lambda}{d_{mirror}}$$

The diameter D' of the beam on return to earth will be:

$$D' \approx 1.22L\frac{\lambda}{d_{mirror}}$$

Letting P'' represent the power intercepted by the telescope, we have:

$$\frac{P''}{P'} = \frac{A_{telescope}}{A_{beam}} = \frac{\frac{\pi}{4}d_{telescope}^2}{\frac{\pi}{4}D'^2}$$

$$= \left(\frac{d_{telescope}}{D'}\right)^2$$

326   Chapter 33

Substitute for D' and simplify:

$$\frac{P''}{P'} = \left(\frac{d_{telescope}d_{mirror}}{1.22L\lambda}\right)^2 \quad (2)$$

Multiply equation (2) by equation (1) and simplify to obtain:

$$\frac{P''}{P'}\frac{P'}{P} = \frac{P''}{P} = \left(\frac{d_{telescope}d_{mirror}}{1.22L\lambda}\right)^2 \left(\frac{d_{mirror}d_{telescope}}{1.22L\lambda}\right)^2 = \left(\frac{d_{mirror}d_{telescope}}{1.22L\lambda}\right)^4$$

Substitute numerical values and evaluate P''/P:

$$\frac{P''}{P} = \left[\frac{(20\,\text{in})(6\,\text{in})\left(\frac{2.54\,\text{cm}}{\text{in}}\right)^2}{1.22(3.82\times 10^8\,\text{m})(500\,\text{nm})}\right]^4$$

$$= \boxed{1.21\times 10^{-14}}$$

Interference-Diffraction Pattern of Two Slits

*43 ••   Suppose that the *central* diffraction maximum for two slits contains 17 interference fringes for some wavelength of light. How many interference fringes would you expect in the first *secondary* diffraction maximum?

Determine the ConceptPicture the Problem There are 8 interference fringes on each side of the central maximum. The secondary diffraction maximum is half as wide as the central one. It follows that it will contain 8 interference maxima.

Using Phasors to Add Harmonic Waves

*46 •   Find the resultant of the two waves $E_1 = 4\sin\omega t$ and $E_2 = 3\sin(\omega t + 60°)$.

Picture the Problem Chose the coordinate system shown in the phasor diagram. We can use the standard methods of vector addition to find the resultant of the two waves.

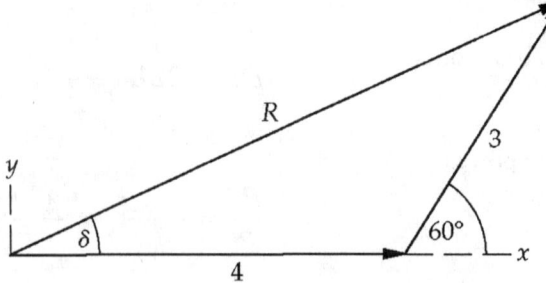

The resultant of the two waves is of the form: $E = R\sin(\omega t + \delta)$

Interference and Diffraction 327

Express the x component of $\vec{R}$: $\quad R_x = 4 + 3\cos 60° = 5.50$

Express the y component of $\vec{R}$: $\quad R_y = 0 + 3\sin 60° = 2.60$

Find the magnitude of $\vec{R}$: $\quad R = \sqrt{(5.50)^2 + (2.60)^2} = 6.08$

Find the phase angle δ between $\vec{R}$ and $\vec{E}_1$:
$$\delta = \tan^{-1}\left(\frac{R_y}{R_x}\right) = \tan^{-1}\left(\frac{2.60}{5.50}\right) = 25.3°$$

Substitute to obtain: $\quad E = \boxed{6.08 \sin(\omega t + 25.3°)}$

Remarks: We could have used the law of cosines to find R and the law of sines to find δ.

*52 ••• For single-slit diffraction, calculate the first three values of ϕ (the total phase difference between rays from each edge of the slit) that produce subsidiary maxima by (*a*) using the phasor model and (*b*) setting $dI/d\phi = 0$, where I is given by Equation 33-19.

Picture the Problem We can use the phasor diagram shown in Figure 33-26 to determine the first three values of ϕ that produce subsidiary maxima. Setting the derivative of Equation 33-19 equal to zero will yield a transcendental equation whose roots are the values of ϕ corresponding to the maxima in the diffraction pattern.

(*a*) Referring to Figure 33-26 we see that the first subsidiary maximum occurs when: $\quad \phi = 3\pi$

A minimum occurs when: $\quad \phi = 4\pi$

Another maximum occurs when: $\quad \phi = 5\pi$

Thus, subsidiary maxima occur when:
$\phi = (2n+1)\pi, \; n = 1, 2, 3, ...$
and the first three subsidiary maxima are at $\phi = 3\pi, 5\pi,$ and 7π.

(*b*) The intensity in the single-slit diffraction pattern is given by:
$$I = I_0 \left(\frac{\sin \tfrac{1}{2}\phi}{\tfrac{1}{2}\phi}\right)^2$$

Set the derivative of this expression equal to zero for extrema:

328   Chapter 33

$$\frac{dI}{d\phi} = 2I_0 \left(\frac{\sin \tfrac{1}{2}\phi}{\tfrac{1}{2}\phi} \right) \left[\frac{\tfrac{1}{4}\phi \cos \tfrac{1}{2}\phi - \tfrac{1}{2}\sin \tfrac{1}{2}\phi}{\left(\tfrac{1}{2}\phi\right)^2} \right] = 0 \text{ for relative maxima and minima}$$

Simplify to obtain the transcendental equation:

$$\tan \tfrac{1}{2}\phi = \tfrac{1}{2}\phi$$

Solve this equation numerically (use the "Solver" function of your calculator) to obtain:

$$\phi = \boxed{2.86\pi, 4.92\pi, \text{ and } 6.94\pi}$$

Remarks: Note that our results in (*b*) are smaller than the approximate values found in (*a*) by 4.80%, 1.63%, and 0.865% and that the agreement improves as *n* increases.

Diffraction and Resolution

*55 •   Two sources of light of wavelength 700 nm are separated by a horizontal distance *x*. They are 5 m from a vertical slit of width 0.5 mm. What is the least value of *x* for which the diffraction pattern of the sources can be resolved by Rayleigh's criterion?

Picture the Problem We can use Rayleigh's criterion for slits and the geometry of the diagram to express *x* in terms of λ, *L*, and the width *a* of the slit.

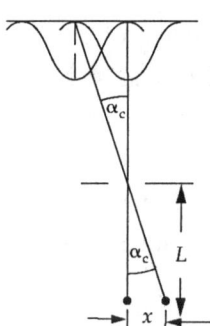

Referring to the diagram, relate α_c, *L*, and *x*:

$$\alpha_c \approx \frac{x}{L}$$

For slits, Rayleigh's criterion is:

$$\alpha_c = \frac{\lambda}{a}$$

Equate these two expressions to obtain:

$$\frac{x}{L} = \frac{\lambda}{a}$$

Solve for *x*:

$$x = \frac{\lambda L}{a}$$

Substitute numerical values and evaluate x:
$$x = \frac{(700\,\text{nm})(5\,\text{m})}{0.5\,\text{mm}} = \boxed{7.00\,\text{mm}}$$

*60 •• The star Mizar in Ursa Major is a binary system of stars of nearly equal magnitudes. The angular separation between the two stars is 14 seconds of arc. What is the minimum diameter of the pupil that allows resolution of the two stars using light of wavelength 550 nm?

Picture the Problem We can use Rayleigh's criterion for circular apertures and the geometry of the diagram to obtain an expression we can solve for the minimum diameter D of the pupil that allows resolution of the binary stars.

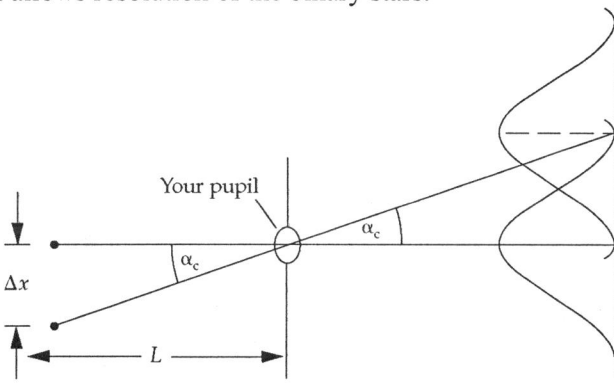

(a) Rayleigh's criterion is satisfied provided:
$$\alpha_c = 1.22 \frac{\lambda}{D}$$

Solve for D:
$$D = 1.22 \frac{\lambda}{\alpha_c}$$

Substitute numerical values and evaluate D:
$$D = 1.22 \frac{550\,\text{nm}}{14'' \times \frac{1°}{3600''} \times \frac{\pi\,\text{rad}}{180°}}$$
$$= \boxed{9.89\,\text{mm}} \approx 1\,\text{cm}$$

Diffraction Gratings

*62 • With the diffraction grating used in Problem 61, two other lines in the first-order hydrogen spectrum are found at angles $\theta_1 = 9.72 \times 10^{-2}$ rad and $\theta_2 = 1.32 \times 10^{-1}$ rad. Find the wavelengths of these lines.

Picture the Problem We can solve $d \sin \theta = m\lambda$ for λ with $m = 1$ to express the location of the first-order maximum as a function of the angles at which the first-order images are found.

330   Chapter 33

The interference maxima in a diffraction pattern are at angles θ given by:	$d \sin \theta = m\lambda$ where d is the separation of the slits and $m = 0, 1, 2, \ldots$
Solve for λ:	$\lambda = \dfrac{d \sin \theta}{m}$
Relate the number of slits N per centimeter to the separation d of the slits:	$N = \dfrac{1}{d}$
Let $m = 1$ and substitute for d to obtain:	$\lambda = \dfrac{d \sin \theta}{N}$
Substitute numerical values and evaluate λ_1 for $\theta_1 = 9.72 \times 10^{-2}$ rad:	$\lambda_1 = \dfrac{\sin(9.72 \times 10^{-2} \text{ rad})}{2000 \text{ cm}^{-1}} = \boxed{485 \text{ nm}}$
Substitute numerical values and evaluate λ_1 for $\theta_2 = 1.32 \times 10^{-1}$ rad:	$\lambda_1 = \dfrac{\sin(1.32 \times 10^{-1} \text{ rad})}{2000 \text{ cm}^{-1}} = \boxed{658 \text{ nm}}$

*67 ••   A diffraction grating with 4800 lines per centimeter is illuminated at normal incidence with white light (wavelength range of 400 nm to 700 nm). For how many orders can one observe the complete spectrum in the transmitted light? Do any of these orders overlap? If so, describe the overlapping regions.

Picture the Problem We can use the grating equation $d \sin \theta = m\lambda$, $m = 1, 2, 3, \ldots$ to express the order number in terms of the slit separation d, the wavelength of the light λ, and the angle θ.

The interference maxima in the diffraction pattern are at angles θ given by:	$d \sin \theta = m\lambda, m = 1, 2, 3, \ldots$
Solve for m:	$m = \dfrac{d \sin \theta}{\lambda}$
If one is to see the complete spectrum:	$\sin \theta \leq 1$ and $m \leq \dfrac{d}{\lambda}$

Evaluate m_{max}:

$$m_{max} \frac{\frac{1}{4800\,cm^{-1}}}{\lambda_{max}} = \frac{\frac{1}{4800\,cm^{-1}}}{700\,nm} = 2.98$$

Because $m_{max} = 2.98$, one can see the complete spectrum only for $m = 1$ and 2.

Express the condition for overlap: $\quad m_1 \lambda_1 \geq m_2 \lambda_2$

Because 700 nm < 2 × 400 nm, there is no overlap of the second - order spectrum into the first - order spectrum; however, there is overlap of long wavelengths in the second order with short wavelengths in the third - order spectrum.

*71 •• Mercury has several stable isotopes, among them ^{198}Hg and ^{202}Hg. The strong spectral line of mercury, at about 546.07 nm, is a composite of spectral lines from the various mercury isotopes. The wavelengths of this line for ^{198}Hg and ^{202}Hg are 546.07532 nm and 546.07355 nm, respectively. What must be the resolving power of a grating capable of resolving these two isotopic lines in the third-order spectrum? If the grating is illuminated over a 2-cm-wide region, what must be the number of lines per centimeter of the grating?

Picture the Problem We can use the expression for the resolving power of a grating to find the resolving power of the grating capable of resolving these two isotopic lines in the third-order spectrum. Because the total number of the slits of the grating N is related to width w of the illuminated region and the number of lines per centimeter of the grating and the resolving power R of the grating, we can use this relationship to find the number of lines per centimeter of the grating

The resolving power of a diffraction grating is given by:

$$R = \frac{\lambda}{|\Delta \lambda|} = mN \qquad (1)$$

Substitute numerical values and evaluate R:

$$R = \frac{546.07532}{|546.07532 - 546.07355|}$$

$$= \boxed{3.09 \times 10^5}$$

Express n, be the number of lines per centimeter of the grating, in terms of the total number of slits N of the grating and the width w of the

$$n = \frac{N}{w}$$

332 Chapter 33

grating:

From equation (1) we have:
$$N = \frac{R}{m}$$

Substitute to obtain:
$$n = \frac{R}{mw}$$

Substitute numerical values and evaluate n:
$$n = \frac{3.09 \times 10^5}{(3)(2\,\text{cm})} = \boxed{5.15 \times 10^4 \text{ cm}^{-1}}$$

General Problems

*77 • A long, narrow horizontal slit lies 1 mm above a plane mirror, which is in the horizontal plane. The interference pattern produced by the slit and its image is viewed on a screen 1 m from the slit. The wavelength of the light is 600 nm. (*a*) Find the distance from the mirror to the first maximum. (*b*) How many dark bands per centimeter are seen on the screen?

Picture the Problem We can apply the condition for constructive interference to find the angular position of the first maximum on the screen. Note that, due to reflection, the wave from the image is 180° out of phase with that from the source.

(*a*) Because $y_0 \ll L$, the distance from the mirror to the first maximum is given by:
$$y_0 = L\theta_0 \qquad (1)$$

Express the condition for constructive interference:
$$d \sin\theta = (m + \tfrac{1}{2})\lambda, \, m = 0, 1, 2, \ldots$$

Solve for θ:
$$\theta = \sin^{-1}\left[(m + \tfrac{1}{2})\frac{\lambda}{d}\right]$$

For the first maximum, $m = 0$ and:
$$\theta_0 = \sin^{-1}\left[(\tfrac{1}{2})\frac{\lambda}{d}\right]$$

Substitute in equation (1) to obtain:
$$y_0 = L\sin^{-1}\left[(\tfrac{1}{2})\frac{\lambda}{d}\right]$$

Because the image of the slit is as far behind the mirror's surface as the slit is in front of it, $d = 2$ mm. Substitute numerical values and evaluate y_0:

$$y_0 = (1\,\text{m})\sin^{-1}\left[\left(\tfrac{1}{2}\right)\frac{600\,\text{nm}}{2\,\text{mm}}\right]$$

$$= \boxed{0.150\,\text{mm}}$$

(b) The separation of the fringes on the screen is given by:

$$\Delta y = \frac{\lambda L}{d}$$

The number of dark bands per centimeter is the reciprocal of the fringe separation:

$$n = \frac{1}{\Delta y} = \frac{d}{\lambda L}$$

Substitute numerical values and evaluate n:

$$n = \frac{2\,\text{mm}}{(600\,\text{nm})(1\,\text{m})} = \boxed{3.33 \times 10^3\,\text{m}^{-1}}$$

*82 •• A thin layer of a transparent material with an index of refraction of 1.30 is used as a nonreflective coating on the surface of glass with an index of refraction of 1.50. What should the thickness of the material be for it to be nonreflecting for light of wavelength 600 nm?

Picture the Problem Note that reflection at both surfaces involves a phase shift of π rad. We can apply the condition for destructive interference to find the thickness t of the nonreflective coating.

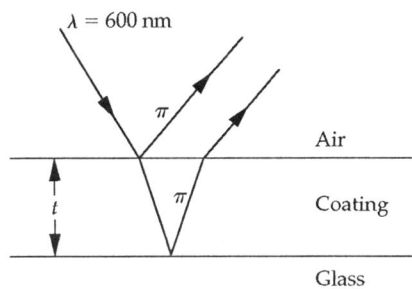

The condition for destructive interference is:

$$2t = \left(m + \tfrac{1}{2}\right)\lambda_{\text{coating}} = \left(m + \tfrac{1}{2}\right)\frac{\lambda_{\text{air}}}{n_{\text{coating}}}$$

Solve for t:

$$t = \left(m + \tfrac{1}{2}\right)\frac{\lambda_{\text{air}}}{2n_{\text{coating}}}$$

Evaluate t for $m = 0$:

$$t = \left(\tfrac{1}{2}\right)\frac{600\,\text{nm}}{2(1.30)} = \boxed{115\,\text{nm}}$$

*87 •• The Impressionist painter Georges Seurat used a technique called *pointillism*, in which his paintings are composed of small, closely spaced dots of pure color, each about 2 mm in diameter. The illusion of the colors blending together smoothly is

334 Chapter 33

produced in the eye of the viewer by diffraction effects. Calculate the minimum viewing distance for this effect to work properly. Use the wavelength of visible light that requires the *greatest* distance, so that you're sure the effect will work for *all* visible wavelengths. Assume the pupil of the eye has a diameter of 3 mm.

Picture the Problem We can use the geometry of the dots and the pupil of the eye and Rayleigh's criterion to find the greatest viewing distance that ensures that the effect will work for all visible wavelengths.

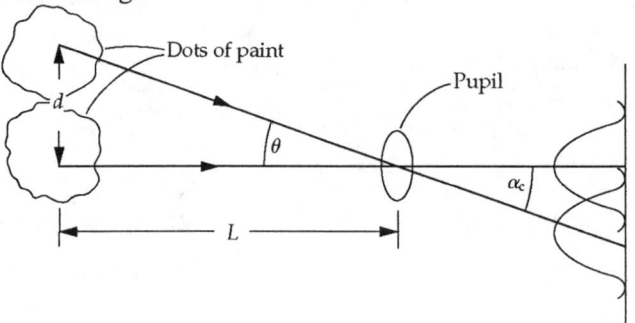

Referring to the diagram, express the angle subtended by the adjacent dots:

$$\theta \approx \frac{d}{L}$$

Letting the diameter of the pupil of the eye be D, apply Rayleigh's criterion to obtain:

$$\alpha_c = 1.22 \frac{\lambda}{D}$$

Set $\theta = \alpha_c$ to obtain:

$$\frac{d}{L} = 1.22 \frac{\lambda}{D}$$

Solve for L:

$$L = \frac{Dd}{1.22\lambda}$$

Evaluate L for the *shortest* wavelength light in the visible portion of the spectrum:

$$L = \frac{(3\,\text{mm})(2\,\text{mm})}{1.22(400\,\text{nm})} = \boxed{12.3\,\text{m}}$$

*88 ••• A *Jamin refractometer* is a device for measuring or for comparing the indexes of refraction of gases. A beam of monochromatic light is split into two parts, each of which is directed along the axis of a separate cylindrical tube before being recombined into a single beam that is viewed through a telescope. Suppose that each tube is 0.4 m long and that sodium light of wavelength 589 nm is used. Both tubes are initially evacuated, and constructive interference is observed in the center of the field of view. As air is slowly allowed to enter one of the tubes, the central field of view changes to dark and back to bright a total of 198 times. (*a*) What is the index of refraction of air? (*b*) If the fringes can be counted to ± 0.25 fringe, where one fringe is equivalent to one complete cycle of intensity variation at the center of the field of view, to what accuracy can the index of refraction of air be determined by this experiment?

Picture the Problem It is given that with one tube evacuated and one full of air at 1-atm pressure, there are 198 more wavelengths of light in the tube full of air than in the evacuated tube of the same length. We can use this condition to obtain an equation that expresses this difference in terms of L, λ_n, and λ_0. We can obtain a second equation relating λ_n, n, and λ_0 ($\lambda_n = \frac{\lambda_0}{n}$) and solve the two equations simultaneously to find n.

(a) The wavelengths are related by:
$$\lambda_n = \frac{\lambda_0}{n}$$

The number of wavelengths in length L is the length L divided by the wavelength. Thus:
$$\frac{L}{\lambda_n} - \frac{L}{\lambda_0} = 198$$

Substitute for λ_n:
$$\frac{nL}{\lambda_0} - \frac{L}{\lambda_0} = 198$$

Solve for λ_n to obtain:
$$n = 1 + \frac{198\lambda_0}{L}$$

Substitute numerical values and evaluate n:
$$n = 1 + 198\left(\frac{589\,\text{nm}}{0.4\,\text{m}}\right) = \boxed{1.0002916}$$

(b) Replace 198 with 198 ± 0.25 and assume that the uncertainties in L and λ_0 are negligible:

$$n = 1 + \frac{\lambda_0}{L}(198 \pm 0.25) = \boxed{1.0002916 \pm 0.0000004}$$

Chapter 34
Wave-Particle Duality and Quantum Physics

Conceptual Problems

*1 •   The quantized character of electromagnetic radiation is revealed by
(a) the Young double-slit experiment. (b) diffraction of light by a small aperture. (c) the photoelectric effect. (d) the J.J. Thomson cathode-ray experiment.

Determine the Concept The Young double-slit experiment, the diffraction of light by a small aperture, and the J.J. Thomson cathode-ray experiment all demonstrated the wave nature of electromagnetic radiation. Only the photoelectric effect requires an explanation based on the quantization of electromagnetic radiation. (c) is correct.

*5 •   The work function of a surface is ϕ. The threshold wavelength for emission of photoelectrons from the surface is (a) hc/ϕ. (b) ϕ/hf. (c) hf/ϕ. (d) none of the answers are correct.

Determine the Concept The threshold wavelength for emission of photoelectrons is related to the work function of a metal through $\phi = hc/\lambda_t$. Hence $\lambda_t = hc/\phi$ and (a) is correct.

*11 •   Explain why the maximum kinetic energy of electrons emitted in the photoelectric effect does not depend on the intensity of the incident light, but the total number of electrons emitted does depend on the intensity of the incident light.

Determine the Concept In the photoelectric effect, an electron absorbs the energy of a single photon. Therefore, $K_{max} = hf - \phi$, independently of the number of photons incident on the surface. However, the number of photons incident on the surface determines the number of electrons that are emitted.

Estimation and Approximation

*16 ••   Students in an advanced physics lab use X rays to measure the Compton wavelength, λ_C. The students obtain the following wavelength shifts $\lambda_2 - \lambda_1$ as a function of scattering angle θ

θ	45°	75°	90°	135°	180°
$\lambda_2 - \lambda_1$	0.647 pm	1.67 pm	2.45 pm	3.98 pm	4.95 pm

337

Use their data to estimate the value for the Compton wavelength. Compare this number with the accepted value.

Picture the Problem From the Compton-scattering equation we have $\lambda_2 - \lambda_1 = \lambda_C(1-\cos\theta)$, where $\lambda_C = h/m_e c$ is the Compton wavelength. Note that this equation is of the form $y = mx + b$ provided we let $y = \lambda_2 - \lambda_1$ and $x = 1 - \cos\theta$. Thus, we can linearize the Compton equation by plotting $\Delta\lambda = \lambda_2 - \lambda_1$ as a function of $1 - \cos\theta$. The slope of the resulting graph will yield an experimental value for the Compton wavelength.

(a) The spreadsheet solution is shown below. The formulas used to calculate the quantities in the columns are as follows:

Cell	Formula/Content	Algebraic Form
A3	45	θ (deg)
B3	1 – cos(A3*PI()/180)	$1 - \cos\theta$
C3	6.47E^–13	$\Delta\lambda = \lambda_2 - \lambda_1$

θ (deg)	$1-\cos\theta$	$\lambda_2-\lambda_1$
45	0.293	6.47E–13
75	0.741	1.67E–12
90	1.000	2.45E–12
135	1.707	3.98E–12
180	2.000	4.95E–12

The following graph was plotted from the data shown in the above table. Excel's "Add Trendline" was used to fit a linear function to the data. The regression equation is
$$\Delta\lambda = 2.48\times10^{-12}(1-\cos\theta) - 1.03\times10^{-13}$$

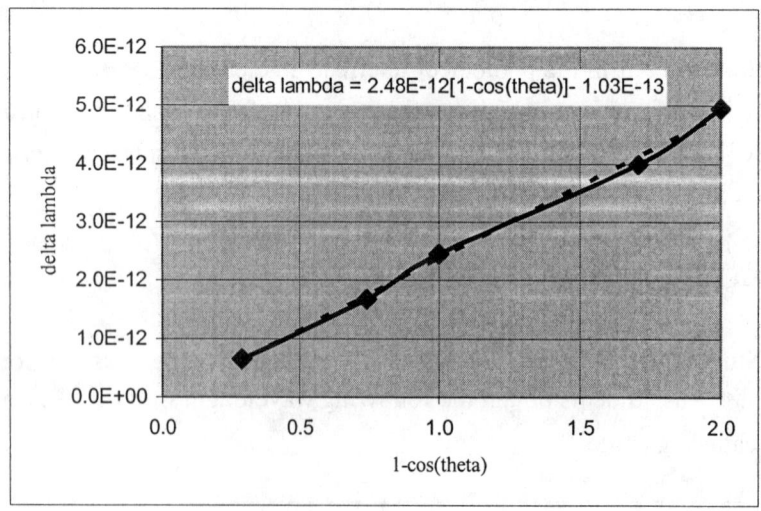

From the regression line we note that the experimental value for the Compton wavelength $\lambda_{C,exp}$ is:

$$\lambda_{C,exp} = \boxed{2.48 \times 10^{-12} \text{ m}}$$

The Compton wavelength is given by:

$$\lambda_C = \frac{h}{m_e c} = \frac{hc}{m_e c^2}$$

Substitute numerical values and evaluate λ_C:

$$\lambda_C = \frac{1240 \text{ eV} \cdot \text{nm}}{5.11 \times 10^5 \text{ eV}} = 2.43 \times 10^{-12} \text{ m}$$

Express the percent difference between λ_C and $\lambda_{C,exp}$:

$$\% \text{diff} = \frac{\lambda_{C,exp} - \lambda_{exp}}{\lambda_{exp}} = \frac{\lambda_{C,exp}}{\lambda_{exp}} - 1$$

$$= \frac{2.48 \times 10^{-12} \text{ m}}{2.43 \times 10^{-12} \text{ m}} - 1 = \boxed{2.06\%}$$

*17 •• Baseball, tennis, golf, and soccer are sports that involve placing a ball in play with a certain speed. Estimate which of these sports has a ball with the smallest de Broglie wavelength when the ball is moving with the highest speed typically created by a professional athlete.

Picture the Problem The de Broglie wavelength of an object is given by $\lambda = h/p$, where p is the momentum of the object.

The de Broglie wavelength of an object, in terms of its mass m and speed v, is:

$$\lambda = \frac{h}{mv}$$

The values in the following table were obtained using the internet:

Type of ball	m (g)	v_{max} (m/s)
Baseball	142	44
Tennis	57	54
Golf	57	42
Soccer	250	31

The de Broglie wavelength of a baseball, moving with its maximum speed, is:

$$\lambda = \frac{6.63 \times 10^{-34} \text{ J} \cdot \text{s}}{(0.142 \text{ kg})(44 \text{ m/s})} = 1.06 \times 10^{-34} \text{ m}$$

Proceed as above to obtain the values shown in the table:

Type of ball	m	v_{max}	λ
	(g)	(m/s)	(m)
Baseball	142	44	1.06×10^{-34}
Tennis	57	54	2.15×10^{-34}
Golf	57	42	2.77×10^{-34}
Soccer	250	31	0.855×10^{-34}

Examination of the table indicates that the soccer ball has the shortest de Broglie wavelength.

The Particle Nature of Light: Photons

*20 • Find the photon energy for light of wavelength (*a*) 450 nm, (*b*) 550 nm, and (*c*) 650 nm.

Picture the Problem We can use $E = hc/\lambda$ to find the photon energy when we are given the wavelength of the radiation.

(*a*) Express the photon energy as a function of wavelength and evaluate E for $\lambda = 450$ nm:

$$E = \frac{hc}{\lambda} = \frac{1240 \, \text{eV} \cdot \text{nm}}{450 \, \text{nm}} = \boxed{2.76 \, \text{eV}}$$

(*b*) For $\lambda = 550$ nm:

$$E = \frac{1240 \, \text{eV} \cdot \text{nm}}{550 \, \text{nm}} = \boxed{2.25 \, \text{eV}}$$

(*c*) For $\lambda = 650$ nm:

$$E = \frac{1240 \, \text{eV} \cdot \text{nm}}{650 \, \text{nm}} = \boxed{1.91 \, \text{eV}}$$

*23 • Lasers used in the telecommunications network typically have a wavelength near 1.55 μm. How many photons per second are being transmitted if such a laser has an output power of 2.5 mW?

Picture the Problem The number of photons per unit volume is, in turn, the ratio of the power of the laser to the energy of the photons and the volume occupied by the photons emitted in one second is the product of the cross-sectional area of the beam and the speed at which the photons travel; i.e., the speed of light.

Relate the number of photons emitted per second to the power of

$$N = \frac{P}{E} = \frac{P\lambda}{hc}$$

the laser and the energy of the photons:

Substitute numerical values and evaluate N:

$$N = \frac{(2.5\,\text{mW})(1.55\,\mu\text{m})}{(6.63 \times 10^{-34}\,\text{J}\cdot\text{s})(3 \times 10^8\,\text{m/s})}$$

$$= \boxed{1.95 \times 10^{16}\,\text{s}^{-1}}$$

The Photoelectric Effect

*28 •• When a surface is illuminated with light of wavelength 780 nm, the maximum kinetic energy of the emitted electrons is 0.37 eV. What is the maximum kinetic energy if the surface is illuminated with light of wavelength 410 nm?

Picture the Problem We can use Einstein's photoelectric equation to find the work function of this surface and then apply it a second time to find the maximum kinetic energy of the photoelectrons when the surface is illuminated with light of wavelength 365 nm.

Use Einstein's photoelectric equation to relate the maximum kinetic energy of the emitted electrons to their total energy and the work function of the surface:

$$K_{max} = \frac{hc}{\lambda} - \phi$$

Using Einstein's photoelectric equation, find the work function of the surface:

$$\phi = E - K_{max} = \frac{hc}{\lambda} - K_{max}$$

$$= \frac{1240\,\text{eV}\cdot\text{nm}}{780\,\text{nm}} - 0.37\,\text{eV}$$

$$= 1.22\,\text{eV}$$

Substitute for ϕ and λ and evaluate K_{max}:

$$K_{max} = \frac{1240\,\text{eV}\cdot\text{nm}}{410\,\text{nm}} - 1.22\,\text{eV}$$

$$= \boxed{1.80\,\text{eV}}$$

Compton Scattering

*32 • Compton used photons of wavelength 0.0711 nm. (*a*) What is the energy of these photons? (*b*) What is the wavelength of the photon scattered at $\theta = 180°$? (*c*) What is the energy of the photon scattered at this angle?

342   Chapter 34

Picture the Problem We can use the Einstein equation for photon energy to find the energy of both the incident and scattered photon and the Compton scattering equation to find the wavelength of the scattered photon.

(a) Use the Einstein equation for photon energy to obtain:
$$E = \frac{hc}{\lambda_1} = \frac{1240 \, \text{eV} \cdot \text{nm}}{0.0711 \, \text{nm}} = \boxed{17.4 \, \text{keV}}$$

(b) Express the wavelength of the scattered photon in terms of its pre-scattering wavelength and the shift in its wavelength during scattering:
$$\lambda_2 = \lambda_1 + \Delta\lambda = \lambda_1 + \frac{h}{m_e c}(1 - \cos\theta)$$

Substitute numerical values and evaluate λ_2:

$$\lambda_2 = 0.0711 \, \text{nm} + \frac{6.63 \times 10^{-34} \, \text{J} \cdot \text{s}}{(9.11 \times 10^{-31} \, \text{kg})(3 \times 10^8 \, \text{m/s})}(1 - \cos 180°) = \boxed{0.0760 \, \text{nm}}$$

(c) Use the Einstein equation for photon energy to obtain:
$$E = \frac{hc}{\lambda_2} = \frac{1240 \, \text{eV} \cdot \text{nm}}{0.0760 \, \text{nm}} = \boxed{16.3 \, \text{keV}}$$

Electrons and Matter Waves

***39 ••** An electron, a proton, and an alpha particle (the nucleus of a helium atom) each have a kinetic energy of 150 keV. Find (a) their momenta and (b) their de Broglie wavelengths.

Picture the Problem The momenta of these particles can be found from their kinetic energies and speeds. Their de Broglie wavelengths are given by $\lambda = h/p$.

(a) The momentum of a particle p, in terms of its kinetic energy K, is given by:
$$p = \sqrt{2mK}$$

Substitute numerical values and evaluate p_e:

$$p_e = \sqrt{2m_e K} = \sqrt{2(9.11 \times 10^{-31} \, \text{kg})\left(150 \, \text{keV} \times \frac{1.6 \times 10^{-19} \, \text{C}}{\text{eV}}\right)}$$
$$= \boxed{2.09 \times 10^{-22} \, \text{N} \cdot \text{s}}$$

Substitute numerical values and evaluate p_p:

$$p_p = \sqrt{2m_p K} = \sqrt{2(1.67 \times 10^{-27} \text{ kg})\left(150 \text{ keV} \times \frac{1.6 \times 10^{-19} \text{ C}}{\text{eV}}\right)}$$
$$= \boxed{8.95 \times 10^{-21} \text{ N} \cdot \text{s}}$$

Substitute numerical values and evaluate p_α:

$$p_\alpha = \sqrt{2m_\alpha K} = \sqrt{2\left(4\text{u} \times \frac{1.66 \times 10^{-27} \text{ kg}}{\text{u}}\right)\left(150 \text{ keV} \times \frac{1.6 \times 10^{-19} \text{ C}}{\text{eV}}\right)}$$
$$= \boxed{1.79 \times 10^{-20} \text{ N} \cdot \text{s}}$$

(b) The de Broglie wavelengths of the particles are given by:
$$\lambda = \frac{h}{p}$$

Substitute numerical values and evaluate λ_p:
$$\lambda_p = \frac{h}{p_p}$$
$$= \frac{6.63 \times 10^{-34} \text{ J} \cdot \text{s}}{8.95 \times 10^{-21} \text{ N} \cdot \text{s}} = \boxed{7.41 \times 10^{-14} \text{ m}}$$

Substitute numerical values and evaluate λ_e:
$$\lambda_e = \frac{h}{p_e}$$
$$= \frac{6.63 \times 10^{-34} \text{ J} \cdot \text{s}}{2.09 \times 10^{-22} \text{ N} \cdot \text{s}} = \boxed{3.17 \times 10^{-12} \text{ m}}$$

Substitute numerical values and evaluate λ_α:
$$\lambda_\alpha = \frac{h}{p_\alpha}$$
$$= \frac{6.63 \times 10^{-34} \text{ J} \cdot \text{s}}{1.79 \times 10^{-20} \text{ N} \cdot \text{s}} = \boxed{3.70 \times 10^{-14} \text{ m}}$$

*42 • A proton is moving at $v = 0.003c$, where c is the speed of light. Find the electron's de Broglie wavelength.

Picture the Problem We can use its definition to calculate the de Broglie wavelength of this proton.

Chapter 34

Use its definition to express the de Broglie wavelength of the proton:

$$\lambda_p = \frac{h}{p_p} = \frac{h}{m_p v_p}$$

Substitute numerical values and evaluate λ_p:

$$\lambda = \frac{6.63 \times 10^{-34}\ \text{J} \cdot \text{s}}{(1.67 \times 10^{-27}\ \text{kg})[0.003(3 \times 10^8\ \text{m/s})]} = \boxed{0.441\ \text{pm}}$$

*47 • An electron microscope uses electrons of energy 70 keV. Find the wavelength of these electrons.

Picture the Problem We can use $\lambda = \dfrac{1.226}{\sqrt{K}}$ nm, where K is in eV, to find the wavelength of 70-keV electrons.

Relate the wavelength of the electrons to their kinetic energy:

$$\lambda = \frac{1.23}{\sqrt{K}}\ \text{nm}$$

Substitute numerical values and evaluate λ:

$$\lambda = \frac{1.226}{\sqrt{70 \times 10^3\ \text{eV}}}\ \text{nm} = \boxed{4.63\ \text{pm}}$$

A Particle in a Box

*52 •• Use a spreadsheet program or graphing calculator to plot $\psi(x)$ and the probability distribution $\psi^2(x)$ of a particle in a box for the states $n = 1, 2,$ and 3.

Picture the Problem The wave function for state n is $\psi_n(x) = \sqrt{\dfrac{2}{L}} \sin \dfrac{n\pi x}{L}$. The following graphs were plotted using a spreadsheet program.

The graph of $\psi(x)$ for $n = 1$ is shown below:

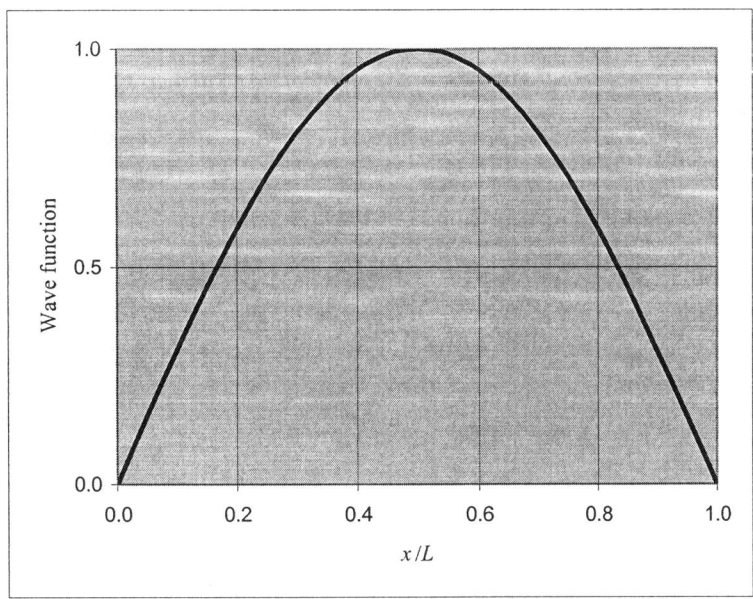

The graph of $\psi^2(x)$ for $n = 1$ is shown below:

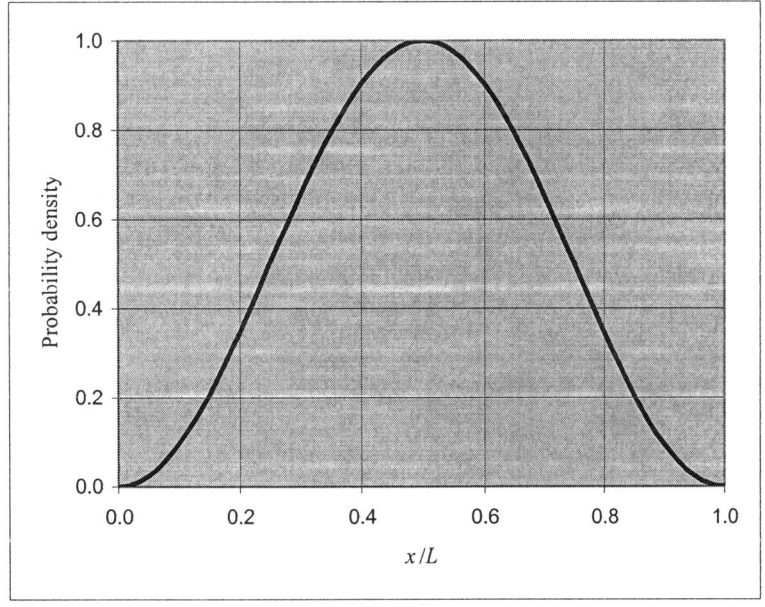

The graph of $\psi(x)$ for $n = 2$ is shown below:

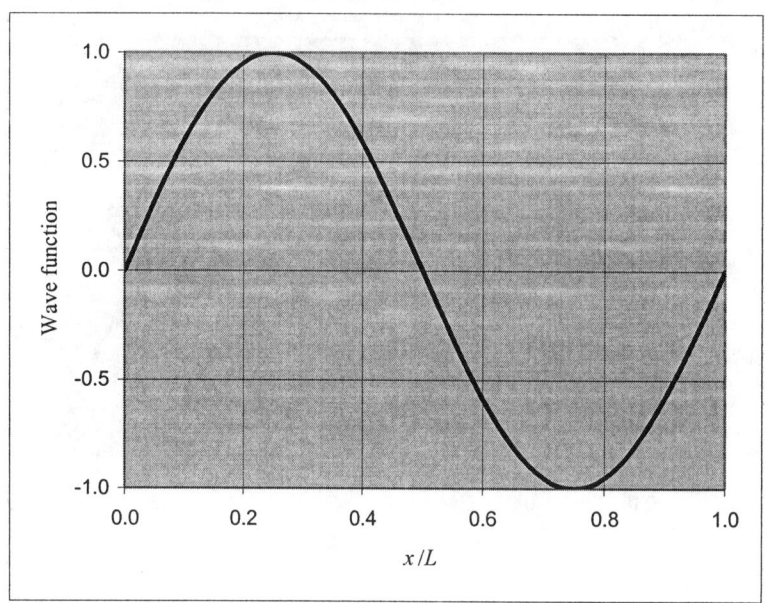

The graph of $\psi^2(x)$ for $n = 2$ is shown below:

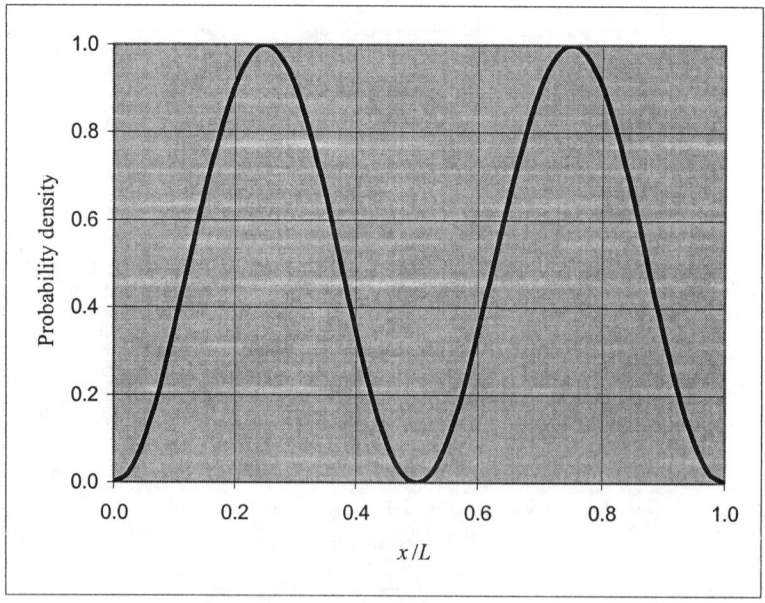

The graph of $\psi(x)$ for $n = 3$ is shown below:

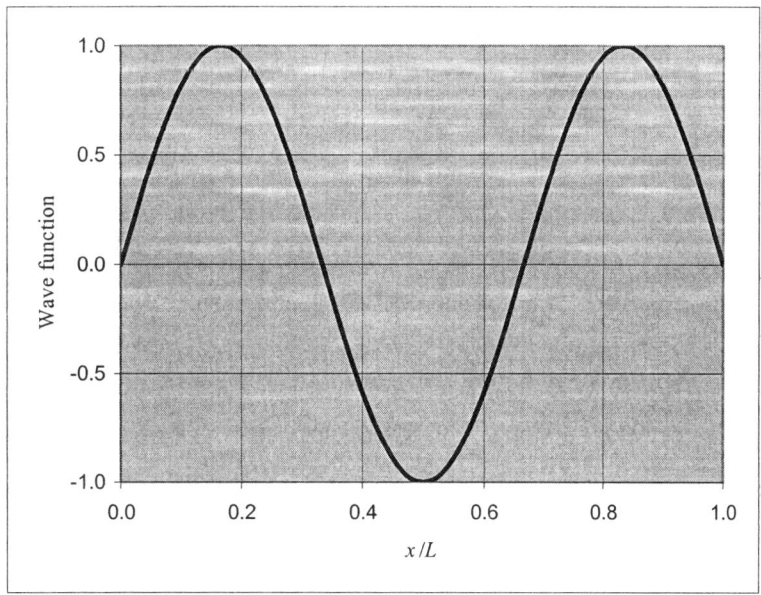

The graph of $\psi^2(x)$ for $n = 3$ is shown below:

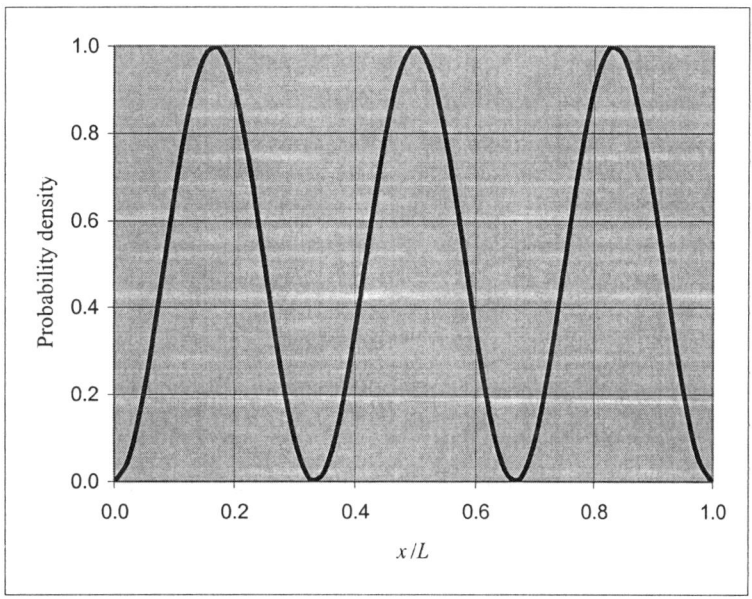

Calculating Probabilities and Expectation Values

*56 •• Repeat Problem 55 for a particle in the first excited state ($n = 2$).

Picture the Problem The probability of finding the particle in some range Δx is $\psi^2 dx$. The interval $\Delta x = 0.002L$ is so small that we can neglect the variation in $\psi(x)$ and just compute $\psi^2 \Delta x$.

Chapter 34

Express the probability of finding the particle in the interval Δx:

$$P = P(x)\Delta x = \psi^2(x)\Delta x$$

Express the wave function for a particle in its first excited state:

$$\psi_2(x) = \sqrt{\frac{2}{L}}\sin\frac{2\pi x}{L}$$

Substitute to obtain:

$$P = \frac{2}{L}\sin^2\frac{2\pi x}{L}\Delta x$$

$$= \frac{2}{L}\left(\sin^2\frac{2\pi x}{L}\right)(0.002L)$$

$$= 0.004\sin^2\frac{2\pi x}{L}$$

(a) Evaluate P at $x = L/2$:

$$P = 0.004\sin^2\frac{2\pi L}{2L} = 0.004\sin^2\pi$$

$$= \boxed{0}$$

(b) Evaluate P at $x = 2L/3$:

$$P = 0.004\sin^2\frac{4\pi L}{3L} = 0.004\sin^2\frac{4\pi}{3}$$

$$= \boxed{0.003}$$

(c) Evaluate P at $x = L$:

$$P = 0.004\sin^2\frac{2\pi L}{L} = 0.004\sin^2 2\pi$$

$$= \boxed{0}$$

*64 •• (a) Use a spreadsheet program or graphing calculator to plot the expectation value for position $\langle x \rangle$ and the square of the position $\langle x^2 \rangle$ as a function of the quantum number n for the particle in the box described in Problem 60, for values of n from 1 to 100. Assume $L = 1$m for your graph. Refer to Problem 63. (b) Comment on the significance of any asymptotic limits that your graph shows.

Picture the Problem From Problem 63 we have $\langle x \rangle = \frac{L}{2}$ and $\langle x^2 \rangle = \frac{L^2}{3} - \frac{L^2}{2n^2\pi^2}$. A spreadsheet program was used to plot the following graphs of <x> and <x²> as a function of n.

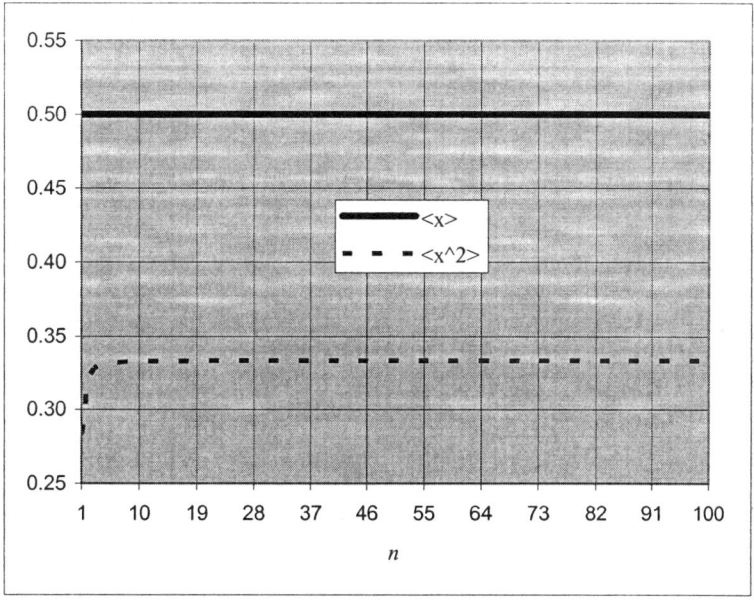

$$\boxed{\text{As } n \to \infty, \langle x^2 \rangle \to \frac{L^2}{3}}$$

General Problems

*67 • A light beam of wavelength 400 nm has an intensity of 100 W/m².
(*a*) What is the energy of each photon in the beam? (*b*) How much energy strikes an area of 1 cm² perpendicular to the beam in 1 s? (*c*) How many photons strike this area in 1 s?

Picture the Problem We can use the Einstein equation for photon energy to find the energy of each photon in the beam. The intensity of the energy incident on the surface is the ratio of the power delivered by the beam to its delivery time. Hence, we can express the energy incident on the surface in terms of the intensity of the beam.

(*a*) Use the Einstein equation for photon energy to express the energy of each photon in the beam:

$$E_{\text{photon}} = hf = \frac{hc}{\lambda}$$

Substitute numerical values and evaluate E_{photon}:

$$E_{\text{photon}} = \frac{1240\,\text{eV}\cdot\text{nm}}{400\,\text{nm}} = \boxed{3.10\,\text{eV}}$$

(*b*) Relate the energy incident on a surface of area A to the intensity of the beam:

$$E = IA\Delta t$$

Substitute numerical values and evaluate E:

$$E = (100 \text{ W/m}^2)(10^{-4} \text{ m}^2)(1\text{s})$$
$$= 0.01 \text{ J} \times \frac{1 \text{ eV}}{1.60 \times 10^{-19} \text{ J}}$$
$$= \boxed{6.25 \times 10^{16} \text{ eV}}$$

(c) Express the number of photons striking this area in 1 s as the ratio of the total energy incident on the surface to the energy delivered by each photon:

$$N = \frac{E}{E_{photon}} = \frac{6.25 \times 10^{16} \text{ ev}}{3.10 \text{ eV}}$$
$$= \boxed{2.02 \times 10^{16}}$$

*73 •• Suppose that a 100-W source radiates light of wavelength 600 nm uniformly in all directions and that the eye can detect this light if only 20 photons per second enter a dark-adapted eye with a pupil 7 mm in diameter. How far from the source can the light be detected under these rather extreme conditions?

Picture the Problem We can relate the fraction of the photons entering the eye to ratio of the area of the pupil to the area of a sphere of radius R. We can find the number of photons emitted by the source from the rate at which it emits and the energy of each photon which we can find using the Einstein equation.

Letting r be the radius of the pupil, $N_{entering\ eye}$ the number of photons per second entering the eye, and $N_{emitted}$ the number of photons emitted by the source per second, express the fraction of the light energy entering the eye at a distance R from the source:

$$\frac{N_{entering\ eye}}{N_{emitted}} = \frac{A_{eye}}{4\pi R^2}$$
$$= \frac{\pi r^2}{4\pi R^2}$$
$$= \frac{r^2}{4R^2}$$

Solve for R to obtain:

$$R = \frac{r}{2}\sqrt{\frac{N_{emitted}}{N_{entering\ eye}}} \qquad (1)$$

Find the number of photons emitted by the source per second:

$$N_{emitted} = \frac{P}{E_{photon}}$$

Using the Einstein equation, express the energy of the photons:

$$E_{photon} = \frac{hc}{\lambda}$$

Wave-Particle Duality and Quantum Physics 351

Substitute numerical values and evaluate E_{photon}:

$$E_{photon} = \frac{1240\,eV\cdot nm}{600\,nm} = 2.07\,eV$$

Substitute and evaluate $N_{emitted}$:

$$N_{emitted} = \frac{100\,W}{(2.07\,eV)(1.60\times 10^{-19}\,J/eV)}$$
$$= 3.02\times 10^{20}\,s^{-1}$$

Substitute for $N_{emitted}$ in equation (1) and evaluate R:

$$R = \frac{3.5\,mm}{2}\sqrt{\frac{3.02\times 10^{20}\,s^{-1}}{20\,s^{-1}}}$$
$$= \boxed{6.80\times 10^3\,km}$$

***79 ••** The Pauli exclusion principle states that no more than one electron may occupy a particular quantum state at a time. Therefore, if we wish to model an atom as a collection of electrons trapped in a one-dimensional box, each electron in the box must have a unique value of the quantum number n. Calculate the energy that the most energetic electron would have for the uranium atom with atomic number 92, assuming the box has a width of 0.05 nm. How does this energy compare to the rest-mass energy of the electron?

Picture the Problem We can use the expression for the energy of a particle in a well to find the energy of the most energetic electron in the uranium atom.

Relate the energy of an electron in the uranium atom to its quantum number n:

$$E_n = n^2\left(\frac{h^2}{8mL^2}\right)$$

Substitute numerical values and evaluate E_{92}:

$$E_{92} = (92)^2\left[\frac{(6.63\times 10^{-34}\,J\cdot s)^2}{8(9.11\times 10^{-31}\,kg)(0.05\,nm)^2}\times \frac{1\,eV}{1.6\times 10^{-19}\,J}\right] = \boxed{1.28\,MeV}$$

The rest energy of an electron is:

$$m_ec^2 = (9.11\times 10^{-31}\,kg)(3\times 10^8\,m/s)^2\left(\frac{1\,eV}{1.6\times 10^{-19}\,J}\right) = 0.512\,MeV$$

Express the ratio of E_{92} to m_ec^2:

$$\frac{E_{92}}{m_ec^2} = \frac{1.28\,MeV}{0.512\,MeV} = 2.50$$

352   Chapter 34

> The energy of the most energetic electron is approximately 2.5 times the rest-mass energy of an electron.

*83 •• (a) Show that for large n, the fractional difference in energy between state n and state $n + 1$ for a particle in a one-dimensional box is given approximately by

$$(E_{n+1} - E_n)/E_n \approx 2/n$$

(b) What is the approximate percentage energy difference between the states $n_1 = 1000$ and $n_2 = 1001$? (c) Comment on how this result is related to Bohr's correspondence principle.

Picture the Problem We can use the fact that the energy of the nth state is related to the energy of the ground state according to $E_n = n^2 E_1$ to express the fractional change in energy in terms of n and then examine this ratio as n grows without bound.

(a) Express the ratio $(E_{n+1} - E_n)/E_n$:

$$\frac{E_{n+1} - E_n}{E_n} = \frac{(n+1)^2 - n^2}{n^2} = \frac{2n+1}{n^2}$$

$$= \frac{2}{n} + \frac{1}{n^2} \approx \boxed{\frac{2}{n}}$$

for $n \gg 1$.

(b) Evaluate $\dfrac{E_{1001} - E_{1000}}{E_{1000}}$:

$$\frac{E_{1001} - E_{1000}}{E_{1000}} \approx \frac{2}{1000} = \boxed{0.2\%}$$

(c) > Classically, the energy is continuous. For very large values of n, the energy difference between adjacent levels is infinitesimal.

Chapter 35
Applications of the Schrödinger Equation

Conceptual Problems

*4 • The Schrödinger Equation could be applied equally well to baseballs as to electrons yet we would never analyze the motion of a baseball with a wave function. Explain why this is the case by estimating the quantum mechanically predicted lowest energy level of a baseball trapped inside a locker. You can treat the locker as if it were a one-dimensional infinite potential well. What value of the quantum number n would you need for a ball rolling around in the locker, after you toss it in, so that the kinetic energy is approximately equal to the quantum mechanically calculated energy?

Picture the Problem Assume a mass of 150 g for the baseball, 30 cm for the width of the locker, and 1 cm/s for the speed of the ball, and equate the kinetic energy of the ball and the quantum-mechanical energy and solve for the quantum number n.

The allowed energy states of a particle of mass m in a 1-dimensional infinite potential well of width L are given by:
$$E_n = n^2 \left(\frac{h^2}{8mL^2} \right)$$

The kinetic energy of the ball is:
$$K = \frac{1}{2}mv^2$$

For $E_n = K$:
$$n^2 \left(\frac{h^2}{8mL^2} \right) = \frac{1}{2}mv^2$$

Solve for the quantum number n:
$$n = \frac{2mvL}{h}$$

Substitute numerical values and evaluate n:
$$n = \frac{2(0.15\,\text{kg})(0.01\,\text{m/s})(0.3\,\text{m})}{6.63 \times 10^{-34}\,\text{J}\cdot\text{s}}$$
$$= 1.36 \times 10^{30} \approx \boxed{10^{30}}$$

The Harmonic Oscillator

*8 •• Use the procedure of Example 35-1 to verify that the energy of the first excited state of the harmonic oscillator is $E_1 = \tfrac{3}{2}\hbar\omega_0$. (*Note*: Rather than solve for a again, use the result $a = m\omega_0/(2\hbar)$ obtained in Example 35-1.)

353

354   Chapter 35

Picture the Problem We can differentiate $\psi(x)$ twice and substitute in the Schrödinger equation for the harmonic oscillator. Substitution of the given value for a will lead us to an expression for E_1.

The wave function for the first excited state of the harmonic oscillator is:
$$\psi_1(x) = A_1 x e^{-ax^2}$$

Compute $d\psi_1(x)/dx$:
$$\frac{d\psi_1(x)}{dx} = \frac{d}{dx}\left[A_1 x e^{-ax^2}\right] = A_1 e^{-ax^2}$$

Compute $d^2\psi_1(x)/dx^2$:

$$\frac{d^2\psi_1(x)}{dx^2} = \frac{d}{dx}\left[A_1 e^{-ax^2}\right] = -2axA_1 e^{-ax^2} - 4axA_1 e^{-ax^2} + 4a^2 x^3 A_1 e^{-ax^2}$$
$$= \left(4a^2 x^3 - 6ax\right)A_1 e^{-ax^2}$$

Substitute in the Schrödinger equation:

$$-\frac{\hbar^2}{2m}\left[\left(4a^2 x^3 - 6ax\right)A_1 e^{-ax^2}\right] + \tfrac{1}{2}m\omega_0^2 x^2 A_1 x e^{-ax^2} = E_1 A_1 x e^{-ax^2}$$

Divide out $A_1 e^{-ax^2}$ to obtain:

$$-\frac{\hbar^2}{2m}\left[\left(4a^2 x^3 - 6ax\right)\right] + \tfrac{1}{2}m\omega_0^2 x^3 = E_1 x$$

or

$$-\frac{\hbar^2}{2m}\left(4a^2 x^3\right) + \frac{\hbar^2}{2m}(6ax) + \tfrac{1}{2}m\omega_0^2 x^3 = E_1 x$$

Substitute for a to obtain:

$$-\frac{\hbar^2}{2m}4\left(\frac{m\omega_0}{2\hbar}\right)^2 x^3 + \frac{\hbar^2}{2m}6\left(\frac{m\omega_0}{2\hbar}\right)x + \tfrac{1}{2}m\omega_0^2 x^3 = E_1 x$$

Solve for E_1 to obtain:
$$E_1 = \boxed{\tfrac{3}{2}\hbar\omega_0} = 3E_0$$

Reflection and Transmission of Electron Waves: Barrier Penetration

*14 ••   A particle of mass m with wave number k_1 is traveling to the right along the negative x axis. The potential energy of the particle is equal to zero everywhere on the negative x axis, and is equal to U_0 everywhere on the positive x axis, $U_0 > 0$. (a) Show that if the total energy is $E = \alpha U_0$, where $\alpha \geq 1$, wave number k_2 in the region $x > 0$ is given by

$$k_2 = \sqrt{\frac{\alpha-1}{\alpha}} k_1$$

(b) Using a spreadsheet program or graphing calculator, graph the reflection coefficient R and transmission coefficient T for $1 \leq \alpha \leq 5$.

Picture the Problem We can use the total energy of the particle in the region $x > 0$ to express k_2 in terms of α and k_1. Knowing k_2 in terms of k_1, we can use $R = \dfrac{(k_1 - k_2)^2}{(k_1 + k_2)^2}$ to find R and $T = 1 - R$ to determine the transmission coefficient T.

(a) Using conservation of energy, express the energy of the particle in the region $x > 0$:

$$\frac{\hbar^2 k_2^2}{2m} + U_0 = \alpha U_0$$

Solve for k_2:

$$k_2 = \frac{\sqrt{2mU_0(\alpha-1)}}{\hbar}$$

From the equation for the total energy of the particle:

$$k_1 = \frac{\sqrt{2m\alpha U_0}}{\hbar}$$

Express the ratio of k_2 to k_1:

$$\frac{k_2}{k_1} = \frac{\dfrac{\sqrt{2mU_0(\alpha-1)}}{\hbar}}{\dfrac{\sqrt{2m\alpha U_0}}{\hbar}} = \sqrt{\frac{\alpha-1}{\alpha}}$$

and $\boxed{k_2 = \sqrt{\dfrac{\alpha-1}{\alpha}} k_1}$

(b) The reflection coefficient R is given by:

$$R = \frac{(k_1 - k_2)^2}{(k_1 + k_2)^2}$$

356   Chapter 35

Factor k_1 from the numerator and denominator to obtain:

$$R = \frac{\left(1 - \frac{k_2}{k_1}\right)^2}{\left(1 + \frac{k_2}{k_1}\right)^2}$$

Substitute our result from (*a*) for k_2/k_1:

$$R = \frac{\left(1 - \sqrt{\frac{\alpha-1}{\alpha}}\right)^2}{\left(1 + \sqrt{\frac{\alpha-1}{\alpha}}\right)^2} = \left(\frac{1 - \sqrt{\frac{\alpha-1}{\alpha}}}{1 + \sqrt{\frac{\alpha-1}{\alpha}}}\right)^2$$

The transmission coefficient is given by:

$$T = 1 - R = 1 - \left(\frac{1 - \sqrt{\frac{\alpha-1}{\alpha}}}{1 + \sqrt{\frac{\alpha-1}{\alpha}}}\right)^2$$

A spreadsheet program to plot R and T as functions of α is shown below. The formulas used to calculate the quantities in the columns are as follows:

Cell	Content/Formula	Algebraic Form
A2	1.0	α
B2	(1−SQRT((A2−1)/A2))/ (1+SQRT((A2−1)/A2))^2	$\left(\dfrac{1 - \sqrt{\frac{\alpha-1}{\alpha}}}{1 + \sqrt{\frac{\alpha-1}{\alpha}}}\right)^2$
C2	1−B2	$1 - \left(\dfrac{1 - \sqrt{\frac{\alpha-1}{\alpha}}}{1 + \sqrt{\frac{\alpha-1}{\alpha}}}\right)^2$

	A	B	C
1	alpha	R	T
2	1.0	1.000	0.000
3	1.2	0.298	0.702
4	1.4	0.198	0.802
5	1.6	0.149	0.851
18	4.2	0.036	0.964
19	4.4	0.034	0.966
20	4.6	0.032	0.968
21	4.8	0.031	0.969

| 22 | 5.0 | 0.029 | 0.971 |

The following graph was plotted using the data in the above table:

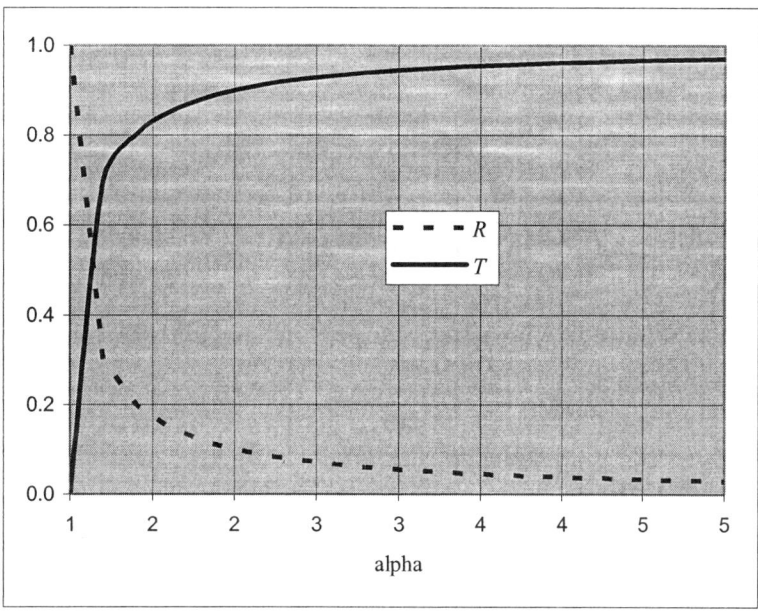

*18   ••   A 10-eV electron is incident on a potential barrier of height 25 eV and width of 1 nm. (a) Use Equation 35-29 to calculate the order of magnitude of the probability that the electron will tunnel through the barrier. (b) Repeat your calculation for a width of 0.1 nm.

Picture the Problem The probability that the electron with a given energy will tunnel through the given barrier is given by Equation 35-29.

(a) Equation 35-29 is:
$$T = e^{-2\alpha a}$$
where
$$\alpha = \sqrt{\frac{2m(U_0 - E)}{\hbar^2}} = \frac{\sqrt{2m(U_0 - E)}}{\hbar}$$

Multiply the numerator and denominator of α by c to obtain:
$$\alpha = \frac{\sqrt{2mc^2(U_0 - E)}}{\hbar c}$$
where
$$\hbar c = 1.974 \times 10^{-13} \text{ MeV} \cdot \text{m}$$

Using $m_e c^2 = 511 \text{ keV}$, evaluate T:

358   Chapter 35

$$T = \exp\left\{-2(10^{-9}\text{ m})\frac{\sqrt{2(511\text{keV})(25\text{ eV} - 10\text{ eV})}}{1.974 \times 10^{-13}\text{ MeV} \cdot \text{m}}\right\} = \boxed{5.91 \times 10^{-18}}$$

(b) Repeat with $a = 10^{-10}$ m:

$$T = \exp\left\{-2(10^{-10}\text{ m})\frac{\sqrt{2(511\text{keV})(25\text{ eV} - 10\text{ eV})}}{1.974 \times 10^{-13}\text{ MeV} \cdot \text{m}}\right\} = \boxed{1.89 \times 10^{-2}}$$

The Schrödinger Equation in Three Dimensions

*23 • Give the wave functions for the lowest ten quantum states of the particle in Problem 22.

Picture the Problem The wave functions are of the form

$$\psi = A \sin\left(\frac{n_1 \pi}{L_1} x\right) \sin\left(\frac{n_2 \pi}{2L_1} y\right) \sin\left(\frac{n_3 \pi}{4L_1} z\right)$$

The Schrödinger Equation for Two Identical Particles

*28 • What is the ground-state energy of ten noninteracting fermions, such as neutrons, in a one-dimensional box of length L? (Because the quantum number associated with spin can have two values, each spatial state can hold two neutrons.)

Picture the Problem For fermions, such as neutrons for which the spin quantum number is ½, two particles can occupy the same spatial state.

The lowest total energy for the 10 fermions is:

$$E = 2E_1(1^2 + 2^2 + 3^2 + 4^2 + 5^2)$$
$$= 2\left(\frac{h^2}{8mL^2}\right)(55) = \boxed{\frac{55h^2}{4mL^2}}$$

General Problems

*33 •• Eight identical noninteracting fermions (e.g., neutrons) are confined to a two-dimensional square box of side length L. Determine the energies of the three lowest states. (See Problem 26.)

Picture the Problem We can determine the energies of the state by identifying the four lowest quantum states that are occupied in the ground state and computing their combined energies. We can then find the energy difference between the ground state and

Applications of the Schrödinger Equation 359

the first excited state and use this information to find the energy of the excited state.

Each n, m state can accommodate only 2 particles. Therefore, in the ground state of the system of 8 fermions, the four lowest quantum states are occupied. These are:

(1,1), (1,2), (2,1) and (2,2)

Note that the states (1,2) and (2,1) are distinctly different states because the x and y directions are distinguishable.

The energies are quantized to the values given by:

$$E_{n_1,n_2} = 2\left(\frac{h^2}{8mL^2}\right)\left(n_1^2 + n_2^2\right)$$

The energy of the ground state is the sum of the energies of the four lowest quantum states:

$$E_0 = E_{1,1} + E_{1,2} + E_{2,1} + E_{2,2}$$

$$= 2\left(\frac{h^2}{8mL^2}\right)\left(1^2 + 1^2\right) + 2\left(\frac{h^2}{8mL^2}\right)\left(1^2 + 2^2\right) + 2\left(\frac{h^2}{8mL^2}\right)\left(2^2 + 1^2\right) + 2\left(\frac{h^2}{8mL^2}\right)\left(2^2 + 2^2\right)$$

$$= 2\left(\frac{h^2}{8mL^2}\right)(2 + 5 + 5 + 8)$$

$$= \frac{5h^2}{mL^2}$$

The next higher state is achieved by taking one fermion from the (2,2) state and raising it to the next higher unoccupied state. That state is the (1,3) state. The energy difference between the ground state and this state is:

$$\Delta E = E_{1,3} - E_{2,2}$$

$$= \frac{h^2}{8mL^2}\left(1^2 + 3^2\right) - \frac{h^2}{8mL^2}\left(2^2 + 2^2\right)$$

$$= \frac{h^2}{8mL^2}(10 - 8) = \frac{h^2}{4mL^2}$$

Hence, the energies of the degenerate states (1,3) and (3,1) are:

$$E_{1,3} = E_{3,1} = E_0 + \Delta E$$

$$= \frac{5h^2}{mL^2} + \frac{h^2}{4mL^2} = \frac{21h^2}{4mL^2}$$

The three lowest energy levels are therefore:

$$E_0 = \boxed{\frac{5h^2}{mL^2}}$$

and two states of energy

$$E_1 = E_2 = \boxed{\frac{21h^2}{4mL^2}}$$

360 Chapter 35

*41 ••• In this problem you will derive the ground-state energy of the harmonic oscillator using the precise form of the uncertainty principle, $\Delta x \Delta p \geq \hbar/2$, where Δx and Δp are defined to be the standard deviations $(\Delta x)^2 = [(x - x_{av})^2]_{av}$ and $(\Delta p)^2 = [(p - p_{av})^2]_{av}$ (see Equation 17-35a). Proceed as follows:

1. Write the total classical energy in terms of the position x and momentum p using $U(x) = m\omega_0^2 x^2$ and $K = p^2/2m$.
2. Use the result of Equation 17-35 to write $(\Delta x)^2 = [(x - x_{av})^2]_{av} = (x^2)_{av} - x_{av}^2$ and $(\Delta p)^2 = [(p - p_{av})^2]_{av} = (p^2)_{av} - p_{av}^2$.
3. Use the symmetry of the potential energy function to argue that x_{av} and p_{av} must be zero, so that $(\Delta x)^2 = (x)_{av}$ and $(\Delta p)^2 = (p^2)_{av}$.
4. Assume that $\Delta p = \hbar/(2\Delta x)$ to eliminate $(p^2)_{av}$ from the average energy $E_{av} = (p^2)_{av}/(2m) + \frac{1}{2}m\omega^2(x^2)_{av}$ and write E_{av} as $E_{av} = \hbar^2/(8mZ) + \frac{1}{2}m\omega^2 Z$, where $Z = (x^2)_{av}$.
5. Set $dE/dZ = 0$ to find the value of Z for which E is a minimum.
6. Show that the minimum energy is given by $(E_{av})_{min} = +\frac{1}{2}\hbar\omega_0$.

Picture the Problem We can follow the step-by-step procedure outlined in the problem statement to show that $(E_{av})_{min} = +\frac{1}{2}\hbar\omega$.

1. The total classical energy is:

$$E_{av} = U_{av} + K_{av}$$
$$= \frac{1}{2}m\omega^2(x^2)_{av} + \frac{(p^2)_{av}}{2m} \quad (1)$$

2. Express the standard deviation of Δp:

$$(\Delta p)^2 = [(p - p_{av})^2]_{av}$$
$$= [p^2 - 2pp_{av} - p_{av}^2]_{av}$$

Because $p_{av} = 0$:

$$(\Delta p)^2 = (p^2)_{av}$$

3. Express the standard deviation of Δx:

$$(\Delta x)^2 = [(x - x_{av})^2]_{av}$$
$$= [x^2 - 2xx_{av} - x_{av}^2]_{av}$$

Because $x_{av} = 0$:

$$(\Delta x)^2 = (x^2)_{av}$$

Applications of the Schrödinger Equation

4. Use the uncertainty principle $\Delta p = \hbar/2\Delta x$ to eliminate $(p^2)_{av}$ from the average energy in equation (1):

$$E_{av} = \tfrac{1}{2}m\omega^2 (x^2)_{av} + \frac{(\Delta p^2)}{2m}$$

$$= \tfrac{1}{2}m\omega^2 (x^2)_{av} + \frac{1}{2m}\left[\frac{\hbar^2}{4(\Delta x)^2}\right]$$

$$= \tfrac{1}{2}m\omega^2 (x^2)_{av} + \frac{\hbar^2}{8m(x^2)_{av}}$$

Let $Z = (x^2)_{av}$ to obtain:

$$E_{av} = \tfrac{1}{2}m\omega^2 Z + \frac{\hbar^2}{8mZ}$$

5. Differentiate E_{av} with respect to Z and set this derivative equal to zero:

$$\frac{dE_{av}}{dZ} = \frac{d}{dZ}\left[\tfrac{1}{2}m\omega^2 Z + \frac{\hbar^2}{8mZ}\right]$$

$$= \tfrac{1}{2}m\omega^2 - \frac{\hbar^2}{8mZ^2} = 0 \text{ for extrema}$$

Solve for Z to find the value of Z that minimizes E_{av} (see the remark below):

$$Z = \frac{\hbar}{2m\omega}$$

6. Evaluate E_{av} when $Z = \hbar/2m\omega$:

$$(E_{av})_{min} = \tfrac{1}{2}m\omega^2 \left(\frac{\hbar}{2m\omega}\right) + \frac{\hbar^2}{8m}\left(\frac{2m\omega}{\hbar}\right)$$

$$= \boxed{\tfrac{1}{2}\hbar\omega}$$

Remarks: All we've shown is that $Z = \hbar/2m\omega$ is an extreme value, i.e., either a *maximum* or a *minimum*. To show that $Z = \hbar/2m\omega$ minimizes E_{av}, we must either 1) show that the second derivative of E_{av} with respect to Z evaluated at $Z = \hbar/2m\omega$ is positive, or 2) confirm that the graph of E_{av} as a function of Z opens upward at $Z = \hbar/2m\omega$.

Chapter 36
Atoms

Conceptual Problems

*1 • As n increases, does the spacing of adjacent energy levels increase or decrease?

Determine the Concept Examination of Figure 35-4 indicates that as n increases, the spacing of adjacent energy levels decreases.

*7 • For the principal quantum number $n = 4$, how many different values can the orbital quantum number ℓ have? (a) 4, (b) 3, (c) 7, (d) 16, or (e) 6.

Determine the Concept We can find the possible values of ℓ by using the constraints on the quantum numbers n and ℓ.

The allowed values for the orbital quantum number ℓ for $n = 1, 2, 3,$ and 4 are summarized in table shown to the right:

n	ℓ
1	0
2	0, 1
3	0, 1, 2
4	0, 1, 2, 3

From the table it is clear that ℓ can have 4 values.

(a) is correct.

*10 •• Why is the energy of the 3s state considerably lower than the energy of the 3p state for sodium, whereas in hydrogen these states have essentially the same energy?

Determine the Concept The s state, with $\ell = 0$, is a "penetrating" state in which the probability density near the nucleus is significant. Consequently, the 3s electron in sodium is in a region of low potential energy for a significant portion of the time. In the state $\ell = 1$, the probability density at the nucleus is zero, so the 2p electron of sodium is shielded from the nuclear charge by the 1s electrons. In hydrogen, the 3s and 2p electrons experience the same nuclear potential.

*14 • For the principal quantum number $n = 3$, what are the possible values of the quantum numbers ℓ and m_ℓ?

Picture the Problem We can apply the constraints on the quantum numbers ℓ and m_ℓ to find the possible values for each when $n = 3$.

Express the constraints on the quantum numbers n, ℓ, and m_ℓ:

$n = 1, 2, 3, ...,$
$\ell = 0, 1, 2, ..., n-1,$
and

364 Chapter 36

$$m_\ell = -\ell, -\ell+1, ..., \ell$$

So, for $n = 3$, the constraints on ℓ limit it to the values:

$$\ell = \boxed{0, 1, \text{and } 2.}$$

m_ℓ can take on the values:

$$m_\ell = \boxed{-2, -1, 0, 1, 2}$$

*18 •• The Ritz combination principle states that for any atom, one can find different spectral lines λ_1, λ_2, λ_3, and λ_4, so that $1/\lambda_1 + 1/\lambda_2 = 1/\lambda_3 + 1/\lambda_4$. Show why this is true using an energy-level diagram.

Determine the Concept The Ritz combination principle is due to the quantization of energy levels in the atom. We can use the relationship between the wavelength of the emitted photon and the difference in energy levels within the atom that results in the emission of the photon to express each of the wavelengths and then the sum of the reciprocals of the first and second wavelengths and the sum of the reciprocals of the third and fourth wavelengths.

Express the wavelengths of the spectral lines λ_1, λ_2, λ_3, and λ_4 in terms of the corresponding energy transitions:

$$\lambda_1 = \frac{hc}{E_3 - E_2}$$

$$\lambda_2 = \frac{hc}{E_2 - E_0}$$

$$\lambda_3 = \frac{hc}{E_3 - E_1}$$

and

$$\lambda_4 = \frac{hc}{E_1 - E_0}$$

Add the reciprocals of λ_1 and λ_2 to obtain:

$$\frac{1}{\lambda_1} + \frac{1}{\lambda_2} = \frac{E_3 - E_2}{hc} + \frac{E_2 - E_0}{hc}$$

$$= \frac{E_3 - E_0}{hc} \quad (1)$$

Add the reciprocals of λ_3 and λ_4 to obtain:

$$\frac{1}{\lambda_3} + \frac{1}{\lambda_4} = \frac{E_3 - E_1}{hc} + \frac{E_1 - E_0}{hc}$$

$$= \frac{E_3 - E_0}{hc} \quad (2)$$

Because the right-hand sides of equations (1) and (2) are equal:

$$\boxed{\frac{1}{\lambda_1} + \frac{1}{\lambda_2} = \frac{1}{\lambda_3} + \frac{1}{\lambda_4}}$$

One possible set of energy levels is shown to the right:

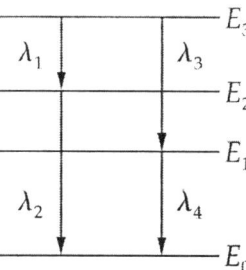

Estimation and Approximation

*20 •• In laser cooling and trapping, atoms in a beam traveling in one direction are slowed by interaction with an intense laser beam in the opposite direction. The photons scatter off the atoms via resonance absorption, a process by which the incident photon is absorbed by the atom, and a short time later a photon of equal energy is emitted in a random direction. The net result of a single such scattering event is a transfer of momentum to the atom in a direction opposite to the motion of the atom, followed by a second transfer of momentum to the atom in a random direction. Thus, during photon absorption the atom loses speed, but during photon emission the change in speed of the atom is, on average, zero (because the directions of the emitted photons are random). An analogy often made to this process is that of slowing down a bowling ball by bouncing ping-pong balls off of it. (a) Given a typical photon energy used in these experiments of about 1 eV and a momentum typical for an atom with a thermal speed appropriate to a temperature of about 500 K (a typical temperature for an atomic beam), estimate the number of photon-atom collisions that are required to bring an atom to rest. (The average kinetic energy of an atom is equal to $\frac{3}{2}kT$, where k is the Boltzmann constant. Use this to estimate the speed of the atoms.) (b) Compare this with the number of ping-pong ball-bowling ball collisions that are required to bring the bowling ball to rest. (Assume the speeds of the incident ping-pong balls are all equal to the initial speed of the bowling ball.) (c) ^{85}Rb is a type of atom often used in cooling experiments. The wavelength of the light resonant with the cooling transition is $\lambda = 780.24$ nm. Estimate the number of photons needed to slow down an ^{85}Rb atom from a typical thermal velocity of 300 m/s to a stop.

Picture the Problem The number of photons need to stop a ^{85}Rb atom traveling at 300 m/s is the ratio of its momentum to that of a typical photon.

(a) The number N of photon-atom collisions needed to bring an atom to rest is the ratio of the change in the momentum of the atom as it stops to the momentum brought to the collision by each photon:

$$N = \frac{\Delta p_{atom}}{p_{photon}} = \frac{mv}{\frac{E}{c}} = \frac{mvc}{E}$$

where m is the mass of the atom.

The kinetic energy of an atom whose temperature is T is:

$$\tfrac{1}{2}mv^2 = \tfrac{3}{2}kT \Rightarrow v = \sqrt{\frac{3kT}{m}}$$

Substitute for v to obtain:

$$N = \frac{mc}{E}\sqrt{\frac{3kT}{m}} = \frac{c}{E}\sqrt{3mkT}$$

366 Chapter 36

For an atom use mass is 50 u:

$$N = \frac{3\times 10^8 \text{ m/s}}{1\text{eV}\times \dfrac{1.6\times 10^{-19}\text{ J}}{\text{eV}}}\sqrt{3\left(50\text{u}\times \frac{1.66\times 10^{-27}\text{ kg}}{\text{u}}\right)(1.38\times 10^{-23}\text{ J/K})(500\text{ K})} \approx \boxed{10^5}$$

(b) The number N of ping-pong ball-bowling ball collisions needed to bring the bowling ball to rest is the ratio of the change in the momentum of the bowling ball as it stops to the momentum brought to the collision by each ping-pong ball:

$$N = \frac{\Delta p_{\text{bowling ball}}}{p_{\text{ping-pong ball}}} = \frac{m_{\text{bb}}v_{\text{bb}}}{m_{\text{ppb}}v_{\text{ppb}}}$$

Provided the speeds of the approaching bowling ball and ping-pong ball are approximately the same:

$$N = \frac{\Delta p_{\text{bowling ball}}}{p_{\text{ping-pong ball}}} \approx \frac{m_{\text{bb}}}{m_{\text{ppb}}} \approx \frac{6\text{ kg}}{4\text{ g}} \approx \boxed{10^3}$$

(c) The number of photons N needed to stop a ^{85}Rb atom is the ratio of the change in the momentum of the atom to the momentum brought to the collision by each photon:

$$N = \frac{\Delta p_{\text{atom}}}{p_{\text{photon}}} = \frac{mv}{\dfrac{h}{\lambda}} = \frac{mv\lambda}{h}$$

Substitute numerical values and evaluate N:

$$N = \frac{85(1.66\times 10^{-27}\text{ kg})(300\text{ m/s})(780.24\text{ nm})}{6.63\times 10^{-34}\text{ J}\cdot\text{s}} = \boxed{4.98\times 10^4}$$

The Bohr Model of the Hydrogen Atom

*26 ••   Repeat Problem 25 for the Brackett series, $n_2 = 4$.

Picture the Problem We can use Bohr's second postulate to relate the photon energy to its frequency and use $\lambda = \dfrac{1240\text{ eV}\cdot\text{nm}}{E_i - E_f}$ to find the wavelengths of the three longest wavelengths in the Brackett series.

(a) Use Bohr's second postulate to express the energy of the photons in the Paschen series:

$$hf = \Delta E = E_i - E_f$$

For the series limit:

$$n = \infty \text{ and } E_i = 0$$

Substitute to obtain:

$$\Delta E = -E_f = -\left(-\frac{E_0}{n_2^2}\right) = \frac{E_0}{n_2^2} \quad (1)$$

Evaluate the photon energy for $n_2 = 4$:

$$hf - \frac{13.6\,\text{eV}}{4^2} = \boxed{0.850\,\text{eV}}$$

Express the wavelength of the radiation resulting from an energy transition $\Delta E = hf$:

$$\lambda = \frac{1240\,\text{eV}\cdot\text{nm}}{\Delta E} \quad (2)$$

provided the energies are expressed in eV.

Evaluate λ_{min} for the transition $n = \infty$ to $n_2 = 4$:

$$\lambda_{min} = \frac{1240\,\text{eV}\cdot\text{nm}}{0.850\,\text{eV}} = \boxed{1459\,\text{nm}}$$

(b) For the three longest wavelengths:

$$n_i = 5, 6, \text{and } 7$$

Equation (1) becomes:

$$\Delta E = E_i - E_f = -\frac{E_0}{n_i^2} - \left(-\frac{E_0}{n_2^2}\right)$$
$$= E_0\left(\frac{1}{n_2^2} - \frac{1}{n_i^2}\right) = E_0\left(\frac{1}{16} - \frac{1}{n_i^2}\right) \quad (3)$$

Evaluate equation (3) for $n = 5$:

$$\Delta E_{5\to 4} = (13.6\,\text{eV})\left(\frac{1}{16} - \frac{1}{25}\right)$$
$$= \boxed{0.306\,\text{eV}}$$

Evaluate equation (2) for $\Delta E = 0.306$ eV:

$$\lambda_{5\to 4} = \frac{1240\,\text{eV}\cdot\text{nm}}{0.306\,\text{eV}} = \boxed{4052\,\text{nm}}$$

Evaluate equation (3) for $n = 6$:

$$\Delta E_{6\to 4} = (13.6\,\text{eV})\left(\frac{1}{16} - \frac{1}{36}\right)$$
$$= \boxed{0.472\,\text{eV}}$$

Evaluate equation (2) for $\Delta E = 0.472$ eV:

$$\lambda_{6\to 4} = \frac{1240\,\text{eV}\cdot\text{nm}}{0.472\,\text{eV}} = \boxed{2627\,\text{nm}}$$

Evaluate equation (3) for $n = 7$:

$$\Delta E_{7\to 4} = (13.6\,\text{eV})\left(\frac{1}{16} - \frac{1}{49}\right)$$
$$= \boxed{0.572\,\text{eV}}$$

Evaluate equation (2) for
$\Delta E = 0.572$ eV:

$$\lambda_{7 \to 4} = \frac{1240 \text{ eV} \cdot \text{nm}}{0.572 \text{ eV}} = \boxed{2168 \text{ nm}}$$

The positions of these lines on a horizontal linear scale are shown below with the wavelengths and transitions indicated.

```
        7→4  6→4                              5→4
        ---|---------|-----------------------------|-----
       2168 nm 2627 nm                          4052 nm
```

*29 ••• In the center-of-mass reference frame of a hydrogen atom, the electron and nucleus have equal and opposite momenta of magnitude p. (a) Show that the total kinetic energy of the electron and nucleus can be written $K = p^2/(2\mu)$ where $\mu = m_e M/(M + m_e)$ is called the reduced mass, m_e is the mass of the electron, and M is the mass of the nucleus. (b) For the equations for the Bohr model of the atom, the motion of the nucleus can be take into account by replacing the mass of the electron with the reduced mass. Use Equation 36-14 to calculate the Rydberg constant for a hydrogen atom with a nucleus of mass $M = m_p$. Find the approximate value of the Rydberg constant by letting M go to infinity in the reduced mass formula. To how many figures does this approximate value agree with the actual value. (c) Find the percentage correction for the ground-state energy of the hydrogen atom by using the reduced mass in Equation 36-16. *Remark: In general, the reduced mass for a two-body problem with masses m_1 and m_2 is given by*

$$\mu = \frac{m_1 m_2}{m_1 + m_2}$$

Picture the Problem We can express the total kinetic energy of the electron-nucleus system as the sum of the kinetic energies of the electron and the nucleus. Rewriting these kinetic energies in terms of the momenta of the electron and nucleus will lead to $K = p^2/2m_r$.

(a) Express the total kinetic energy of the electron-nucleus system:

$$K = K_e + K_n$$

Express the kinetic energies of the electron and the nucleus in terms of their momenta:

$$K_e = \frac{p^2}{2m_e} \text{ and } K_n = \frac{p^2}{2M}$$

Substitute to obtain:

$$K = \frac{p^2}{2m_e} + \frac{p^2}{2M} = \frac{p^2}{2}\left(\frac{1}{m_e} + \frac{1}{M}\right)$$

$$= \frac{p^2}{2}\left(\frac{M+m_e}{m_e M}\right) = \frac{p^2}{2\left(\frac{m_e M}{M+m_e}\right)}$$

$$= \boxed{\frac{p^2}{2m_r}}$$

provided we define $\mu = m_e M/(M+m_e)$.

(b) From Equation 36-14 we have:

$$R = \frac{m_r k^2 e^4}{4\pi c \hbar^3} = C\left(\frac{m_e}{1+\frac{m_e}{M}}\right) \quad (1)$$

where

$$C = \frac{k^2 e^4}{4\pi c \hbar^3}$$

Use the Table of Physical Constants at the end of the text to obtain:

$$C = 1.204663 \times 10^{37} \text{ m}^{-1}/\text{kg}$$

For H:

$$R_H = C\left(\frac{m_e}{1+\frac{m_e}{m_p}}\right)$$

Substitute numerical values and evaluate R_H:

$$R_H = \left(1.204663 \times 10^{37} \text{ m}^{-1}/\text{kg}\right)\left(\frac{9.11 \times 10^{-31} \text{ kg}}{1+\frac{9.11 \times 10^{-31} \text{ kg}}{1.67 \times 10^{-27} \text{ kg}}}\right) = \boxed{1.096850 \times 10^7 \text{ m}^{-1}}$$

Let $M \to \infty$ in equation (1) to obtain $R_{H,\text{approx}}$:

$$R_{H,\text{approx}} = C m_e$$

Substitute numerical values and evaluate $R_{H,\text{approx}}$:

$$R_{H,\text{approx}} = \left(1.204663 \times 10^{37} \text{ m}^{-1}/\text{kg}\right)\left(9.11 \times 10^{-31} \text{ kg}\right) = \boxed{1.097448 \times 10^7 \text{ m}^{-1}}$$

370   Chapter 36

$\boxed{R_\text{H} \text{ and } R_\text{H,approx} \text{ agree to three significant figures.}}$

(c) Express the ratio of the kinetic energy K of the electron in its orbit about a stationary nucleus to the kinetic energy of the reduced-mass system K':

$$\frac{K}{K'} = \frac{\frac{p^2}{2m_\text{e}}}{\frac{p^2}{2m_\text{r}}} = \frac{\mu}{m_\text{e}} = \frac{1}{m_\text{e}}\left(\frac{m_\text{p}m_\text{e}}{m_\text{p}+m_\text{e}}\right)$$

$$= \frac{m_\text{p}}{m_\text{p}+m_\text{e}} = \frac{1}{1+\frac{m_\text{e}}{m_\text{p}}}$$

Substitute numerical values and evaluate the ratio of the kinetic energies:

$$\frac{K}{K'} = \frac{1}{1+\frac{9.11\times 10^{-31}\text{ kg}}{1.67\times 10^{-27}\text{ kg}}}$$

$$= 0.999455$$

or

$K = 0.999455 K'$

and the correction factor is the ratio of the masses or $\boxed{0.0545\%}$

Remarks: The correct energy is slightly less than that calculated neglecting the motion of the nucleus.

*30   ••   The Pickering series of the spectrum of He⁺ (singly-ionized helium) consists of spectral lines due to transitions to the $n = 4$ state of He⁺. Experimentally, every other line of the Pickering series is very close to a spectral line in the Balmer series for hydrogen transitions to $n = 2$. (a) Show that this is true. (b) Calculate the wavelength of a transition from the $n = 6$ level to the $n = 4$ level of He⁺, and show that it corresponds to one of the Balmer lines.

Picture the Problem We can use Equation 36-15 with $Z = 2$ to explain how it is that every other line of the Pickering series is very close to a line in the Balmer series. We can use the relationship between the energy difference between two quantum states and the wavelength of the photon emitted during a transition from the higher state to the lower state to find the wavelength of the photon corresponding to a transition from the $n = 6$ to the $n = 4$ level of He⁺.

(a) From Equation 36-15, the energy levels of an atom are given by:

$$E_n = -Z^2\frac{E_0}{n^2}$$

where E_0 is the Rydberg constant (13.6 eV).

For He⁺, $Z = 2$ and:

$$E_n = -4\frac{E_0}{n^2}$$

Atoms 371

Because of this, an energy level with even principal quantum number n in He^+ will have the same energy as a level with quantum number $n/2$ in H. Therefore, a transition between levels with principal quantum numbers $2m$ and $2p$ in He^+ will have almost the same energy as a transition between level m and p in H. In particular, transitions from $2m$ to $2p = 4$ in He^+ will have the same energy as transitions from m to $n = 2$ in H (the Balmer series).

(b) Transitions between these energy levels result in the emission or absorption of a photon whose wavelength is given by:

$$\lambda = \frac{hc}{E_6 - E_4} \quad (1)$$

Evaluate E_6 and E_4:

$$E_6 = -4\left(\frac{13.6\,\text{eV}}{6^2}\right) = -1.51\,\text{eV}$$

and

$$E_4 = -4\left(\frac{13.6\,\text{eV}}{4^2}\right) = -3.40\,\text{eV}$$

Substitute for E_6 and E_4 in equation (1) and evaluate λ:

$$\lambda = \frac{1240\,\text{eV}\cdot\text{nm}}{-1.51\,\text{eV} - (-3.40\,\text{eV})} = \boxed{656\,\text{nm}}$$

which is the same as the $n = 3$ to $n = 2$ transition in H.

Quantum Numbers in Spherical Coordinates

*35 •• Find the minimum value of the angle θ between $\vec{L}$ and the z axis for (a) $\ell = 1$, (b) $\ell = 4$, and (c) $\ell = 50$.

Picture the Problem The minimum angle between the z axis and $\vec{L}$ is the angle between the $\vec{L}$ vector for $m = \ell$ and the z axis.

Express the angle θ as a function of L_z and L:

$$\theta = \cos^{-1}\left(\frac{L_z}{L}\right)$$

Relate the z component of $\vec{L}$ to m_ℓ and ℓ:

$$L_z = m_\ell \hbar = \ell \hbar$$

Express the angular momentum L:

$$L = \sqrt{\ell(\ell+1)}\,\hbar$$

Substitute to obtain:

$$\theta = \cos^{-1}\left(\frac{\ell\hbar}{\sqrt{\ell(\ell+1)}\,\hbar}\right) = \cos^{-1}\left(\sqrt{\frac{\ell}{\ell+1}}\right)$$

372   Chapter 36

(a) For $\ell = 1$:
$$\theta = \cos^{-1}\left(\sqrt{\frac{1}{1+1}}\right) = \boxed{45.0°}$$

(b) For $\ell = 4$:
$$\theta = \cos^{-1}\left(\sqrt{\frac{4}{4+1}}\right) = \boxed{26.6°}$$

(c) For $\ell = 50$:
$$\theta = \cos^{-1}\left(\sqrt{\frac{50}{50+1}}\right) = \boxed{8.05°}$$

Quantum Theory of the Hydrogen Atom

*39 • (a) If spin is not included, how many different wave functions are there corresponding to the first excited energy level $n = 2$ for hydrogen? (b) List these functions by giving the quantum numbers for each state.

Picture the Problem We can use the constraints on n, ℓ, and m to determine the number of different wave functions, excluding spin, corresponding to the first excited energy state of hydrogen.

For $n = 2$: $\ell = 0$ or 1

(a) For $\ell = 0$, $m_\ell = 0$ and we have: 1 state

For $\ell = 1$, $m_\ell = -1, 0, +1$ and we have: 3 states

Hence, for $n = 2$ we have: $\boxed{4 \text{ states}}$

(b) The four wave functions are summarized to the right.

n	ℓ	m_ℓ	(n, ℓ, m_ℓ)
2	0	0	(2,0,0)
2	1	−1	(2,1,−1)
2	1	0	(2,1,0)
2	1	1	(2,1,1)

*44 •• Show that the ground-state hydrogen wave function (Equation 36-33) is a solution to Schrödinger's equation (Equation 36-21) and the potential energy function (Equation 36-26).

Atoms 373

Picture the Problem We wish to show that $\psi_{1,0,0} = \dfrac{1}{\sqrt{\pi}}\left(\dfrac{Z}{a_0}\right)^{3/2} e^{-Zr/a_0} = Ce^{-Zr/a_0}$ is a solution to $-\dfrac{\hbar^2}{2mr^2}\dfrac{\partial}{\partial r}\left(r^2\dfrac{\partial \psi}{\partial r}\right) + U(r)\psi = E\psi$, where $U(r) = -\dfrac{kZe^2}{r}$. Because the ground state is spherically symmetric, we do not need to consider the angular partial derivatives in Equation 36-21.

The normalized ground-state wave function is:

$$\psi_{1,0,0} = \dfrac{1}{\sqrt{\pi}}\left(\dfrac{Z}{a_0}\right)^{3/2} e^{-Zr/a_0} = Ce^{-Zr/a_0}$$

Differentiate this expression with respect to r to obtain:

$$\dfrac{\partial \psi_{1,0,0}}{\partial r} = C\dfrac{\partial}{\partial r}\left[e^{-Zr/a_0}\right] = -C\dfrac{Z}{a_0}e^{-Zr/a_0}$$

Multiply both sides of this equation by r^2:

$$r^2\dfrac{\partial \psi_{1,0,0}}{\partial r} = -C\dfrac{Z}{a_0}r^2 e^{-Zr/a_0}$$

Differentiate this expression with respect to r to obtain:

$$\dfrac{\partial}{\partial r}\left(r^2\dfrac{\partial \psi_{1,0,0}}{\partial r}\right) = -C\dfrac{Z}{a_0}\dfrac{\partial}{\partial r}\left(r^2 e^{-Zr/a_0}\right) = \left[-\dfrac{2Zr}{a_0} + r^2\left(\dfrac{Z}{a_0}\right)^2\right]Ce^{-Zr/a_0}$$

Substitute in Schrödinger's equation to obtain:

$$-\dfrac{\hbar^2}{2mr^2}\left[-\dfrac{2Zr}{a_0} + r^2\left(\dfrac{Z}{a_0}\right)^2\right]Ce^{-Zr/a_0} - \dfrac{kZe^2}{r}Ce^{-Zr/a_0} = ECe^{-Zr/a_0}$$

Solve for E:

$$E = -\dfrac{\hbar^2}{2mr^2}\left[-\dfrac{2Zr}{a_0} + r^2\left(\dfrac{Z}{a_0}\right)^2\right] - \dfrac{kZe^2}{r}$$

Because $a_0 = \dfrac{\hbar^2}{mke^2}$:

$$E = -\dfrac{\hbar^2}{2mr^2}\left[-\dfrac{2mke^2 Zr}{\hbar^2} + r^2\left(\dfrac{Zmke^2}{\hbar^2}\right)^2\right] - \dfrac{kZe^2}{r} = \dfrac{kZe^2}{r} - \dfrac{Z^2k^2e^4m}{2\hbar^2} - \dfrac{kZe^2}{r}$$

$$= \boxed{-\dfrac{Z^2k^2e^4m}{2\hbar^2}}$$

Because this is the correct ground state energy, we have shown that Equation 36-33, is a solution to Schrödinger's Equation 36-21 with the potential energy function Equation 36-26.

The Spin-Orbit Effect and Fine Structure

*50 • The potential energy of a magnetic moment in an external magnetic field is given by $U = -\vec{\mu} \cdot \vec{B}$. (a) Calculate the difference in energy between the two possible orientations of an electron in a magnetic field $\vec{B} = 1.50\,\text{T}\hat{k}$. (b) If these electrons are bombarded with photons of energy equal to this energy difference, "spin flip" transitions can be induced. Find the wavelength of the photons needed for such transitions. This phenomenon is called *electron spin resonance*.

Picture the Problem The energy difference between the two possible orientations of an electron in a magnetic field is $2\mu B$ and the wavelength of the photons required to induce a spin-flip transition can be found from $hc/\Delta E$. The magnetic moment μ_B associated with the spin of an electron is 5.79×10^{-5} eV/T.

(a) Relate the difference in energy between the two spin orientations in terms of the difference in the potential energies of the two states:

$$\Delta E = 2\mu B$$
$$= 2(5.79 \times 10^{-5}\,\text{eV/T})(0.6\,\text{T})$$
$$= \boxed{6.95 \times 10^{-5}\,\text{eV}}$$

(b) Relate the wavelength of the photon needed to induce such a transition to the energy required:

$$\lambda = \frac{hc}{\Delta E}$$

Substitute numerical values and evaluate λ:

$$\lambda = \frac{1240\,\text{eV} \cdot \text{nm}}{6.95 \times 10^{-5}\,\text{eV}} = 1.78 \times 10^{7}\,\text{nm}$$
$$= \boxed{1.78\,\text{cm}}$$

The Periodic Table

*56 • Write the electron configuration of (a) carbon and (b) oxygen.

Determine the Concept We can use the atomic numbers of carbon and oxygen to determine the sum of the exponents in their electronic configurations and then use the rules for the filling of the shells to find their electronic configurations.

(a) The atomic number Z of carbon is 6. So we must fill the subshells of the electronic configuration until we have placed its 6 electrons. This is accomplished by writing $\boxed{1s^2\,2s^2\,2p^2}$

(b) The atomic number Z of oxygen is 8. So we must fill the subshells of the electronic configuration until we have placed its 8 electrons. This is accomplished by writing

$\boxed{1s^2 2s^2 2p^4}$

Optical Spectra and X-Ray Spectra

*61 • (*a*) Calculate the next two longest wavelengths in the K series (after the K_α line) of molybdenum. (*b*) What is the wavelength of the shortest wavelength in this series?

Picture the Problem When an electron from state *n* drops into a vacated state in the $n = 1$ shell, a photon of energy $\Delta E = E_n - E_1$ is emitted. We can find the wavelength of this photon using $\lambda = hc/\Delta E$. The second and third longest wavelengths in the K series correspond to transitions from $n = 3$ to $n = 1$ and $n = 4$ to $n = 1$ and the shortest wavelength to the transition from $n = \infty$ to $n = 1$.

Express the wavelength of the emitted photon in terms of the energy transition within the atom:
$$\lambda = \frac{hc}{E_n - E_1} = \frac{1240\,\text{eV}\cdot\text{nm}}{E_n - E_1}$$

Express the energy of the *n*th energy state:
$$E_n = -(Z-1)^2 \frac{E_0}{n^2}$$
where $n = 1, 2, \ldots$

Substitute to obtain:
$$\lambda = \frac{hc}{E_n - E_1}$$
$$= \frac{1240\,\text{eV}\cdot\text{nm}}{-(Z-1)^2 \frac{E_0}{n^2} - \left(-(Z-1)^2 \frac{E_0}{1^2}\right)}$$
$$= \frac{1240\,\text{eV}\cdot\text{nm}}{(Z-1)^2 E_0 \left(1 - \frac{1}{n^2}\right)}$$

(*a*) Evaluate this expression with $n = 3$ and $Z = 42$ to obtain:
$$\lambda_3 = \frac{1240\,\text{eV}\cdot\text{nm}}{(42-1)^2 (13.6\,\text{eV})\left(1 - \frac{1}{3^2}\right)}$$
$$= \boxed{0.0610\,\text{nm}}$$

Use $n = 4$ and $Z = 42$ to obtain:
$$\lambda_4 = \frac{1240\,\text{eV}\cdot\text{nm}}{(42-1)^2(13.6\,\text{eV})\left(1-\frac{1}{4^2}\right)}$$
$$= \boxed{0.0578\,\text{nm}}$$

(b) The shortest wavelength in the series corresponds to the largest energy difference between the initial and final states. Repeat the calculation in part (a) with $n = \infty$ to obtain:
$$\lambda_\infty = \frac{1240\,\text{eV}\cdot\text{nm}}{(42-1)^2(13.6\,\text{eV})(1-0)}$$
$$= \boxed{0.0542\,\text{nm}}$$

General Problems

*68 ••   We are often interested in finding the quantity ke^2/r in electron volts when r is given in nanometers. Show that $ke^2 = 1.44$ eV·nm.

Picture the Problem We can show that $ke^2 = 1.44$ eV·nm by solving the equation for the ground state energy of an atom for ke^2.

Express the ground state energy of an atom as a function of k, e, and a_0:
$$E_0 = \frac{ke^2}{2a_0}$$

Solve for ke^2:
$$ke^2 = 2E_0 a_0$$

Substitute for E_0 and a_0 to obtain:
$$ke^2 = 2(13.6\,\text{eV})(0.0529\,\text{nm})$$
$$= \boxed{1.44\,\text{eV}\cdot\text{nm}}$$

*71 ••   The combination of physical constants $\alpha = e^2k/\hbar c$, where k is the Coulomb constant, is known as the *fine-structure constant*. It appears in numerous relations in atomic physics. (a) Show that α is dimensionless. (b) Show that in the Bohr model of hydrogen $v_n = c\alpha/n$, where v_n is the speed of the electron in the stationary state of quantum number n.

Picture the Problem We can show that α is dimensionless by showing that it has no units. In part (b) we can use Bohr's 3rd postulate and the expression for the radii of the Bohr orbits, together with the definition of α, to show that the speed of the electron in a stationary state of quantum number n is related to α according to $v_n = c\alpha/n$.

(a) Express the units of α:

$$\frac{C^2\left(\dfrac{N\cdot m^2}{C^2}\right)}{(J\cdot s)\dfrac{m}{s}} = \frac{N\cdot m^2}{J\cdot m} = 1$$

Because α is unitless, it is also dimensionless.

(b) Apply the quantization of angular momentum postulate to obtain:

$$v_n = \frac{n\hbar}{mr_n}$$

The radii of the Bohr orbits are given by:

$$r_n = n^2 \frac{\hbar^2}{mkZe^2}$$

or, because $Z = 1$ for hydrogen,

$$r_n = n^2 \frac{\hbar^2}{mke^2}$$

Substitute and simplify to obtain:

$$v_n = \frac{n\hbar}{mn^2 \dfrac{\hbar^2}{mke^2}} = \frac{ke^2}{n\hbar}$$

Divide this expression by the definition of α to obtain:

$$\frac{v_n}{\alpha} = \frac{\dfrac{ke^2}{n\hbar}}{\dfrac{e^2 k}{\hbar c}} = \frac{c}{n}$$

Solve for v_n:

$$v_n = \boxed{\frac{\alpha c}{n}}$$

*74 • A Rydberg atom is one in which an outer-shell electron is placed into a *very* high excited state ($n \approx 40$ or higher.) Such atoms are useful for experiments which probe the transition from quantum mechanical behavior to classical. Furthermore, these excited states have extremely long lifetimes (i.e., the electron will stay in this high excited state for a very long time.) A hydrogen atom is in the $n = 45$ state. (a) What is the ionization energy of the atom when it is in this state? (b) What is the energy level separation (in eV) between this state and the $n = 44$ state? (c) What is the wavelength of a photon resonant with this transition? (d) How large is the atom in the $n = 45$ state?

Picture the Problem The ionization energy of the electron is the magnitude of the energy of the atom in the given state. We can use $E = -E_0/n^2$, where E_0 is the ground-state energy, to find the energy levels in the 44th and 45th states and, hence, the energy level separation between the states. The wavelength of a photon resonant with this transition can be found from $\lambda = hc/\Delta E$. We'll approximate the size of the atom in the $n = 45$ state by finding the radius of the outer-shell electron.

(a) The energy of the atom in its nth state is:
$$E_n = -\frac{E_0}{n^2}$$

The energy of the atom in the $n = 45$ state is:
$$E_{45} = -\frac{13.6\,\text{eV}}{(45)^2} = -6.72\,\text{meV}$$

The ionization energy is the negative of the energy in the state $n = 45$:
$$E_{\text{ionizing}} = -E_{45} = \boxed{6.72\,\text{meV}}$$

(b) The energy level separation between the $n = 45$ and $n = 44$ state is:
$$E_{45 \to 44} = -\left(\frac{13.6\,\text{eV}}{(45)^2} - \frac{13.6\,\text{eV}}{(44)^2}\right)$$
$$= \boxed{3.09 \times 10^{-4}\,\text{eV}}$$

(c) The photon wavelength is:
$$\lambda = \frac{hc}{\Delta E}$$

Substitute numerical values and evaluate λ:
$$\lambda = \frac{1240\,\text{eV}\cdot\text{nm}}{3.09 \times 10^{-4}\,\text{eV}} = \boxed{4.01 \times 10^6\,\text{nm}}$$

(d) The radii of the Bohr orbits are given by:
$$r = n^2 \frac{a_0}{Z}$$

Substitute numerical values and evaluate the radius of the 45th Bohr orbit:
$$r = (45)^2 \frac{0.0529\,\text{nm}}{1} = \boxed{107\,\text{nm}}$$

Chapter 37
Molecules

Conceptual Problems

*1 • Would you expect the NaCl molecule to be polar or nonpolar?

Determine the Concept Yes. Because the center of charge of the positive Na ion does not coincide with the center of charge for the negative Cl ion, the NaCl molecule has a permanent dipole moment. Hence, it is a polar molecule.

5 •• The elements on the far right column of the periodic table are sometimes called noble gases because they virtually never react with other atoms to form molecules. However, this behavior is sometimes modified if the resulting molecule is formed in an electronic excited state. An example is ArF. When it is formed in the excited state, it is written ArF and is called an excimer (for <u>ex</u>cited di<u>mer</u>). Refer to Figure 37-13 and discuss how this diagram would look for ArF in which the ArF ground state is unstable but the ArF* excited state is stable. *Remark: Excimers are used in certain kinds of lasers.*

Determine the Concept The diagram would consist of a non-bonding ground state with no vibrational or rotational states for ArF (similar to the upper curve in Figure 37-4) but for ArF* there should be a bonding excited state with a definite minimum with respect to inter-nuclear separation and several vibrational states as in the excited state curve of Figure 37-13.

Estimation and Approximation

*14 •• Repeat Problem 13, finding the quantum number v and spacing between adjacent energy levels for a 5-kg mass attached to 1500-N/m spring vibrating with an amplitude of 2 cm. *Hint: Pick v so that the quantum energy formula (Equation 37-18) gives the correct energy for the given system. Then find the energy increase for the next highest energy level.*

Picture the Problem We can solve Equation 37-18 for v and substitute for the frequency of the mass-and-spring oscillator to estimate the quantum number v and spacing between adjacent energy levels for this system.

The vibrational energy levels are given by Equation 37-18:
$$E_v = (v + \tfrac{1}{2})hf$$
where $v = 0, 1, 2, \ldots$

Solve for v:
$$v = \frac{E_v}{hf} - \frac{1}{2}$$

379

or, because $v \gg 1$,

$$v \approx \frac{E_v}{hf}$$

The vibrational energy of the object attached to the spring is:

$$E_v = \tfrac{1}{2}kA^2$$

where A is the amplitude of its motion.

Substitute for E_v in the expression for v to obtain:

$$v = \frac{kA^2}{2hf}$$

The frequency of oscillation f of the mass-and-spring oscillator is given by:

$$f = \frac{1}{2\pi}\sqrt{\frac{k}{m}}$$

$$v = \frac{\pi k A^2}{h}\sqrt{\frac{m}{k}} = \frac{\pi A^2}{h}\sqrt{mk}$$

Substitute numerical values and evaluate v:

$$v = \frac{\pi(0.02\,\text{m})^2}{6.63\times 10^{-34}\,\text{J}\cdot\text{s}}\sqrt{(5\,\text{kg})(1500\,\text{N/m})} = \boxed{1.64\times 10^{32}}$$

Set $v = 0$ in Equation 37-18 to express the spacing between adjacent energy levels:

$$E_{0v} = \frac{1}{2}hf = \frac{h}{4\pi}\sqrt{\frac{k}{m}}$$

Substitute numerical values and evaluate E_{0v}:

$$E_{0v} = \frac{6.63\times 10^{-34}\,\text{J}\cdot\text{s}}{4\pi}\sqrt{\frac{1500\,\text{N/m}}{5\,\text{kg}}}$$

$$= \boxed{9.14\times 10^{-34}\,\text{J}}$$

Remarks: Note that our value for v justifies our assumption that $v \gg 1$.

Molecular Bonding

*18 • The equilibrium separation of the HF molecule is 0.0917 nm, and its measured electric dipole moment is 6.40×10^{-30} C·m. What percentage of the bonding is ionic?

Picture the Problem The percentage of the bonding that is ionic is given by $100\left(\dfrac{p_{\text{meas}}}{p_{100}}\right)$.

Express the percentage of the bonding that is ionic:

$$\text{Percent ionic bonding} = 100\left(\frac{p_{\text{meas}}}{p_{100}}\right)$$

Express the dipole moment for 100% ionic bonding:

$$p_{100} = er$$

Substitute to obtain:

$$\text{Percent ionic bonding} = 100\left(\frac{p_{\text{meas}}}{er}\right)$$

Substitute numerical values and evaluate the percent ionic bonding:

$$\text{Percent ionic bonding} = 100\left[\frac{6.40\times10^{-30}\,\text{C}\cdot\text{m}}{(1.60\times10^{-19}\,\text{C})(0.0917\,\text{nm})}\right] = \boxed{43.6\%}$$

*24 ••• Assume that the potential energy associated with the core repulsion of the two ions of a diatomic molecule with ionic bonding can be represented by a potential energy of the form $U_{\text{rep}} = C/r^n$, so the total potential energy is $U = U_e + U_{\text{rep}} + \Delta E$, where $U_e = -ke^2/r$. ΔE is the energy of the two ions at infinite separation less the energy of the two neutral atoms at infinite separation (see Figure 37-1). Use $dU/dr = 0$ at $r = r_0$ to show that $n = \dfrac{|U_e(r_0)|}{U_{\text{rep}}(r_0)}$

Picture the Problem $U(r)$ is the potential energy of the two ions as a function of separation distance r. $U(r)$ is chosen so $U(\infty) = -\Delta E$, where ΔE is the negative of the energy required to form two ions at infinite separation from two neutral atoms also at infinite separation. $U_{\text{rep}}(r)$ is the potential energy of the two ions due to the repulsion of the two closed-shell cores. E_d is the disassociation energy, the energy required to separate the two ions plus the energy ΔE required to form two neutral atoms from the two ions at infinite separation. The net force acting on the ions is the sum of F_{rep} and F_e. We can find F_{rep} from U_{rep} and F_e from Coulomb's law and then use $dU/dr = F_{\text{net}} = 0$ at $r = r_0$ to solve for n.

Express the net force acting on the ions:

$$F_{\text{net}} = F_{\text{rep}} + F_e \qquad (1)$$

Find F_{rep} from U_{rep}:

$$F_{\text{rep}} = \frac{dU_{\text{rep}}}{dr} = \frac{d}{dr}\left[Cr^{-n}\right] = -nCr^{-n-1}$$

$$= -\frac{nC}{r^{n+1}}$$

382   Chapter 37

The electrostatic potential energy of the two ions as a function of separation distance is given by:

$$U_e = -\frac{ke^2}{r}$$

Find the electrostatic force of attraction F_e from U_e:

$$F_e = \frac{dU_e}{dr} = \frac{d}{dr}\left[-\frac{ke^2}{r}\right] = \frac{ke^2}{r^2}$$

Substitute for F_{rep} and F_e in equation (1) to obtain:

$$F_{net} = -\frac{nC}{r^{n+1}} + \frac{ke^2}{r^2}$$

Because $dU/dr = F_{net} = 0$ at $r = r_0$:

$$0 = -\frac{nC}{r_0^{n+1}} + \frac{ke^2}{r_0^2}$$

Multiply both sides of this equation by r_0 to obtain:

$$0 = -\frac{nC}{r_0^n} + \frac{ke^2}{r_0} = -nU_{rep}(r_0) + |U_e(r_0)|$$

Solve for n to obtain:

$$\boxed{n = \frac{|U_e(r_0)|}{U_{rep}(r_0)}}$$

Energy Levels of Spectra of Diatomic Molecules

*27 • The separation of the two oxygen atoms in a molecule of O_2 is actually slightly greater than the 0.1 nm used in Example 37-3, and the characteristic energy of rotation E_{0r} is 1.78×10^{-4} eV, rather than the result obtained in that example. Use this value to calculate the separation distance of the two oxygen atoms.

Picture the Problem We can relate the characteristic rotational energy E_{0r} to the moment of inertia of the molecule and model the moment of inertia of the O_2 molecule as two point objects separated by a distance r.

The characteristic rotational energy of a molecule is given by:

$$E_{0r} = \frac{\hbar^2}{2I}$$

Express the moment of inertia of the molecule:

$$I = 2M_O\left(\frac{r}{2}\right)^2 = \tfrac{1}{2}M_O r^2$$

Substitute for I to obtain:

$$E_{0r} = \frac{\hbar^2}{2(\tfrac{1}{2}M_O r^2)} = \frac{\hbar^2}{M_O r^2} = \frac{\hbar^2}{16 m_p r^2}$$

Solve for r:

$$r = \frac{\hbar}{4}\sqrt{\frac{1}{E_{0r}m_p}}$$

Substitute numerical values and evaluate r:

$$r = \left(\frac{1.055\times10^{-34}\text{ J}\cdot\text{s}}{4}\right)\sqrt{\frac{1}{(1.78\times10^{-4}\text{ eV})(1.6\times10^{-19}\text{ J/eV})(1.67\times10^{-27}\text{ kg})}}$$
$$= \boxed{0.121\text{ nm}}$$

*30 •• The equilibrium separation between the nuclei of the LiH molecule is 0.16 nm. Determine the energy separation between the $\ell = 3$ and $\ell = 2$ rotational levels of this diatomic molecule.

Picture the Problem We can use the expression for the rotational energy levels of the diatomic molecule to express the energy separation ΔE between the $\ell = 3$ and $\ell = 2$ rotational levels and model the moment of inertia of the LiH molecule as two point objects separated by a distance r_0.

The energy separation between the $\ell = 3$ and $\ell = 2$ rotational levels of this diatomic molecule is given by:

$$\Delta E = E_{\ell=3} - E_{\ell=2}$$

Express the rotational energy levels $E_{\ell=3}$ and $E_{\ell=2}$ in terms of E_{0r}:

$$E_{\ell=3} = 3(3+1)E_{0r} = 12E_{0r}$$
and
$$E_{\ell=2} = 2(2+1)E_{0r} = 6E_{0r}$$

Substitute for $E_{\ell=3}$ and $E_{\ell=2}$ to obtain:

$$\Delta E = 12E_{0r} - 6E_{0r} = 6E_{0r}$$
or
$$E_{0r} = \tfrac{1}{6}\Delta E$$

The characteristic rotational energy of a molecule is given by:

$$E_{0r} = \frac{\hbar^2}{2I} = \tfrac{1}{6}\Delta E \Rightarrow \Delta E = \frac{3\hbar^2}{I}$$

Express the moment of inertia of the molecule:

$$I = \mu r_0^2$$
where μ is the reduced mass of the molecule.

384   Chapter 37

Substitute for I to obtain:

$$\Delta E = \frac{3\hbar^2}{\mu r_0^2} = \frac{3\hbar^2}{\dfrac{m_{Li} m_H}{m_{Li} + m_H} r_0^2}$$

$$= \frac{3\hbar^2 (m_{Li} + m_H)}{m_{Li} m_H r_0^2}$$

Substitute numerical values and evaluate ΔE:

$$\Delta E = \frac{3(1.055 \times 10^{-34}\,\text{J}\cdot\text{s})^2 (6.94\,\text{u} + 1\,\text{u})}{(6.94\,\text{u})(1\,\text{u})(0.16\,\text{nm})^2 (1.602 \times 10^{-19}\,\text{J/eV})(1.660 \times 10^{-27}\,\text{kg/u})} = \boxed{5.61\,\text{meV}}$$

*31 ••   Derive Equations 37-14 and 37-15 for the moment of inertia in terms of the reduced mass of a diatomic molecule.

Picture the Problem Let the origin of coordinates be at the point mass m_1 and point mass m_2 be at a distance r_0 from the origin. We can express the moment of inertia of a diatomic molecule with respect to its center of mass using the definitions of the center of mass and the moment of inertia of point particles.

Express the moment of inertia of a diatomic molecule:

$$I = m_1 r_1^2 + m_2 r_2^2 \qquad (1)$$

The r coordinate of the center of mass is:

$$r_{CM} = \frac{m_2}{m_1 + m_2} r_0$$

The distances of m_1 and m_2 from the center of mass are:

$$r_1 = r_{CM}$$

and

$$r_2 = r_0 - r_{CM} = r_0 - \frac{m_2}{m_1 + m_2} r_0$$

$$= \frac{m_1}{m_1 + m_2} r_0$$

Substitute for r_1 and r_2 in equation (1) to obtain:

$$I = m_1 \left(\frac{m_2}{m_1 + m_2} r_0 \right)^2 + m_2 \left(\frac{m_1}{m_1 + m_2} r_0 \right)^2$$

Simplifying this expression leads to:

$$I = \frac{m_1 m_2}{m_1 + m_2} r_0^2$$

or

Molecules 385

$$I = \mu r_0^2 \qquad \text{36-14}$$

where

$$\mu = \frac{m_1 m_2}{m_1 + m_2} \qquad \text{36-15}$$

*36 •• Two objects of mass m_1 and m_2 are attached to a spring of force constant k and equilibrium length r_0. (*a*) Show that when m_1 is moved a distance Δr_1 from the center of mass, the force exerted by the spring is

$$F = -k\left(\frac{m_1 + m_2}{m_2}\right)\Delta r_1$$

(*b*) Show that the frequency of oscillation is $f = \frac{1}{2\pi}\sqrt{k/\mu}$, where μ is the reduced mass.

Picture the Problem For a two-mass and spring system on which no external forces are acting, the center of mass must remain fixed. We can use this condition to express the net force acting on either object. Because this force is a linear restoring force, we can conclude that the motion of the object whose mass is m_1 will be simple harmonic with an angular frequency given by $\omega = \sqrt{\dfrac{k_{\text{eff}}}{m_1}}$. Substitution for k_{eff} will lead us to the result given in (*b*).

(*a*) If the particle whose mass is m_1 moves a distance Δr_1 from (or toward) the center of mass, then the particle whose mass is m_2 must move a distance:

$$\Delta r_2 = \frac{m_1}{m_2}\Delta r_1 \text{ from (or toward) the center of mass.}$$

Express the force exerted by the spring:

$$F = -k\Delta r = -k(\Delta r_1 + \Delta r_2)$$

Substitute for Δr_2 to obtain:

$$F = -k\left(\Delta r_1 + \frac{m_1}{m_2}\Delta r_1\right)$$

$$= \boxed{-k\left(\frac{m_1 + m_2}{m_2}\right)\Delta r_1}$$

(*b*) A displacement Δr_1 of m_1 results in a restoring force:

$$F = -k\left(\frac{m_1 + m_2}{m_2}\right)\Delta r_1 = -k_{\text{eff}}\Delta r_1$$

Chapter 37

where $k_{\text{eff}} = k\left(\dfrac{m_1 + m_2}{m_2}\right)$

Because this is a linear restoring force, we know that the motion will be simple harmonic with:

$$\omega = \sqrt{\dfrac{k_{\text{eff}}}{m_1}}$$

or

$$f = \dfrac{\omega}{2\pi} = \dfrac{1}{2\pi}\sqrt{\dfrac{k_{\text{eff}}}{m_1}}$$

Substitute for k_{eff} and simplify to obtain:

$$f = \dfrac{\omega}{2\pi} = \dfrac{1}{2\pi}\sqrt{k\left(\dfrac{m_1 + m_2}{m_1 m_2}\right)}$$

or, because $\mu = \dfrac{m_1 m_2}{m_1 + m_2}$ is the reduced mass of the two-particle system,

$$\boxed{f = \dfrac{1}{2\pi}\sqrt{\dfrac{k}{\mu}}}.$$

General Problems

*40 •• The effective force constant for the HF molecule is 970 N/m. Find the frequency of vibration for this molecule.

Picture the Problem We can use the result of Problem 36 to find the frequency of vibration of the HF molecule.

In Problem 36 it was established that:

$$f = \dfrac{1}{2\pi}\sqrt{\dfrac{k}{\mu}}$$

The reduced mass is:

$$\mu = \dfrac{m_H m_F}{m_H + m_F}$$

Substitute for μ to obtain:

$$f = \dfrac{1}{2\pi}\sqrt{\dfrac{k}{\dfrac{m_H m_F}{m_H + m_F}}} = \dfrac{1}{2\pi}\sqrt{\dfrac{k(m_H + m_F)}{m_H m_F}}$$

Substitute numerical values and evaluate f:

$$f = \frac{1}{2\pi}\sqrt{\frac{(970\,\text{N/m})(1\,\text{u}+19\,\text{u})}{(1\,\text{u})(19\,\text{u})(1.66\times10^{-27}\,\text{kg/u})}} = \boxed{1.25\times10^{14}\,\text{Hz}}$$

***47 ••** For a molecule such as CO, which has a permanent electric dipole moment, radiative transitions obeying the selection rule $\Delta\ell = \pm 1$ between two rotational energy levels of the same vibrational level are allowed. (That is, the selection rule $\Delta v = \pm 1$ does not hold.) (*a*) Find the moment of inertia of CO and calculate the characteristic rotational energy E_{0r} (in eV). (*b*) Make an energy level diagram for the rotational levels for $\ell = 0$ to $\ell = 5$ for some vibrational level. Label the energies in electron volts starting with $E = 0$ for $\ell = 0$. Indicate on your diagram transitions that obey $\Delta\ell = -1$ and calculate the energy of the photon emitted (*c*) Find the wavelength of the photons emitted for each transition in (*b*). In what region of the electromagnetic spectrum are these photons?

Picture the Problem We can find the reduced mass of CO and the moment of inertia of a CO molecule from their definitions. The energy level diagram for the rotational levels for $\ell = 0$ to $\ell = 5$ can be found using $\Delta E_{\ell,\ell-1} = 2\ell E_{0r}$. Finally, we can find the wavelength of the photons emitted for each transition using $\lambda_{\ell,\ell-1} = \dfrac{hc}{\Delta E_{\ell,\ell-1}} = \dfrac{hc}{2\ell\Delta E_{0r}}$.

(*a*) Express the moment of inertia of CO:

$$I = \mu r_0^2$$

where μ is the reduced mass of the CO molecule.

Find μ:

$$\mu = \frac{m_C m_O}{m_C + m_O} = \frac{(12\,\text{u})(16\,\text{u})}{12\,\text{u}+16\,\text{u}} = 6.86\,\text{u}$$

In Problem 39 it was established that $r_0 = 0.113$ nm. Use this result to evaluate *I*:

$$I = (6.86\,\text{u})(1.66\times10^{-27}\,\text{kg/u})(0.113\,\text{nm})^2 = \boxed{1.45\times10^{-46}\,\text{kg}\cdot\text{m}^2}$$

The characteristic rotational energy E_{0r} is given by:

$$E_{0r} = \frac{\hbar^2}{2I}$$

Substitute numerical values and evaluate E_{0r}:

$$E_{0r} = \frac{(6.58\times10^{-16}\,\text{eV}\cdot\text{s})^2(1.6\times10^{-19}\,\text{J/eV})}{2(1.45\times10^{-46}\,\text{kg}\cdot\text{m}^2)} = \boxed{0.239\,\text{meV}}$$

(b) The energy level diagram is shown to the right. Note that $\Delta E_{\ell,\ell-1}$, the energy difference between adjacent levels for $\Delta\ell = -1$, is
$$\Delta E_{\ell,\ell-1} = 2\ell E_{0r}.$$

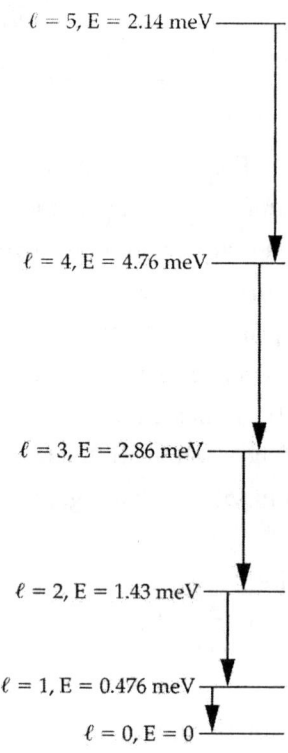

(c) Express the energy difference $\Delta E_{\ell,\ell-1}$ between energy levels in terms of the frequency of the emitted radiation:

$$\Delta E_{\ell,\ell-1} = hf_{\ell,\ell-1}$$

Because $c = f_{\ell,\ell-1}\lambda_{\ell,\ell-1}$:

$$\lambda_{\ell,\ell-1} = \frac{hc}{\Delta E_{\ell,\ell-1}} = \frac{hc}{2\ell\Delta E_{0r}}$$

Substitute numerical values to obtain:

$$\lambda_{\ell,\ell-1} = \frac{(4.136\times 10^{-15}\,\text{eV}\cdot\text{s})(3\times 10^8\,\text{m/s})}{2\ell(0.239\,\text{meV})} = \frac{2596\,\mu\text{m}}{\ell}$$

For $\ell = 1$:
$$\lambda_{1,0} = \frac{2596\,\mu\text{m}}{1} = \boxed{2596\,\mu\text{m}}$$

For $\ell = 2$:
$$\lambda_{2,1} = \frac{2596\,\mu\text{m}}{2} = \boxed{1298\,\mu\text{m}}$$

For $\ell = 3$:
$$\lambda_{3,2} = \frac{2596\,\mu\text{m}}{3} = \boxed{865\,\mu\text{m}}$$

For $\ell = 4$:
$$\lambda_{4,3} = \frac{2596 \, \mu m}{4} = \boxed{649 \, \mu m}$$

For $\ell = 5$:
$$\lambda_{5,4} = \frac{2596 \, \mu m}{5} = \boxed{519 \, \mu m}$$

$\boxed{\text{These wavelengths fall in the microwave region of the spectrum.}}$

*48 ••• Use the results of Problem 24 to calculate the vibrational frequency of the LiCl molecule. The dissociation energy of LiCl is 4.86 eV, and the equilibrium separation is 0.202 nm. The electron affinity of chlorine is 3.62 eV, and the ionization energy of lithium is 5.39 eV. To do this, expand the potential about $r = r_0$, where r_0 is the equilibrium separation, in a Taylor series. Retain only the term proportional to $(r - r_0)^2$. Recall that the potential energy of a simple harmonic oscillator is given by $U_{SHO} = \frac{1}{2} m \omega^2 x^2$. What is the wavelength resulting from transitions between adjacent harmonic oscillator levels of this molecule?

Picture the Problem The wavelength resulting from transitions between adjacent harmonic oscillator levels of a LiCl molecule is given by $\lambda = \frac{2\pi c}{\omega}$. We can find an expression for ω by following the procedure outlined in the problem statement.

The wavelength resulting from transitions between adjacent harmonic oscillator levels of this molecule is given by:

$$\lambda = \frac{hc}{\Delta E} = \frac{hc}{\hbar \omega} = \frac{2\pi c}{\omega} \quad (1)$$

From Problem 24 we have:

$$U(r) = -\frac{ke^2}{r} + \frac{C}{r^n}, \text{ where } \Delta E \text{ is constant.}$$

The Taylor expansion of $U(r)$ about $r = r_0$ is:

$$U(r) = U(r_0) + \left(\frac{dU}{dr}\right)_{r_0} (r - r_0)$$
$$+ \frac{1}{2}\left(\frac{d^2U}{dr^2}\right)_{r_0} (r - r_0)^2 + \ldots$$

Because $U(r_0)$ is a constant, it can be dropped without affecting the physical results and because

$$U(r) \approx \frac{1}{2}\left(\frac{d^2U}{dr^2}\right)_{r_0} (r - r_0)^2 \quad (2)$$

$\left(\dfrac{dU}{dr}\right)_{r_0} = 0$:

Differentiate $U(r)$ twice to obtain:
$$\dfrac{d^2U}{dr^2} = -2\dfrac{ke^2}{r^3} + n(n-1)\dfrac{C}{r^{n+2}}$$

Because $dU/dr = F_{net} = 0$ at $r = r_0$:
$$0 = -\dfrac{nC}{r_0^{n+1}} + \dfrac{ke^2}{r_0^2}$$

Solving for C yields:
$$C = \dfrac{ke^2 r_0^{n+1}}{nr_0^2} = \dfrac{ke^2 r_0^{n-1}}{n}$$

Substitute for C and evaluate $\left(\dfrac{d^2U}{dr^2}\right)_{r_0}$ to obtain:
$$\left(\dfrac{d^2U}{dr^2}\right)_{r_0} = -2\dfrac{ke^2}{r_0^3}(n-1) + \dfrac{n(n-1)}{r_0^{n+2}}\dfrac{ke^2 r_0^{n-1}}{n}$$
$$= \dfrac{ke^2}{r_0^3}(n-1)$$

Substitute for $\left(\dfrac{d^2U}{dr^2}\right)_{r_0}$ in equation (2):
$$U(r) \approx \dfrac{1}{2}\left[\dfrac{ke^2}{r_0^3}(n-1)\right](r-r_0)^2$$

Because the potential energy of a simple harmonic oscillator is given by $U_{SHO} = \tfrac{1}{2}m\omega^2(r-r_0)^2$:
$$\dfrac{1}{2}m\omega^2(r-r_0)^2 \approx \dfrac{1}{2}\left[\dfrac{ke^2}{r_0^3}(n-1)\right](r-r_0)^2$$

Solve for ω to obtain:
$$\omega \approx \sqrt{\dfrac{(n-1)ke^2}{mr_0^3}}$$

Substitute μ_{LiCl} for m to obtain:
$$\omega \approx \sqrt{\dfrac{(n-1)ke^2}{\dfrac{m_{Li}m_{Cl}}{m_{Li}+m_{Cl}}r_0^3}} \quad (3)$$
$$= \sqrt{\dfrac{(n-1)(m_{Li}+m_{Cl})ke^2}{m_{Li}m_{Cl}r_0^3}}$$

From Problem 24:
$$n = \dfrac{|U_e(r_0)|}{U_{rep}(r_0)} \quad (4)$$

U_{rep} is related to U_e, E_d, and ΔE
$$U_{rep} = -(U_e + E_d + \Delta E) \quad (5)$$

according to:

The energy needed to form Li⁺ and Cl⁻ from neutral lithium and chlorine atoms is:

$$\Delta E = E_{\text{ionization}} - E_{\text{electron affinity}}$$
$$= 5.39\,\text{eV} - 3.62\,\text{eV} = 1.77\,\text{eV}$$

$U_e(r_0)$ is given by:

$$U_e = -\frac{ke^2}{r_0} = -\frac{1.44\,\text{eV}\cdot\text{nm}}{r_0}$$

Substitute r_0 and evaluate U_e:

$$U_e = -\frac{1.44\,\text{eV}\cdot\text{nm}}{0.202\,\text{nm}} = -7.13\,\text{eV}$$

Substitute numerical values in equation (5) and evaluate U_{rep}:

$$U_{\text{rep}} = -(-7.13\,\text{eV} + 4.86\,\text{eV} + 1.77\,\text{eV})$$
$$= 0.500\,\text{eV}$$

Substitute for $U_{\text{rep}}(r_0)$ and $U_e(r_0)$ in equation (4) and evaluate n:

$$n = \frac{|-7.13\,\text{eV}|}{0.500\,\text{eV}} = 14.3$$

Substitute numerical values in equation (3) and evaluate ω:

$$\omega \approx \sqrt{\frac{(14.3-1)(6.941\,\text{u} + 35.453\,\text{u})(1.44\,\text{eV}\cdot\text{nm})(1.60\times10^{-19}\,\text{J/eV})}{(6.941\,\text{u})(35.453\,\text{u})(1.66\times10^{-27}\,\text{kg/u})(0.202\,\text{nm})^3}}$$
$$= \boxed{1.96\times10^{14}\,\text{s}^{-1}}$$

Substitute numerical values in equation (1) and evaluate λ:

$$\lambda = \frac{2\pi(3\times10^8\,\text{m/s})}{1.96\times10^{14}\,\text{s}^{-1}} = \boxed{9.62\,\mu\text{m}}$$

Chapter 38
Solids and the Theory of Conduction

Conceptual Problems

*2 • When the temperature of pure copper is lowered from 300 K to 4 K, its resistivity drops by a much greater factor than that of brass when it is cooled in the same way. Why?

Determine the Concept The resistivity of brass at 4 K is almost entirely due to the "residual resistance," the resistance due to impurities and other imperfections of the crystal lattice. In brass, the zinc ions act as impurities in copper. In pure copper, the resistivity at 4 K is due to its residual resistance, which is very low if the copper is very pure.

*8 • How does the change in the resistivity of copper compare with that of silicon when the temperature increases?

Determine the Concept The resistivity of copper increases with increasing temperature; the resistivity of (pure) silicon decreases with increasing temperature because the number of charge carriers increases.

The Structure of Solids

*18 • Find the value of n in Equation 38-6 that gives the measured dissociation energy of 741 kJ/mol for LiCl, which has the same structure as NaCl and for which $r_0 = 0.257$ nm.

Picture the Problem We can solve Equation 38-6 for n.

Equation 38-6 is:
$$U(r_0) = -\alpha \frac{ke^2}{r_0}\left(1 - \frac{1}{n}\right)$$
and
$$|U(r_0)| = \alpha \frac{ke^2}{r_0}\left(1 - \frac{1}{n}\right)$$

Solve for n to obtain:
$$n = \frac{1}{1 - \dfrac{|U(r_0)|r_0}{\alpha ke^2}}$$

393

Substitute numerical values and evaluate n:

$$n = \cfrac{1}{1 - \cfrac{(741\,\text{kJ/mol})\left(\cfrac{1\,\text{eV/ion pair}}{96.47\,\text{kJ/mol}}\right)(0.257\,\text{nm})}{(1.7476)(1.44\,\text{eV}\cdot\text{nm})}} = \boxed{4.64}$$

A Microscopic Picture of Conduction

*22 •• Silicon has an atomic weight of 28.09 and a density of 2.41×10^3 kg/m^3. Each atom of silicon has 2 valence electrons and the Fermi energy of the material is 4.88 eV. (*a*) Given that the electron mean free path at room temperature is $\lambda = 27.0$ nm, estimate the resistivity. (*b*) The accepted value for the resistivity of Silicon is 640 Ω·m (at room temperature). How does this accepted value compare to the value calculated in part (*a*)?

Picture the Problem We can use Equation 38-14 to estimate the resistivity of silicon.

(*a*) From Equation 38-14:

$$\rho = \frac{m_e v_{av}}{n_e e^2 \lambda} \qquad (1)$$

The speed of the electrons is given by:

$$v_{av} = v_F = \sqrt{\frac{2E_F}{m_e}}$$

Substitute numerical values and evaluate v_{av}:

$$v_{av} = \sqrt{\frac{2(4.88\,\text{eV})}{(9.11 \times 10^{-31}\,\text{kg})} \frac{1.60 \times 10^{-19}\,\text{J}}{1\,\text{eV}}}$$

$$= 1.31 \times 10^6\,\text{m/s}$$

The electron density of Si is given by:

$$n_e = M N_A N_{\text{atom}}$$

where N_{atom} is the number of electrons per atom.

Substitute numerical values and evaluate n_e:

$$n_e = \left(2.41 \times 10^3\,\frac{\text{kg}}{\text{m}^3}\right)\left(\frac{6.02 \times 10^{23}\,\text{atoms}}{0.02809\,\text{kg}}\right)\left(\frac{2\,\text{e}}{\text{atom}}\right) = 1.03 \times 10^{29}\,\text{e/m}^3$$

Substitute numerical values in equation (1) and evaluate ρ:

$$\rho = \frac{(9.11 \times 10^{-31}\,\text{kg})(1.31 \times 10^6\,\text{m/s})}{(1.60 \times 10^{-19}\,\text{C})^2 (1.03 \times 10^{29}\,\text{e/m}^3)(27.0 \times 10^{-9}\,\text{m})} = \boxed{1.66 \times 10^{-8}\,\Omega\cdot\text{m}}$$

(*b*) The accepted resistivity of 640 Ω·m is much greater than the calculated value. We assume that valence electrons will produce conduction in the material. Silicon is a

Solids and the Theory of Conduction 395

semiconductor and a gap between the valence band and conduction band exists. Only electrons with sufficient energies will be found in the conduction band.

The Fermi Electron Gas

*26 • Calculate the Fermi temperature for (a) Al, (b) K, and (c) Sn.

Picture the Problem The Fermi temperature T_F is defined by $kT_F = E_F$, where E_F is the Fermi energy.

The Fermi temperature is given by:
$$T_F = \frac{E_F}{k}$$

(a) For Al:
$$T_F = \frac{11.7\,\text{eV}}{8.62\times10^{-5}\,\text{eV/K}} = \boxed{1.36\times10^5\,\text{K}}$$

(b) For K:
$$T_F = \frac{2.11\,\text{eV}}{8.62\times10^{-5}\,\text{eV/K}} = \boxed{2.45\times10^4\,\text{K}}$$

(c) For Sn:
$$T_F = \frac{10.2\,\text{eV}}{8.62\times10^{-5}\,\text{eV/K}} = \boxed{1.18\times10^5\,\text{K}}$$

*31 •• (a) Assuming that gold contributes one free electron per atom to the metal, calculate the electron density in gold knowing that its atomic weight is 196.97 and its mass density is 19.3×10^3 kg/m^3. (b) If the Fermi speed for gold is 1.39×10^6 m/s, what is the Fermi energy in electron volts? (c) By what factor is the Fermi energy higher than the kT energy at room temperature? (d) Explain the difference between the Fermi energy and kT energy.

Picture the Problem We can use $n_e = \rho V = \dfrac{\rho N_A N_{atom}}{m}$, where N_{atom} is the number of electrons per atom, to calculate the electron density of gold. The Fermi energy is given by $E_F = \tfrac{1}{2}m_e v_F^2$.

(a) The electron density of gold is given by:
$$n_e = \rho V = \frac{\rho N_A N_{atom}}{m}$$

Substitute numerical values and evaluate n_e:

$$n_e = \frac{\left(19.3\times10^3\,\dfrac{\text{kg}}{\text{m}^3}\right)(6.02\times10^{23}\,\text{atoms})\left(\dfrac{1\text{e}}{1\,\text{atom}}\right)}{0.197\,\text{kg}} = \boxed{5.90\times10^{28}\,\text{e/m}^3}$$

(b) The Fermi energy is given by:
$$E_F = \tfrac{1}{2}m_e v_F^2$$

Substitute numerical values and evaluate E_F:

$$E_F = \frac{1}{2}(9.11\times10^{-31}\,\text{kg})(1.39\times10^6\,\text{m/s})^2\left(\frac{1\,\text{eV}}{1.60\times10^{-19}\,\text{J}}\right) = \boxed{5.50\,\text{eV}}$$

(c) The factor by which the Fermi energy is higher than the kT energy at room temperature is:

$$f = \frac{E_F}{kT}$$

At room temperature $kT = 0.026$ eV. Substitute numerical values and evaluate f:

$$f = \frac{5.50\,\text{eV}}{0.026\,\text{eV}} = \boxed{212}$$

(d) E_F is 212 times kT at room temperature. There are so many free electrons present that most of them are crowded, as described by the Pauli exclusion principle, up to energies far higher than they would be according to the classical model.

*32 ••   The pressure of an ideal gas is related to the average energy of the gas particles by $PV = \frac{2}{3}NE_{av}$, where N is the number of particles and E_{av} is the average energy. Use this to calculate the pressure of the Fermi electron gas in copper in newtons per square meter, and compare your result with atmospheric pressure, which is about 10^5 N/m². (*Note:* The units are most easily handled by using the conversion factors 1 N/m² = 1 J/m³ and 1 eV = 1.6×10⁻¹⁹ J.)

Picture the Problem We can solve $PV = \frac{2}{3}NE_{av}$ for P and substitute for E_{av} in order to express P in terms of N/V and E_F.

Solve $PV = \frac{2}{3}NE_{av}$ for P:

$$P = \frac{2}{3}\left(\frac{N}{V}\right)E_{av}$$

Because $E_{av} = \frac{3}{5}E_F$:

$$P = \frac{2}{3}\left(\frac{N}{V}\right)\left(\frac{3}{5}E_F\right) = \frac{2}{5}\left(\frac{N}{V}\right)E_F$$

Substitute numerical values (see Table 38-1) and evaluate P:

$$P = \frac{2}{5}(8.47\times10^{22}\,\text{electrons/cm}^3)(7.04\,\text{eV})(1.60\times10^{-19}\,\text{J/eV})$$

$$= \boxed{3.82\times10^{10}\,\text{N/m}^2} = 3.82\times10^{10}\,\text{N/m}^2 \times \frac{1\,\text{atm}}{101.325\times10^3\,\text{N/m}^2}$$

$$= \boxed{3.77\times10^5\,\text{atm}}$$

Heat Capacity Due to Electrons in a Metal

*35 •• Gold has a Fermi energy of 5.53 eV. Determine the molar specific heat at constant volume and room temperature for gold.

Picture the Problem We can use Equation 38-29 to find the molar specific heat of gold at constant volume and room temperature.

The molar specific heat is given by Equation 38-29:
$$c'_V = \frac{\pi^2 RT}{2T_F}$$

The Fermi energy is given by:
$$E_F = kT_F \Rightarrow T_F = \frac{E_F}{k}$$

Substitute for T_F to obtain:
$$c'_V = \frac{\pi^2 RkT}{2E_F}$$

Substitute numerical values and evaluate c'_V:

$$c'_V = \frac{\pi^2 (8.31\,\text{J/mol K})(1.38\times 10^{-23}\,\text{J/mol})\left(\frac{1\,\text{eV}}{1.60\times 10^{-19}\,\text{J}}\right)(300\,\text{K})}{2(5.53\,\text{eV})}$$

$$= \boxed{0.192\,\text{J/mol}\cdot\text{K}}$$

Remarks: The value 0.192 J/mol K is for a mole of gold atoms. Since each gold atom contributes one electron to the metal, a mole of gold corresponds to a mole of electrons.

Quantum Theory of Electrical Conduction

*37 •• The resistivity of pure copper is increased approximately 1×10^{-8} $\Omega\cdot$m by the addition of 1 percent (by number of atoms) of an impurity throughout the metal. The mean free path depends on both the impurity and the oscillations of the lattice ions according to the equation $1/\lambda = 1/\lambda_t + 1/\lambda_i$. (*a*) Estimate λ_i from data given in Table 38-1. (*b*) If *r* is the effective radius of an impurity lattice ion seen by an electron, the scattering cross section is πr^2. Estimate this area, using the fact that *r* is related to λ_i by Equation 38-16.

Picture the Problem We can solve the resistivity equation for the mean free path and then substitute the Fermi speed for the average speed to express the mean free path as a function of the Fermi energy.

(a) In terms of the mean free path and the mean speed, the resistivity is:

$$\rho_i = \frac{m_e v_{av}}{ne^2 \lambda_i} = \frac{m_e u_F}{ne^2 \lambda_i}$$

Solve for λ to obtain:

$$\lambda_i = \frac{m_e u_F}{ne^2 \rho_i}$$

Express the Fermi speed u_F in terms of the Fermi energy E_F:

$$u_F = \sqrt{\frac{2E_F}{m_e}}$$

Substitute to obtain:

$$\lambda_i = \frac{\sqrt{2 m_e E_F}}{ne^2 \rho_i}$$

Substitute numerical values (see Table 38-1) and evaluate λ_i:

$$\lambda_i = \frac{\sqrt{2(9.11 \times 10^{-31} \text{ kg})(7.04 \text{ eV})(1.60 \times 10^{-19} \text{ J/eV})}}{(8.47 \times 10^{28} \text{ electrons/m}^3)(1.60 \times 10^{-19} \text{ C})^2 (10^{-8} \Omega \cdot \text{m})} = \boxed{66.1 \text{ nm}}$$

(b) From Equation 38-16 we have:

$$\lambda = \frac{1}{n \pi r^2}$$

Solve for πr^2:

$$\pi r^2 = \frac{1}{n \lambda}$$

Substitute numerical values and evaluate πr^2:

$$\pi r^2 = \frac{1}{(8.47 \times 10^{28} \text{ m}^{-3})(66.1 \text{ nm})}$$
$$= 1.79 \times 10^{-22} \text{ m}^2 = \boxed{1.79 \times 10^{-4} \text{ nm}^2}$$

Band Theory of Solids

*39 • You are an electron sitting at the top of the valence band in a silicon atom, longing to jump across the 1.14-eV energy gap that separates you from the bottom of the conduction band and all of the adventures that it may contain. What you need, of course, is a photon. What is the maximum photon wavelength that will get you across the gap?

Picture the Problem We can relate the maximum photon wavelength to the energy gag using $\Delta E = hf = hc/\lambda$.

Express the energy gap as a function of the wavelength of the photon:

$$\Delta E = hf = \frac{hc}{\lambda}$$

Solids and the Theory of Conduction 399

Solve for λ:
$$\lambda = \frac{hc}{\Delta E}$$

Substitute numerical values and evaluate λ:
$$\lambda = \frac{1240\,\text{eV}\cdot\text{nm}}{1.14\,\text{eV}} = \boxed{1.09\,\mu\text{m}}$$

Semiconductors

*44 •• When a thin slab of semiconducting material is illuminated with monochromatic electromagnetic radiation, most of the radiation is transmitted through the slab if the wavelength is greater than 1.85 μm. For wavelengths less than 1.85 μm, most of the incident radiation is absorbed. Determine the energy gap of this semiconductor.

Picture the Problem We can use $E = hf$ to find the energy gap of this semiconductor.

The energy gap of the semiconductor is given by:
$$E_g = hf = \frac{hc}{\lambda}$$
where
$hc = 1240$ eV·nm

Substitute numerical values and evaluate E_g:
$$E_g = \frac{1240\,\text{eV}\cdot\text{nm}}{1.85\,\mu\text{m}} = \boxed{0.670\,\text{eV}}$$

*46 •• The ground-state energy of the hydrogen atom is given by
$$E_1 = -\frac{mk^2 e^4}{2\hbar^2} = -\frac{e^2 m_e}{8\epsilon_0^2 h^2}$$
Modify this equation in the spirit of Problem 45 by replacing ϵ_0 by $\kappa\epsilon_0$ and m_e by an effective mass for the electron to estimate the binding energy of the extra electron of an impurity arsenic atom in (*a*) silicon and (*b*) germanium.

Picture the Problem We can make the same substitutions we made in Problem 45 in the expression for E_1 (= 13.6 eV) to obtain an expression that we can use to estimate the binding energy of the extra electron of an impurity arsenic atom in silicon and germanium.

Make the indicated substitutions in the expression for E_1 to obtain:
$$E_1 = -\frac{e^2 m_{\text{eff}}}{8(\kappa\epsilon_0)^2 h^2} = -\frac{e^2 m_e m_{\text{eff}}}{8 m_e \kappa^2 \epsilon_0^2 h^2}$$
$$= -\frac{m_{\text{eff}}}{m_e \kappa^2 \epsilon_0^2} \frac{e^2 m_e}{8\epsilon_0^2 h^2}$$
$$= -\frac{m_{\text{eff}}}{m_e \kappa^2 \epsilon_0^2} E_1$$

(a) For silicon:
$$E_1 = -\frac{0.2m_e}{m_e(12)^2}(13.6\,\text{eV}) = \boxed{18.9\,\text{meV}}$$

(b) For germanium:
$$E_1 = -\frac{0.1m_e}{m_e(16)^2}(13.6\,\text{eV}) = \boxed{5.31\,\text{meV}}$$

Semiconductor Junctions and Devices

*51 •• In Figure 38-27 for the *pnp*-transistor amplifier, suppose $R_b = 2\,\text{k}\Omega$ and $R_L = 10\,\text{k}\Omega$. Suppose further that a 10-μA ac base current generates a 0.5-mA ac collector current. What is the voltage gain of the amplifier?

Picture the Problem We can use its definition to compute the voltage gain of the amplifier.

The voltage gain of the amplifier is given by:
$$\text{Voltage gain} = \frac{I_c R_L}{I_b R_b}$$

Substitute numerical values and evaluate the voltage gain:
$$\text{Voltage gain} = \frac{(0.5\,\text{mA})(10\,\text{k}\Omega)}{(10\,\mu\text{A})(2\,\text{k}\Omega)}$$
$$= \boxed{250}$$

*54 •• A "good" silicon diode has the current–voltage characteristic given in Problem 49. Let $kT = 0.025$ eV (room temperature) and the saturation current $I_0 = 1$ nA. (a) Show that for small reverse-bias voltages, the resistance is 25 MΩ. (*Hint:* Do a Taylor expansion of the exponential function or use your calculator and enter small values for V_b). (b) Find the dc resistance for a reverse bias of 0.5 V. (c) Find the dc resistance for a 0.5-V forward bias. What is the current in this case? (d) Calculate the ac resistance dV/dI for a 0.5-V forward bias.

Picture the Problem We can use Ohm's law and the expression for the current from Problem 49 to find the resistance for small reverse-and-forward bias voltages.

(a) Use Ohm's law to express the resistance:
$$R = \frac{V_b}{I} \quad (1)$$

From Problem 47, the current across a *pn* junction is given by:
$$I = I_0\left(e^{eV_b/kT} - 1\right) \quad (2)$$

For $eV_b \ll kT$:
$$e^{eV_b/kT} - 1 \approx 1 + \frac{eV_b}{kT} - 1 = \frac{eV_b}{kT}$$

Substitute to obtain:
$$I = I_0 \frac{eV_b}{kT}$$

Solids and the Theory of Conduction 401

Substitute for I in equation (1) and simplify:
$$R = \frac{V_b}{I_0 \frac{eV_b}{kT}} = \frac{kT}{eI_0}$$

Substitute numerical values and evaluate R:
$$R = \frac{(0.025\,\text{eV})(1.60\times10^{-19}\,\text{J/eV})}{(1.60\times10^{-19}\,\text{C})(10^{-9}\,\text{A})}$$
$$= \boxed{25.0\,\text{M}\Omega}$$

(b) Substitute equation (2) in equation (1) to obtain:
$$R = \frac{V_b}{I_0\left(e^{eV_b/kT} - 1\right)} \quad (3)$$

Evaluate $\dfrac{eV_b}{kT}$ for $V_b = -0.5$ V:
$$\frac{eV_b}{kT} = \frac{(1.60\times10^{-19}\,\text{C})(-0.5\,\text{V})}{(1.38\times10^{-23}\,\text{J/K})(293\,\text{K})} = -19.8$$

Evaluate equation (3) for $V_b = -0.5$ V:
$$R = \frac{-0.5\,\text{V}}{(10^{-9}\,\text{A})(e^{-19.8} - 1)} = \boxed{500\,\text{M}\Omega}$$

(c) Evaluate $\dfrac{eV_b}{kT}$ for $V_b = +0.5$ V:
$$\frac{eV_b}{kT} = \frac{(1.60\times10^{-19}\,\text{C})(0.5\,\text{V})}{(1.38\times10^{-23}\,\text{J/K})(293\,\text{K})} = 19.8$$

Evaluate equation (3) for $V_b = +0.5$ V:
$$R = \frac{0.5\,\text{V}}{(10^{-9}\,\text{A})(e^{19.8} - 1)} = \boxed{1.26\,\Omega}$$

(d) Evaluate $R_{ac} = dV/dI$ to obtain:
$$R_{ac} = \frac{dV}{dI} = \left(\frac{dI}{dV}\right)^{-1}$$
$$= \left\{\frac{d}{dV}\left[I_0\left(e^{eV_b/kT} - 1\right)\right]\right\}^{-1}$$
$$= \left\{\frac{eI_0}{kT}e^{eV_b/kT}\right\}^{-1} = \frac{kT}{eI_0}e^{-eV_b/kT}$$

Substitute numerical values and evaluate R_{ac}:
$$R_{ac} = (25\,\text{M}\Omega)e^{-19.8} = \boxed{0.0629\,\Omega}$$

The BCS Theory

*57 • Repeat Problem 56 for lead ($T_c = 7.19$ K), which has a measured energy gap of 2.73×10^{-3} eV.

Picture the Problem We can calculate E_g using $E_g = 3.5kT_c$ and find the wavelength of a photon having sufficient energy to break up Cooper pairs in tin at $T = 0$ using

$\lambda = hc/E_g$.

(a) From Equation 38-24 we have:

$$E_g = 3.5kT_c$$

Substitute numerical values and evaluate E_g:

$$E_g = 3.5(8.62 \times 10^{-5} \text{ eV/K})(7.19 \text{ K})$$
$$= \boxed{2.17 \text{ meV}}$$

Express the ratio of E_g to $E_{g,\text{measured}}$:

$$\frac{E_g}{E_{g,\text{measured}}} = \frac{2.17 \text{ meV}}{2.73 \times 10^{-3} \text{ eV}} = 0.795$$

or

$$E_g \approx \boxed{0.8 E_{g,\text{measured}}}$$

(b) The wavelength of a photon having sufficient energy to break up Cooper pairs in tin at $T = 0$ is given by:

$$\lambda = \frac{hc}{E_g}$$

Substitute numerical values and evaluate λ:

$$\lambda = \frac{1240 \text{ eV} \cdot \text{nm}}{2.73 \times 10^{-3} \text{ eV}} = 4.54 \times 10^5 \text{ nm}$$
$$= \boxed{0.454 \text{ mm}}$$

The Fermi-Dirac Distribution

*60 •• (a) Use Equation 38-22a to calculate the Fermi energy for silver. (b) Determine the average energy of a free electron and (c) find the Fermi speed for silver.

Picture the Problem Equation 38-22a expresses the dependence of the Fermi energy E_F on the number density of free electrons. Once we've determined the Fermi energy for silver, we can find the average electron energy from the Fermi energy for silver and then use the average electron energy to find the Fermi speed for silver.

(a) From Equation 38-22a we have:

$$E_F = \frac{h^2}{8m_e}\left(\frac{3N}{\pi V}\right)^{3/2}$$

Use Table 27-1 to find the free-electron number density N/V for silver:

$$\frac{N}{V} = 5.86 \times 10^{22} \frac{\text{electrons}}{\text{cm}^3}$$
$$= 5.86 \times 10^{28} \frac{\text{electrons}}{\text{m}^3}$$

Substitute numerical values and evaluate E_F:

$$E_F = \frac{(6.63 \times 10^{-34}\,\text{J}\cdot\text{s})^2}{8(9.11 \times 10^{-31}\,\text{kg})} \left[\frac{3(5.86 \times 10^{28}\,\text{electrons/m}^3)}{\pi}\right]^{2/3} \left(\frac{1\,\text{eV}}{1.60 \times 10^{-19}\,\text{J}}\right)$$

$$= \boxed{5.51\,\text{eV}}$$

(b) The average electron energy is given by:

$$E_{av} = \frac{3}{5} E_F$$

Substitute numerical values and evaluate E_{av}:

$$E_{av} = \frac{3}{5}(5.51\,\text{eV}) = \boxed{3.31\,\text{eV}}$$

(c) Express the Fermi energy in terms of the Fermi speed of the electrons:

$$E_F = \tfrac{1}{2} m_e v_F^2$$

Solve for v_F:

$$v_F = \sqrt{\frac{2 E_F}{m_e}}$$

Substitute numerical values and evaluate v_F:

$$v_F = \sqrt{\frac{2(3.31\,\text{eV})}{9.11 \times 10^{-31}\,\text{kg}} \left(\frac{1.60 \times 10^{-19}\,\text{J}}{1\,\text{eV}}\right)}$$

$$= \boxed{1.08 \times 10^6\,\text{m/s}}$$

*63 •• What is the probability that a conduction electron in silver will have a kinetic energy of 4.9 eV at $T = 300$ K?

Picture the Problem The probability that a conduction electron will have a given kinetic energy is given by the Fermi factor.

The Fermi factor is:

$$f(E) = \frac{1}{e^{(E - E_F)/kT} + 1}$$

Because $E_F - 4.9\,\text{eV} \gg 300k$:

$$f(4.9\,\text{eV}) = \frac{1}{0 + 1} = \boxed{1}$$

*70 ••• (a) Show that for $E \geq 0$, the Fermi factor may be written as $f(E) = 1/(Ce^{E/kT} + 1)$. (b) Show that if $C \gg e^{-E/kT}$, $f(E) = Ae^{-E/kT} \ll 1$; in other words, show that the Fermi factor is a constant times the classical Boltzmann factor if $A \ll 1$. (c) Use $\int n(E)\,dE = N$ and Equation 38-41 to determine the constant A. (d) Using the result obtained in Part (c), show that the classical approximation is applicable when the electron concentration is very small and/or the temperature is very high. (e) Most semiconductors have impurities added in a process called doping, which increases the free electron concentration so that it is about $10^{17}/\text{cm}^3$ at room temperature. Show that for these systems, the classical distribution function is applicable.

Picture the Problem We can follow the step-by-step procedure outlined in the problem statement to obtain the indicated results.

(a) The Fermi factor is:
$$f(E) = \frac{1}{e^{(E-E_F)/kT}+1} = \frac{1}{e^{-E_F/kT}e^{E/kT}+1}$$
$$= \boxed{\frac{1}{Ce^{E/kT}+1}}$$
provided $C = e^{-E_F/kT}$

(b) If $C \gg e^{-E/kT}$:
$$f(E) = \frac{1}{Ce^{E/kT}+1} \approx \frac{1}{Ce^{E/kT}} = \boxed{Ae^{-E/kT}}$$
where $A = 1/C$

(c) The energy distribution function is:
$$n(E)dE = g(E)dE\, f(E)$$
where
$$g(E) = \frac{8\pi\sqrt{2}\,m_e^{3/2}V}{h^3}E^{1/2}$$

Substitute for $g(E)dE$ and $f(E)$ in the expression for N to obtain:
$$N = A\frac{8\pi\sqrt{2}\,m_e^{3/2}V}{h^3}\int_0^\infty E^{1/2}e^{-E/kT}dE$$

The definite integral has the value:
$$\int_0^\infty E^{1/2}e^{-E/kT}dE = \frac{(kT)^{3/2}}{2}\sqrt{\pi}$$

Substitute to obtain:
$$N = A\frac{8\pi\sqrt{2}\,m_e^{3/2}V}{h^3}\frac{(kT)^{3/2}}{2}\sqrt{\pi}$$

Solve for A:
$$A = \boxed{\frac{\sqrt{2}\,h^3}{8\pi^{3/2}m_e^{3/2}}\left(\frac{N}{V}\right)\frac{1}{(kT)^{3/2}}}$$

(d) Evaluate A at $T = 300$ K:

$$A = \frac{\sqrt{2}(6.63\times 10^{-34}\,\text{J}\cdot\text{s})^3\,n}{8\pi^{3/2}(9.11\times 10^{-31}\,\text{kg})^{3/2}[(1.38\times 10^{-23}\,\text{J/K})(300\,\text{K})]^{3/2}} \approx 4\times 10^{-26}\,n$$

where the units are SI.

> The valence electron concentration is typically about 10^{39} m^{-3}. To satisfy the condition that $A \ll 1$ at room temperature, n should be less than 10^{23} m^{-3}, or about one millionth of the valence electron concentration. Because A depends on $T^{-3/2}$, the electron concentration may be greater the higher the temperature.

(e) > 10^{17} cm^{-3} = 10^{23} m^{-3}. So, according to the criterion in (d), the classical approximation is applicable.

General Problems

*76 •• Determine the energy that has 10 percent free electron occupancy probability for manganese at $T = 1300$ K.

Picture the Problem The Fermi factor gives the probability of an energy state being occupied as a function of the energy of the state E, the Fermi energy E_F for the particular material, and the temperature T.

The Fermi factor is:
$$f(E) = \frac{1}{e^{(E-E_F)/kT} + 1}$$

For 10 percent probability:
$$0.1 = \frac{1}{e^{(E-E_F)/kT} + 1}$$
or
$$e^{(E-E_F)/kT} = 9$$

Take the natural logarithm of both sides of the equation to obtain:
$$\frac{E - E_F}{kT} = \ln 9$$

Solve for E to obtain:
$$E = E_F + kT \ln 9$$

From Table 37-1, E_F(Mn) = 11.0 eV. Substitute numerical values and evaluate E:

$$E = 11.0\,\text{eV} + (1.38 \times 10^{-23}\,\text{J/K})(1300\,\text{K})\left(\frac{1\,\text{eV}}{1.60 \times 10^{-19}\,\text{J}}\right) \ln 9 = \boxed{11.2\,\text{eV}}$$

*78 ••• A 2-cm^2 wafer of pure silicon is irradiated with light having a wavelength of 775 nm. The intensity of the light beam is 4.0 W/m^2 and every photon that strikes the sample is absorbed and creates an electron–hole pair.
(a) How many electron–hole pairs are produced in one second? (b) If the number of electron–hole pairs in the sample is 6.25×10^{11} in the steady state, at what rate do the electron–hole pairs recombine? (c) If every recombination event results in the radiation of one photon, at what rate is energy radiated by the sample?

Picture the Problem The rate of production of electron-hole pairs is the ratio of the incident energy to the energy required to produce an electron-hole pair.

(a) The number of electron-hole pairs N produced in one second is:

$$N = \frac{IA}{\frac{hc}{\lambda}} = \frac{IA\lambda}{hc}$$

Substitute numerical values and evaluate N:

$$N = \frac{(4.0\,\text{W/m}^2)(2\times 10^{-4}\,\text{m}^2)(775\,\text{nm})}{(1240\,\text{eV}\cdot\text{nm})(1.60\times 10^{-19}\,\text{J/eV})} = \boxed{3.12\times 10^{15}\,\text{s}^{-1}}$$

(b) In the steady state, the rate of recombination equals the rate of generation. Therefore:

$$N = \boxed{3.12\times 10^{15}\,\text{s}^{-1}}$$

(c) The power radiated equals the power absorbed:

$$P_\text{rad} = IA$$

Substitute numerical values and evaluate P_rad:

$$P_\text{rad} = (4.0\,\text{W/m}^2)(2\times 10^{-4}\,\text{m}^2) = \boxed{0.800\,\text{mJ/s}}$$

Chapter 39
Relativity

Conceptual Problems

*1 • The approximate total energy of a particle of mass m moving at speed $u \ll c$ is (a) $mc^2 + \frac{1}{2}mu^2$. (b) $\frac{1}{2}mu^2$. (c) cmu. (d) $\frac{1}{2}mc^2$. (e) $\frac{1}{2}cmu$.

Picture the Problem The total relativistic energy E of a particle is defined to be the sum of its kinetic and rest energies.

The total relativistic energy of a particle is given by:
$$E = K + mc^2 = \tfrac{1}{2}mu^2 + mc^2$$
and $\boxed{(a) \text{ is correct.}}$

*2 • A set of twins work in an office building. One twin works on the top floor and the other twin works in the basement. Considering general relativity, which one will age more quickly? (a) They will age at the same rate. (b) The twin who works on the top floor will age more quickly. (c) The twin who works in the basement will age more quickly. (d) It depends on the speed of the office building. (e) None of these is correct.

Determine the Concept The gravitational field of the earth is slightly greater in the basement of the office building than it is at the top floor. Because clocks run more slowly in regions of low gravitational potential, clocks in the basement will run more slowly than clocks on the top floor. Hence, the twin who works on the top floor will age more quickly. $\boxed{(b) \text{ is correct.}}$

Estimation and Approximation

*7 •• The most distant galaxies which can be seen by the Hubble telescope are moving away from us with a redshift parameter of about $z = 5$. (See Problem 30 for a definition of z). (a) What is the relative velocity of these galaxies with respect to us (expressed as a fraction of the speed of light)? (b) Hubble's law states that the recession velocity is given by the expression $v = Hx$, where v is the velocity of recession, x is the distance, and H is the Hubble constant, $H = 75$ km/s/Mpc. (1 pc = 3.26 light-year). Estimate the distance of such a galaxy using the information given.

Picture the Problem We can use the result from Problem 30, for light that is Doppler-shifted with respect to an observer, $v = c\left(\dfrac{u^2-1}{u^2+1}\right)$, where $u = z + 1$ and z is the red-shift parameter, to find the ratio of v to c. In (b) we can solve Hubble's law for x and substitute our result from (a) to estimate the distance to the galaxy.

408 Chapter 39

(a) Use the result of Problem 30 to express v/c as a function of z:

$$\frac{v}{c} = \frac{(z+1)^2 - 1}{(z+1)^2 + 1}$$

Substitute for z and evaluate v/c:

$$\frac{v}{c} = \frac{(5+1)^2 - 1}{(5+1)^2 + 1} = \boxed{0.946}$$

(b) Solve Hubble's law for x:

$$x = \frac{v}{H}$$

Substitute numerical values and evaluate x:

$$x = \frac{0.946c}{H} = \frac{0.946(3 \times 10^5 \text{ km/s})}{75 \frac{\text{km/s}}{\text{Mpc}}}$$

$$= 3.78 \times 10^3 \text{ Mpc} \times \frac{3.26 \times 10^6 \ c \cdot y}{\text{Mpc}}$$

$$= \boxed{12.3 \text{ Gc} \cdot y}$$

Time Dilation and Length Contraction

*10 •• Unobtainium (Un) is an unstable particle that decays into Normalium (Nr) and Standardium (St) particles. (a) An accelerator produces a beam of Un which travels to a detector located 1000 m away from the accelerator. The particles travel with a velocity of $v = 0.866c$. How long do the particles take (in the laboratory frame) to get to the detector? (b) By the time the particles get to the detector, half of them have decayed. What is the half-life of Un? (*Note*: Half-life as it would be measured in a frame moving with the particles). (c) A new detector is going to be used, which is located 10,000 m away from the accelerator. How fast should the particles be moving if half of them are to make it to the new detector?

Picture the Problem The time required for the particles to reach the detector, as measured in the laboratory frame of reference is the ratio of the distance they must travel to their speed. The half life of the particles is the trip time as measured in a frame traveling with the particles. We can find the speed at which the particles must move if they are to reach the more distant detector by equating their half life to the ratio of the distance to the detector in the particle's frame of reference to their speed.

(a) The time required to reach the detector is the ratio of the distance to the detector and the speed with which the particles are traveling:

$$\Delta t = \frac{\Delta x}{v} = \frac{\Delta x}{0.866c}$$

Substitute numerical values and evaluate Δt:

$$\Delta t = \frac{1000 \text{ m}}{0.866(3 \times 10^8 \text{ m/s})} = \boxed{3.85 \ \mu s}$$

(b) The half life is the trip time as measured in a frame traveling with the particles:

$$\Delta t' = \frac{\Delta t}{\gamma} = \Delta t \sqrt{1-\left(\frac{v}{c}\right)^2}$$

Substitute numerical values and evaluate $\Delta t'$:

$$\Delta t' = 3.85\,\mu s\sqrt{1-\left(\frac{0.866c}{c}\right)^2}$$

$$= \boxed{1.93\,\mu s}$$

(c) In order for half the particles to reach the detector:

$$\Delta t' = \frac{\Delta x'}{\gamma v} = \frac{\Delta x'\sqrt{1-\left(\frac{v}{c}\right)^2}}{v}$$

where $\Delta x'$ is the distance to the new detector.

Rewrite this expression to obtain:

$$\frac{v}{\sqrt{1-\left(\frac{v}{c}\right)^2}} = \frac{\Delta x'}{\Delta t'}$$

Squaring both sides of the equation yields:

$$\frac{v^2}{1-\left(\frac{v}{c}\right)^2} = \left(\frac{\Delta x'}{\Delta t'}\right)^2$$

Substitute numerical values for $\Delta x'$ and $\Delta t'$ and simplify to obtain:

$$\frac{v^2}{1-\left(\frac{v}{c}\right)^2} = \left(\frac{10^4\,\text{m}}{1.93\,\mu s}\right)^2 = (17.3c)^2$$

Divide both sides of the equation by c^2 to obtain:

$$\frac{\frac{v^2}{c^2}}{1-\left(\frac{v}{c}\right)^2} = (17.3)^2$$

Solve this equation for v^2/c^2:

$$\frac{v^2}{c^2} = \frac{(17.3)^2}{1+(17.3)^2} = 0.9967$$

Finally, solving for v yields:

$$v = \boxed{0.998c}$$

The Lorentz Transformation, Clock Synchronization, and Simultaneity

*17 •• A spaceship of proper length $L_p = 400$ m moves past a transmitting station at a speed of $0.76c$. At the instant that the nose of the spaceship passes the transmitter, clocks at the transmitter and in the nose of the spaceship are synchronized to $t = t' = 0$. The instant that the tail of the spaceship passes the transmitter a signal is sent and subsequently detected by the receiver in the nose of the spaceship. (*a*) When, according to the clock in the spaceship, is the signal sent? (*b*) When, according to the clock at the transmitter, is the signal received by the spaceship? (*c*) When, according to the clock in the spaceship, is the signal received? (*d*) Where, according to an observer at the transmitter, is the nose of the spaceship when the signal is received?

Picture the Problem Let S be the reference frame of the spaceship and S' be that of the earth (transmitter station). Let event A be the emission of the light pulse and event B the reception of the light pulse at the nose of the spaceship. In (*a*) and (*c*) we can use the classical distance, rate, and time relationship and in (*b*) and (*d*) we can apply the inverse Lorentz transformations.

(*a*) In both S and S' the pulse travels at the speed c. Thus:

$$t_A = \frac{L_p}{v} = \frac{400\,\text{m}}{0.76c} = \boxed{1.76\,\mu s}$$

(*c*) The elapsed time, according to the clock on the ship is:

$$t_B = t_{\text{pulse to travel length of ship}} + t_A$$

Find the time of travel of the pulse to the nose of the ship:

$$t_{\text{pulse to travel length of ship}} = \frac{400\,\text{m}}{2.998 \times 10^8\,\text{m/s}} = 1.33\,\mu s$$

Substitute numerical values and evaluate t_B:

$$t_B = 1.33\,\mu s + 1.76\,\mu s = \boxed{3.09\,\mu s}$$

(*b*) The inverse time transformation is:

$$t_B' = \gamma\left(t - \frac{vx}{c^2}\right)$$

where

$$\gamma = \frac{1}{\sqrt{1 - \frac{v^2}{c^2}}} = \frac{1}{\sqrt{1 - \frac{(0.76c)^2}{c^2}}} = 1.54$$

Substitute numerical values and evaluate t'_B:

$$t_B' = (1.54)\left(3.09\,\mu s - \frac{(-0.76c)(400\,m)}{c^2}\right)$$

$$= (1.54)\left(3.09\,\mu s - \frac{(-0.76)(400\,m)}{3\times10^8\,m/s}\right)$$

$$= \boxed{6.32\,\mu s}$$

(d) The inverse transformation for x is:

$$x' = \gamma(x - vt)$$

Substitute numerical values and evaluate x':

$$x' = (1.54)[400\,m - (-0.76)(3\times10^8\,m/s)(3.09\times10^{-6}\,s)] = \boxed{1.70\,km}$$

*22 ••• An observer in frame S standing at the origin observes two flashes of colored light separated spatially by $\Delta x = 2400$ m. A blue flash occurs first, followed by a red flash 5 μs later. An observer in S' moving along the x axis at speed v relative to S also observes the flashes 5 μs apart and with a separation of 2400 m, but the red flash is observed first. Find the magnitude and direction of v.

Picture the Problem We can use the inverse time dilation equation to derive an expression for the elapsed time between the flashes in S' in terms of the elapsed time between the flashes in S, their separation in space, and the speed v with which S' is moving.

From the inverse time transformation we have:

$$\Delta t' = \gamma\left[\Delta t - \frac{v}{c^2}\Delta x\right]$$

where $\Delta t'$ is the time between the flashes in S' and Δt and Δx are the elapsed time between the flashes and their separation in S.

Set $\Delta t' = -\Delta t$ to obtain:

$$\frac{-\Delta t}{\gamma} = \Delta t - \frac{v}{c^2}\Delta x$$

or

$$-\Delta t\sqrt{1-\frac{v^2}{c^2}} = \Delta t - \frac{v}{c^2}\Delta x$$

Square both sides of the equation:

$$(\Delta t)^2 - \frac{v^2}{c^2}(\Delta t)^2 = (\Delta t)^2 - 2\frac{v}{c^2}\Delta x\Delta t + \frac{v^2}{c^4}(\Delta x)^2$$

Simplify to obtain:

$$-v(\Delta t)^2 = -2\Delta x\Delta t + \frac{v}{c^2}(\Delta x)^2$$

Solve for v:

$$v = \frac{2\dfrac{\Delta x}{\Delta t}}{1+\left(\dfrac{1}{c}\dfrac{\Delta x}{\Delta t}\right)^2}$$

Substitute numerical values and evaluate v:

$$v = \frac{2\left(\dfrac{2400\,\text{m}}{5\,\mu\text{s}}\right)}{1+\left[\dfrac{1}{3\times 10^8\,\text{m/s}}\left(\dfrac{2400\,\text{m}}{5\,\mu\text{s}}\right)\right]^2}$$

$$= 2.697\times 10^8\,\text{m/s} = \boxed{0.899c}$$

Because v is positive, S' is moving in the positive x direction.

The Velocity Transformation

*24 •• A spaceship, at rest in a certain reference frame S, is given a speed increment of $0.50c$ (call this boost 1). Relative to its new rest frame, the spaceship is given a further $0.50c$ increment 10 seconds later (as measured in its new rest frame; call this boost 2). This process is continued indefinitely, at 10-second intervals, as measured in the rest frame of the ship. (Assume that the boost itself takes a very short time compared to 10 s.) (*a*) Using a spreadsheet program, calculate and graph the velocity of the spaceship in reference frame S as a function of the boost number for boost 1 to boost 10. (*b*) Graph the gamma factor the same way. (*c*) How many boosts does it take until the velocity of the ship in S is greater than $0.999c$? (*d*) How far has the spaceship moved after 5 boosts, as measured in reference frame S? What is the average speed of the spaceship (between boost 1 and boost 5) as measured in S?

Picture the Problem We'll let the velocity (in S) of the spaceship after the ith boost be v_i and derive an expression for the ratio of v to c after the spaceship's $(i+1)$th boost as a function of N. We can use the definition of γ, in terms of v/c to plot γ as a function of N.

(*a*) and (*b*) The velocity of the spaceship after the $(i+1)$th boost is given by relativistic velocity addition equation:

$$v_{i+1} = \frac{v_i + 0.5c}{1+\dfrac{(0.5c)v_i}{c^2}}$$

Factor c from both the numerator and denominator to obtain:

$$v_{i+1} = \frac{\dfrac{v_i}{c} + 0.5}{1 + 0.5\dfrac{v_i}{c}}$$

γ_i is given by:

$$\gamma_i = \frac{1}{\sqrt{1 - \left(\dfrac{v_i}{c}\right)}}$$

A spreadsheet program to calculate v/c and γ as functions of the number of boosts N is shown below. The formulas used to calculate the quantities in the columns are as follows:

Cell	Content/Formula	Algebraic Form
A3	0	N
B2	0	v_0
B3	(B2+0.5)/(1+0.5*B2)	v_{i+1}
C1	1/(1−B2^2)^0.5	γ

	A	B	C
1	boost	v/c	gamma
2	0	0.000	1.00
3	1	0.500	1.15
4	2	0.800	1.67
5	3	0.929	2.69
6	4	0.976	4.56
7	5	0.992	7.83
8	6	0.997	13.52
9	7	0.999	23.39
10	8	1.000	40.51
11	9	1.000	70.15
12	10	1.000	121.50

A graph of v/c as a function of N is shown below:

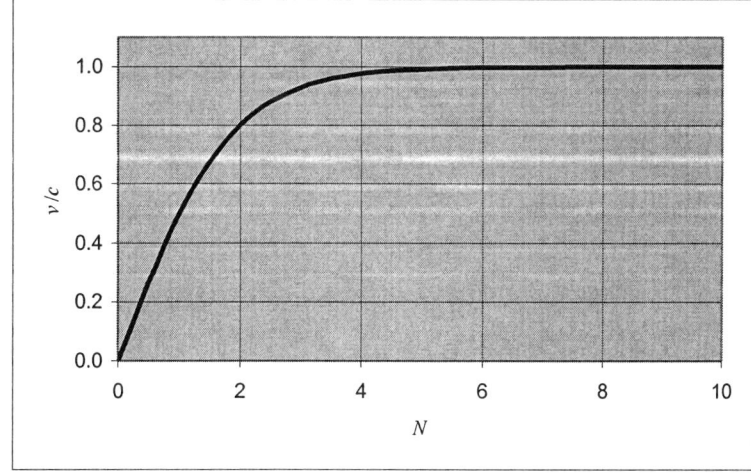

A graph of γ as a function of N is shown below:

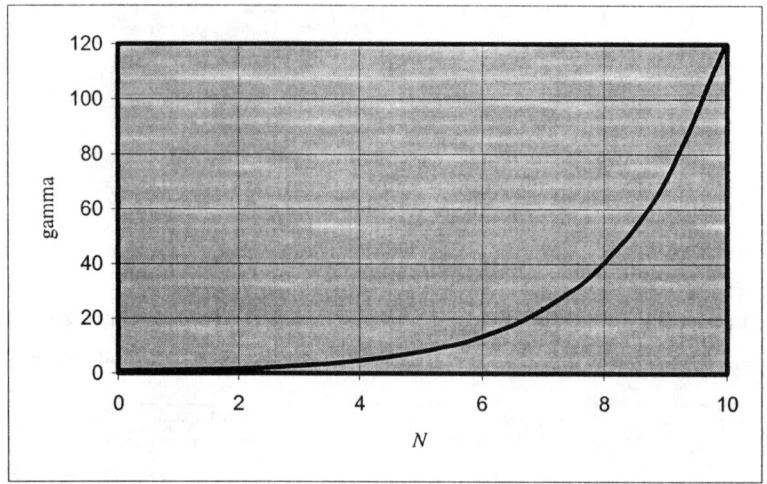

(c) Examination of the spreadsheet or of the graph of v/c as a function of N indicates that, after 8 boosts, the velocity of the spaceship is greater than $0.999c$.

(d) After 5 boosts, the spaceship has traveled a distance Δx, measured in the earth frame of reference (S), given by:

$$\Delta x = \Delta x_{1\to 2} + \Delta x_{2\to 3} + \Delta x_{3\to 4} + \Delta x_{4\to 5}$$
$$= (0.5c)(10\,\text{s})\gamma_{1\to 2} + (0.8c)(10\,\text{s})\gamma_{2\to 3} + (0.929c)(10\,\text{s})\gamma_{3\to 4} + (0.976c)(10\,\text{s})\gamma_{4\to 5}$$
$$\quad + (0.992c)(10\,\text{s})\gamma_{5\to 6}$$
$$= (0.5c)(10\,\text{s})(1.15) + (0.8c)(10\,\text{s})(1.67) + (0.929c)(10\,\text{s})(2.69)$$
$$\quad + (0.976c)(10\,\text{s})(4.56) + (0.992c)(10\,\text{s})(7.83)$$
$$= \boxed{166\,c\cdot\text{s}}$$

The average speed of the spaceship, between boost 1 and boost 5, as measured in S is given by:

$$v_{av} = \frac{\Delta x}{\Delta t}$$

where Δt is the travel time as measured in the earth frame of reference.

Express Δt as the sum of the times the spaceship travels during each 10-s interval following a boost in its speed:

$$\Delta t = \Delta t_{1\to 2} + \Delta t_{2\to 3} + \Delta t_{3\to 4} + \Delta t_{4\to 5}$$
$$= (10\,\text{s})\gamma_{1\to 2} + (10\,\text{s})\gamma_{2\to 3} + (10\,\text{s})\gamma_{3\to 4} + (10\,\text{s})\gamma_{4\to 5} + (10\,\text{s})\gamma_{5\to 6}$$
$$= (10\,\text{s})(\gamma_{1\to 2} + \gamma_{2\to 3} + \gamma_{3\to 4} + \gamma_{4\to 5} + \gamma_{5\to 6})$$

Substitute numerical values and evaluate Δt:

$$\Delta t = (10\,\text{s})(1.15 + 1.67 + 2.69 + 4.56 + 7.83) = 179\,\text{s}$$

Substitute for Δx and Δt and evaluate v_{av}:

$$v_{av} = \frac{166 c \cdot s}{179 s} = \boxed{0.927c}$$

Remarks: This result seems to be reasonable. Relativistic time dilation implies that the spacecraft will be spending larger amounts of time at high speed (as seen in reference frame S).

The Relativistic Doppler Shift

*29 •• A clock is placed in a satellite that orbits the earth with a period of 90 min. By what time interval will this clock differ from an identical clock on earth after 1 y? (Assume that special relativity applies and neglect general relativity.)

Picture the Problem Due to its motion, the orbiting clock will run more slowly than the earth-bound clock. We can use Kepler's third law to find the radius of the satellite's orbit in terms of its period, the definition of speed to find the orbital speed of the satellite from the radius of its orbit, and the time dilation equation to find the difference δ in the readings of the two clocks.

Express the time δ lost by the clock:

$$\delta = \Delta t - \Delta t_p = \Delta t - \frac{\Delta t}{\gamma} = \Delta t \left(1 - \frac{1}{\gamma}\right)$$

Because $v \ll c$, we can use part (*b*) of Problem 13:

$$\frac{1}{\gamma} \approx 1 - \frac{1}{2}\frac{v^2}{c^2}$$

Substitute to obtain:

$$\delta = \Delta t \left[1 - \left(1 - \frac{1}{2}\frac{v^2}{c^2}\right)\right] = \frac{1}{2}\frac{v^2}{c^2} \Delta t \quad (1)$$

Express the square of the speed of the satellite in its orbit:

$$v^2 = \left(\frac{2\pi r}{T}\right)^2 = \frac{4\pi^2 r^2}{T^2} \quad (2)$$

where T is its period and r is the radius of its (assumed) circular orbit.

Use Kepler's third law to relate the period of the satellite to the radius of its orbit about the earth:

$$T^2 = \frac{4\pi^2}{GM_e} r^3 = \frac{4\pi^2}{gR_e^2} r^3$$

Solve for r:

$$r = \sqrt[3]{\frac{gR_e^2 T^2}{4\pi^2}}$$

Substitute numerical values and evaluate r:

$$r = \sqrt[3]{\frac{(9.81\,\text{m/s}^2)(6370\,\text{km})^2(90\,\text{min} \times 60\,\text{s/min})^2}{4\pi^2}} = 6.65 \times 10^6\,\text{m}$$

Substitute numerical values in equation (2) and evaluate v^2:

$$v^2 = \frac{4\pi^2(6.65 \times 10^6\,\text{m})^2}{(90\,\text{min} \times 60\,\text{s/min})^2}$$
$$= 5.99 \times 10^7\,\text{m}^2/\text{s}^2$$

Finally, substitute for v^2 in equation (1) and evaluate δ:

$$\delta = \frac{1}{2}\frac{(5.99 \times 10^7\,\text{m}^2/\text{s}^2)(1\,\text{y} \times 31.56\,\text{Ms/y})}{(3 \times 10^8\,\text{m/s})^2} = \boxed{10.5\,\text{ms}}$$

A particle moves with speed $0.8c$ along the x'' axis of frame S'', which moves with speed $0.8c$ along the x' axis relative to frame S'. Frame S' moves with speed $0.8c$ along the x axis relative to frame S. (a) Find the speed of the particle relative to frame S'. (b) Find the speed of the particle relative to frame S.

Picture the Problem We can apply the inverse velocity transformation equation to express the speed of the particle relative to both frames of reference.

(a) Express u_x' in terms of u_x'':

$$u_x' = \frac{u_x'' + v}{1 + \frac{v u_x''}{c^2}}$$

where V of S', relative to S'', is $0.8c$.

Substitute numerical values and evaluate u_x':

$$u_x' = \frac{0.8c + 0.8c}{1 + \frac{(0.8c)^2}{c^2}} = \frac{1.6c}{1.64} = \boxed{0.976c}$$

(b) Express u_x in terms of u_x':

$$u_x = \frac{u_x' + v}{1 + \frac{v u_x'}{c^2}}$$

where v of S, relative to S', is $0.8c$.

Substitute numerical values and evaluate u_x:

$$u_x = \frac{0.976c + 0.8c}{1 + \frac{(0.8c)(0.976c)}{c^2}} = \frac{1.776c}{1.781}$$
$$= \boxed{0.997c}$$

Relativistic Momentum and Relativistic Energy

*34 • A proton (rest energy 938 MeV) has a total energy of 2200 MeV. (*a*) What is its speed? (*b*) What is its momentum?

Picture the Problem We can use the relation for the total energy, momentum, and rest energy to find the momentum of the proton and Equation 39-26 to relate the speed of the proton to its energy and momentum.

Relate the energy of the proton to its momentum:

$$E^2 = p^2c^2 + (mc^2)^2$$

(*b*) Solve for *p* to obtain:

$$p = \sqrt{\frac{E^2 - (mc^2)^2}{c^2}}$$

Substitute numerical values and evaluate *p*:

$$p = \frac{\sqrt{(2200\,\text{MeV})^2 - (938\,\text{MeV})^2}}{c}$$

$$= \boxed{1.99\,\frac{\text{GeV}}{c}}$$

(*a*) From Equation 39-26 we have:

$$\frac{v}{c} = \frac{pc}{E}$$

Solve for *v* to obtain:

$$v = \frac{pc}{E}c$$

Substitute numerical values and evaluate *v*:

$$v = \frac{1.99\,\text{GeV}}{2200\,\text{MeV}}c = \boxed{0.905c}$$

*39 •• Two protons approach each other head-on at 0.5*c* relative to reference frame *S'*. (*a*) Calculate the total kinetic energy of the two protons as seen in frame *S'*. (*b*) Calculate the total kinetic energy of the protons as seen in reference frame *S*, which is moving with speed 0.5*c* relative to *S'* so that one of the protons is at rest.

Picture the Problem The total kinetic energy of the two protons in part (*a*) is the sum of their kinetic energies and is given by $K = 2(\gamma - 1)E_0$. Part (*b*) differs from part (*a*) in that we need to find the speed of the moving proton relative to frame *S*.

(*a*) The total kinetic energy of the protons in frame *S'* is given by:

$$K = 2(\gamma - 1)E_0$$

418   Chapter 39

Substitute for γ and E_0 and evaluate K:

$$K = 2\left(\frac{1}{\sqrt{1-\frac{(0.5c)^2}{c^2}}} - 1\right)(938.28\,\text{MeV})$$

$$= \boxed{290\,\text{MeV}}$$

(b) The kinetic energy of the moving proton in frame S is given by:

$$K = (\gamma - 1)E_0 \qquad (1)$$

where

$$\gamma = \frac{1}{\sqrt{1-\frac{uv}{c^2}}}$$

Express the speed u of the proton in frame S:

$$u = \frac{u_x' + v}{1 + \frac{vu_x'}{c^2}}$$

Substitute numerical values and evaluate u:

$$u = \frac{0.5c + 0.5c}{1 + \frac{(0.5c)(0.5c)}{c^2}} = 0.800c$$

Evaluate γ:

$$\gamma = \frac{1}{\sqrt{1-\frac{(0.8c)(0.8c)}{c^2}}} = 1.67$$

Substitute numerical values in equation (1) and evaluate K:

$$K = (1.67 - 1)(938.28\,\text{MeV})$$

$$= \boxed{629\,\text{MeV}}$$

General Relativity

*42 ••   Light traveling in the direction of increasing gravitational potential undergoes a frequency redshift. Calculate the shift in wavelength if a beam of light of wavelength $\lambda = 632.8$ nm is sent up a vertical shaft of height $L = 100$ m.

Picture the Problem Let m represent the mass equivalent of a photon. We can equate the change in the gravitational potential energy of a photon as it rises a distance L in the gravitational field to $h\Delta f$ and then express the wavelength shift in terms of the frequency shift.

The speed of the photons in the light beam are related to their frequency and wavelength:

$$c = f\lambda \Rightarrow f = \frac{c}{\lambda}$$

Differentiate this expression with respect to λ to obtain:	$\dfrac{df}{d\lambda} = -c\lambda^{-2} = -\dfrac{c}{\lambda^2}$	
Approximate $df/d\lambda$ by $\Delta f/\Delta\lambda$ and solve for Δf:	$\Delta f = -\dfrac{c}{\lambda^2}\Delta\lambda$	
Divide both sides of this equation by f to obtain:	$\dfrac{\Delta f}{f} = \dfrac{-\dfrac{c}{\lambda^2}\Delta\lambda}{\dfrac{c}{\lambda}} = -\dfrac{\Delta\lambda}{\lambda}$	
Solve for $\Delta\lambda$:	$\Delta\lambda = -\lambda\dfrac{\Delta f}{f}$	(1)
The change in the energy of the photon as it rises a distance L in a gravitational field is given by:	$\Delta E = \Delta U = mgL$	
Because $\Delta E = h\Delta f$:	$h\Delta f = mgL$	(2)
Letting m represent the mass equivalent of the photon:	$E = mc^2 = hf$	(3)
Divide equation (2) by equation (3) to obtain:	$\dfrac{h\Delta f}{hf} = \dfrac{mgL}{mc^2} \Rightarrow \dfrac{\Delta f}{f} = \dfrac{gL}{c^2}$	
Substitute for $\Delta f/f$ in equation (1):	$\Delta\lambda = -\dfrac{gL\lambda}{c^2}$	
Substitute numerical values and evaluate $\Delta\lambda$:	$\Delta\lambda = -\dfrac{(9.81\,\text{m/s}^2)(100\,\text{m})(632.8\,\text{nm})}{(3\times10^8\,\text{m/s})^2}$ $= \boxed{-6.90\times10^{-12}\,\text{nm}}$	

General Problems

*47 •• Frames S and S' are moving relative to each other along the x and x' axes. Observers in the two frames set their clocks to $t = 0$ when the origins coincide. In frame S, event 1 occurs at $x_1 = 1.0\ c\cdot y$ and $t_1 = 1$ y and event 2 occurs at $x_2 = 2.0\ c\cdot y$ and $t_2 = 0.5$ y. These events occur simultaneously in frame S'. (*a*) Find the magnitude and direction of the velocity of S' relative to S. (*b*) At what time do both these events occur as measured in S'?

420   Chapter 39

Picture the Problem We can use Equation 39-12, the inverse time transformation equation, to relate the elapsed times and separations of the events in the two systems to the velocity of S' relative to S. We can use this same relationship in (b) to find the time at which these events occur as measured in S'.

(a) Use Equation 39-12 to obtain:

$$\Delta t' = t_2' - t_1' = \gamma\left[(t_2 - t_1) - \frac{v}{c^2}(x_2 - x_1)\right]$$

$$= \gamma\left[\Delta t - \frac{v}{c^2}\Delta x\right]$$

Because the events occur simultaneously in frame S', $\Delta t' = 0$ and:

$$0 = \Delta t - \frac{v}{c^2}\Delta x$$

Solve for v to obtain:

$$v = \frac{c^2 \Delta t}{\Delta x}$$

Substitute for Δt and Δx and evaluate V:

$$v = \frac{c^2(0.5\,y - 1\,y)}{2.0c \cdot y - 1.0c \cdot y} = \boxed{-0.5c}$$

Because $\Delta t = t_2 - t_1 = -0.5\,y$:

$\boxed{S' \text{ moves in the negative } x \text{ direction.}}$

(b) Use the inverse time transformation to obtain:

$$t_2' = \gamma\left(t_2 - \frac{vx_2}{c^2}\right) = \frac{t_2 - \frac{vx_2}{c^2}}{\sqrt{1 - \frac{v^2}{c^2}}}$$

Substitute numerical values and evaluate t_2' and t_1':

$$t_2' = t_1' = \frac{0.5\,y - \frac{(-0.5c)(2.0c \cdot y)}{c^2}}{\sqrt{1 - \frac{(-0.5c)^2}{c^2}}}$$

$$= \boxed{1.73\,y}$$

*51 •••   In a simple thought experiment, Einstein showed that there is mass associated with electromagnetic radiation. Consider a box of length L and mass M resting on a frictionless surface. At the left wall of the box is a light source that emits radiation of energy E, which is absorbed at the right wall of the box. According to classical electromagnetic theory, this radiation carries momentum of magnitude $p = E/c$ (Equation 30-13). (a) Find the recoil velocity of the box so that momentum is conserved when the light is emitted. (Since p is small and M is large, you may use classical mechanics.) (b)

When the light is absorbed at the right wall of the box the box stops, so the total momentum remains zero. If we neglect the very small velocity of the box, the time it takes for the radiation to travel across the box is $\Delta t = L/c$. Find the distance moved by the box in this time. (*c*) Show that if the center of mass of the system is to remain at the same place, the radiation must carry mass $m = E/c^2$.

Picture the Problem We can use conservation of energy to express the recoil velocity of the box and the relationship between distance, speed, and time to find the distance traveled by the box in time $\Delta t = L/c$. Equating the initial and final locations of the center of mass will allow us to show that the radiation must carry mass $m = E/c^2$.

(*a*) Apply conservation of momentum to obtain:
$$\frac{E}{c} + Mv = p_i = 0$$

Solve for *v*:
$$v = \boxed{-\frac{E}{Mc}}$$

(*b*) The distance traveled by the box in time $\Delta t = L/c$ is:
$$d = v\Delta t = \frac{vL}{c}$$

Substitute for *v* from (*a*):
$$d = \frac{L}{c}\left(-\frac{E}{Mc}\right) = \boxed{-\frac{LE}{Mc^2}}$$

(*c*) Let $x = 0$ be at the center of the box and let the mass of the photon be *m*. Then initially the center of mass is at:
$$x_{CM} = \frac{-\frac{1}{2}mL}{M+m}$$

When the photon is absorbed at the other end of the box, the center of mass is at:
$$x_{CM} = \frac{\left[-\frac{MEL}{Mc^2} + m\left(\frac{1}{2}L - \frac{EL}{Mc^2}\right)\right]}{M+m}$$

Because no external forces act on the system, these expressions for x_{CM} must be equal:
$$\frac{-\frac{1}{2}mL}{M+m} = \frac{\left[-\frac{MEL}{Mc^2} + m\left(\frac{1}{2}L - \frac{EL}{Mc^2}\right)\right]}{M+m}$$

Solve for *m* to obtain:
$$m = \frac{E}{c^2\left(1 - \frac{E}{Mc^2}\right)}$$

422   Chapter 39

Because Mc^2 is of the order of 10^{16} J and $E = hf$ is of the order of 1 J for reasonable values of f, $E/Mc^2 \ll 1$ and:

$$m = \boxed{\dfrac{E}{c^2}}$$

*55 ••• When a projectile particle with kinetic energy greater than the threshold kinetic energy K_{th} strikes a stationary target particle, one or more particles may be created in the inelastic collision. Show that the threshold kinetic energy of the projectile is given by

$$K_{th} = \frac{\left(\sum m_{in} + \sum m_{fin}\right)\left(\sum m_{fin} - \sum m_{in}\right)c^2}{2m_{target}}$$

Here $\sum m_{in}$ is the sum of the masses of the projectile and target particles, $\sum m_{fin}$ is the sum of the masses of the final particles, and m_{target} is the mass of the target particle. Use this expression to determine the threshold kinetic energy of protons incident on a stationary proton target for the production of a proton–antiproton pair; compare your result with the result of Problem 40.

Picture the Problem Let m_i denote the mass of the incident (projectile) particle. Then $\sum m_{in} = m_i + m_{target}$ and we can use this expression to determine the threshold kinetic energy of protons incident on a stationary proton target for the production of a proton–antiproton pair.

Consider the situation in the center of mass reference frame. At threshold we have:

$$E^2 - p^2c^2 = \sum m_{fin}c^2$$

Note that this is a relativistically invariant expression.

In the laboratory frame, the target is at rest so:

$$E_{target} = E_t = E_{t,0}$$

We can, therefore, write:

$$(E_i + E_{t,0})^2 - p_i^2c^2 = \left(\sum m_{fin}c^2\right)^2$$

For the incident particle:

$$E_i^2 - p_i^2c^2 = E_{i,0}^2$$

and

$$E_i = E_{i,0} + K_{th}$$

where K_{th} is the threshold kinetic energy of the incident particle in the laboratory frame.

Express K_{th} in terms of the rest energies:

$$(E_{t,0} + E_{i,0})^2 + 2K_{th}E_{t,0} = \left(\sum m_{fin}c^2\right)^2$$

where

$$E_{t,0} + E_{i,0} = \sum m_{\text{fin}} c^2$$

and

$$E_{t,0} = m_{\text{target}} c^2$$

Substitute to obtain:

$$\left(\sum m_{\text{fin}} c^2\right)^2 + 2K_{\text{th}} m_{\text{target}} c^2 = \left(\sum m_{\text{fin}} c^2\right)^2$$

Solve for K_{th} to obtain:

$$\boxed{K_{\text{th}} = \frac{\left(\sum m_{\text{in}} + \sum m_{\text{fin}}\right)\left(\sum m_{\text{fin}} - \sum m_{\text{in}}\right) c^2}{2 m_{\text{target}}}}$$

For the creation of a proton-antiproton pair in a proton-proton collision:

$$\sum m_{\text{in}} = 2 m_{\text{p}}$$
$$\sum m_{\text{fin}} = 4 m_{\text{p}}$$

and

$$m_{\text{target}} = m_{\text{p}}$$

Substitute to obtain:

$$K_{\text{th}} = \frac{(2m_{\text{p}} + 4m_{\text{p}})(4m_{\text{p}} - 2m_{\text{p}}) c^2}{2 m_{\text{p}}}$$

$$= \frac{(6 m_{\text{p}})(2 m_{\text{p}}) c^2}{2 m_{\text{p}}} = \boxed{6 m_{\text{p}} c^2}$$

in agreement with Problem 40.

*59 ••• For the special case of a particle moving with speed u along the y axis in frame S, show that its momentum and energy in frame S', a frame that is moving along the x axis with velocity v, are related to its momentum and energy in S by the transformation equations

$$p_x' = \gamma\left(p_x - \frac{vE}{c^2}\right), \quad p_y' = p_y, \quad p_z' = p_z$$

$$\frac{E'}{c} = \gamma\left(\frac{E}{c} - \frac{v p_x}{c}\right).$$

Compare these equations with the Lorentz transformation for x', y', z', and t'. These equations show that the quantities p_x, p_y, p_z, and E/c transform in the same way as do x, y, z, and ct.

Picture the Problem We can use the expressions for $\vec{p}$ and E in S together with the relations we wish to verify and the inverse velocity transformation equations to establish the condition $u'^2 = (u_x')^2 + (u_y')^2 + (u_z')^2 = v^2 + \dfrac{u^2}{\gamma^2}$ and then use this result to verify the given expressions for p_x', p_y', p_z' and E'/c.

In any inertial frame the momentum and energy are given by:
$$\vec{p} = \dfrac{m\vec{u}}{\sqrt{1-\dfrac{u^2}{c^2}}} \text{ and } E = \dfrac{mc^2}{\sqrt{1-\dfrac{u^2}{c^2}}}$$

where $\vec{u}$ is the velocity of the particle and u is its speed.

The components of $\vec{p}$ in S are:
$$p_x = \dfrac{mu_x}{\sqrt{1-\dfrac{u^2}{c^2}}}, \; p_y = \dfrac{mu_y}{\sqrt{1-\dfrac{u^2}{c^2}}}, \text{ and}$$

$$p_z = \dfrac{mu_z}{\sqrt{1-\dfrac{u^2}{c^2}}}$$

Because $u_x = u_z = 0$ and $u_y = u$:
$$p_x = p_z = 0$$
and
$$p_y = \dfrac{mu}{\sqrt{1-\dfrac{u^2}{c^2}}}$$

Substituting zeros for p_x and p_z in the relations we are trying to show yields:
$$p_x' = \gamma\left(0 - \dfrac{vE}{c^2}\right) = -\gamma\dfrac{vE}{c^2}, \; p_y' = p_y,$$
$$p_z' = 0, \text{ and}$$
$$\dfrac{E'}{c} = \gamma\left(\dfrac{E}{c} - 0\right) = \gamma\dfrac{E}{c}$$

In S' the momentum components are:
$$p_x' = \dfrac{mu_x'}{\sqrt{1-\dfrac{u'^2}{c^2}}}, \; p_y' = \dfrac{mu_y'}{\sqrt{1-\dfrac{u'^2}{c^2}}}, \text{ and}$$

$$p_z' = \dfrac{mu_z'}{\sqrt{1-\dfrac{u'^2}{c^2}}}$$

The inverse velocity transformations are:
$$u_x' = \dfrac{u_x - v}{\sqrt{1-\dfrac{vu_x}{c^2}}}, \; u_y' = \dfrac{u_y}{\sqrt{1-\dfrac{vu_y}{c^2}}}, \text{ and}$$

$$u_z' = \frac{u_z}{\sqrt{1-\frac{vu_z}{c^2}}}$$

Substitute $u_x = u_z = 0$ and $u_y = u$ to obtain:

$u_x' = -v$, $u_y' = \gamma u$, and $u_z' = 0$

Thus:

$$u'^2 = (u_x')^2 + (u_y')^2 + (u_z')^2$$
$$= v^2 + \frac{u^2}{\gamma^2}$$

First we verify that $p_z' = p_z = 0$:

$$p_z' = \frac{m(0)}{\sqrt{1-\frac{u'^2}{c^2}}} = p_z = \boxed{0}$$

Next we verify that $p_y' = p_y$:

$$p_y' = \frac{mu_y'}{\sqrt{1-\frac{u'^2}{c^2}}} = \frac{mu}{\gamma\sqrt{1-\frac{v^2}{c^2}-\frac{u^2}{\gamma^2 c^2}}} = \frac{mu}{\sqrt{1-\frac{u^2}{c^2}}}\frac{\sqrt{1-\frac{u^2}{c^2}}}{\gamma\sqrt{1-\frac{v^2}{c^2}-\frac{u^2}{\gamma^2 c^2}}}$$

$$= \frac{mu}{\sqrt{1-\frac{u^2}{c^2}}}\sqrt{\frac{\left(1-\frac{u^2}{c^2}\right)\left(1-\frac{v^2}{c^2}\right)}{1-\frac{v^2}{c^2}-\frac{u^2}{c^2}\left(1-\frac{v^2}{c^2}\right)}} = p_y\sqrt{\frac{1-\frac{v^2}{c^2}-\frac{u^2}{c^2}\left(1-\frac{v^2}{c^2}\right)}{1-\frac{v^2}{c^2}-\frac{u^2}{c^2}\left(1-\frac{v^2}{c^2}\right)}}$$

$$= \boxed{p_y}$$

Next we verify that $p_x' = \gamma\left(p_x - \frac{vE}{c^2}\right)$:

$$p_x' = \frac{mu_x'}{\sqrt{1-\frac{u'^2}{c^2}}} = \frac{-mv}{\gamma\sqrt{1-\frac{v^2}{c^2}-\frac{u^2}{\gamma^2 c^2}}} = -\frac{\gamma v}{c^2}\frac{mc^2}{\sqrt{1-\frac{u^2}{c^2}}}\frac{\gamma^{-1}\sqrt{1-\frac{u^2}{c^2}}}{\sqrt{1-\frac{v^2}{c^2}-\frac{u^2}{\gamma^2 c^2}}}$$

$$= -\frac{\gamma v}{c^2} E \sqrt{\frac{\left(1-\frac{u^2}{c^2}\right)\left(1-\frac{v^2}{c^2}\right)}{1-\frac{v^2}{c^2}-\frac{u^2}{c^2}\left(1-\frac{v^2}{c^2}\right)}} = -\frac{\gamma v}{c^2} E \sqrt{\frac{1-\frac{v^2}{c^2}-\frac{u^2}{c^2}\left(1-\frac{v^2}{c^2}\right)}{1-\frac{v^2}{c^2}-\frac{u^2}{c^2}\left(1-\frac{v^2}{c^2}\right)}}$$

$$= \boxed{-\frac{\gamma v}{c^2} E}$$

Finally, we verify that $\dfrac{E'}{c} = \gamma\left(\dfrac{E}{c} - \dfrac{vp_x}{c}\right) = \gamma\dfrac{E}{c}$, or $E' = \gamma E$:

$$E' = \frac{mc^2}{\sqrt{1-\frac{u'^2}{c^2}}} = \frac{\gamma mc^2}{\sqrt{1-\frac{u^2}{c^2}}}\frac{\gamma^{-1}\sqrt{1-\frac{u^2}{c^2}}}{\sqrt{1-\frac{u'^2}{c^2}}} = \gamma E \frac{\gamma^{-1}\sqrt{1-\frac{u^2}{c^2}}}{\sqrt{1-\frac{v^2}{c^2}-\frac{u^2}{\gamma^2 c^2}}}$$

$$= \gamma E \sqrt{\frac{\left(1-\frac{u^2}{c^2}\right)\left(1-\frac{v^2}{c^2}\right)}{1-\frac{v^2}{c^2}-\frac{u^2}{c^2}\left(1-\frac{v^2}{c^2}\right)}} = \gamma E \sqrt{\frac{1-\frac{v^2}{c^2}-\frac{u^2}{c^2}\left(1-\frac{v^2}{c^2}\right)}{1-\frac{v^2}{c^2}-\frac{u^2}{c^2}\left(1-\frac{v^2}{c^2}\right)}}$$

$$= \boxed{\gamma E}$$

The x, y, z, and t transformation equations are:

$$x' = \gamma(x - vt)$$
$$y' = y$$
$$z' = z$$

and

$$t' = \gamma\left(t - \frac{vx}{c^2}\right)$$

The x, y, z, and ct transformation equations are:

$$x' = \gamma\left(x - \frac{v}{c}ct\right)$$
$$y' = y$$
$$z' = z$$

and

$$ct' = \gamma\left(ct - \frac{v}{c}x\right)$$

The p_x, p_y, p_z, and E/c transformation equations are:

$$p_x' = \gamma\left(p_x - \frac{v}{c}\frac{E}{c}\right)$$

$$p_y' = p_y$$

$$p_z' = p_z$$

and

$$\frac{E'}{c} = \gamma\left(\frac{E}{c} - \frac{v}{c}p_x\right)$$

Note that the transformation equations for x, y, z, and ct and the transformation equations for p_x, p_y, p_z, and E/c are identical.

*62 ••• A long rod that is parallel to the x axis is in free fall with acceleration g parallel to the $-y$ axis. An observer in a rocket moving with speed v parallel to the x axis passes by and watches the rod falling. Using the Lorentz transformations, show that the observer will measure the rod to be bent into a parabolic shape. Is the parabola concave upward or concave downward?

Picture the Problem We can use the inverse Lorentz transformation for time to show that the observer will conclude that the rod is bent into a parabolic shape.

In frame S where the rod is not moving along the x axis, the height of the rod at time t is:

$$y(t) = -\tfrac{1}{2}gt^2$$

The inverse Lorentz time transformation is:

$$t = \gamma\left(t' + \frac{vx}{c^2}\right)$$

Express $y'(t)$ in the moving frame of reference:

$$y'(t) = -\tfrac{1}{2}g\gamma\left(t' + \frac{vx}{c^2}\right)^2$$

Evaluate $y'(t)$ at $t' = 0$ to obtain:

$$y'(t) = -\frac{g\gamma v^2}{2c^2}x^2 \qquad (1)$$

> Because equation (1) is the equation of a parabola, we've shown that the moving observer will conclude that the rod is bent into a parabolic shape. Because the coefficient of x^2 is negative, the parabola is concave downward.

Chapter 40
Nuclear Physics

Conceptual Problems

*4 • The half-life of ^{14}C is much less than the age of the universe, yet ^{14}C is found in nature. Why?

Determine the Concept ^{14}C is found on earth because it is constantly being formed by cosmic rays in the upper atmosphere in the reaction $^{14}N + n \rightarrow {}^{14}C + {}^{1}H$.

*13 • Write and balance equations for each of the following nuclear decays:
(a) beta decay of ^{16}N, (b) alpha decay of ^{248}Fm, (c) positron decay of ^{12}N, (d) beta decay of ^{81}Se, (e) positron decay of ^{61}Cu, and (f) alpha decay of ^{228Th}.

Determine the Concept Knowing the parent nucleus and one of the decay products, we can use the conservation of charge, the conservation of energy, and the conservation of the number of nucleons to identify the participants in the decay.

(a) beta decay of ^{16}N

$$\boxed{{}^{16}_{7}N \rightarrow {}^{16}_{8}O + {}^{0}_{-1}\beta + {}^{0}_{0}\overline{\nu} + Q}$$

(b) alpha decay of ^{248}Fm

$$\boxed{{}^{248}_{100}Fm \rightarrow {}^{244}_{98}Cf + {}^{4}_{2}He + Q}$$

(c) positron decay of ^{12}N

$$\boxed{{}^{12}_{7}N \rightarrow {}^{12}_{6}C + {}^{0}_{+1}\beta + {}^{0}_{0}\nu + Q}$$

(d) beta decay of ^{81}Se

$$\boxed{{}^{81}_{34}Se \rightarrow {}^{81}_{35}Br + {}^{0}_{-1}\beta + {}^{0}_{0}\overline{\nu} + Q}$$

(e) positron decay of ^{61}Cu

$$\boxed{{}^{61}_{29}Cu \rightarrow {}^{61}_{28}Ni + {}^{0}_{+1}\beta + {}^{0}_{0}\nu + Q}$$

(f) alpha decay of ^{228Th}

$$\boxed{{}^{228}_{90}Th \rightarrow {}^{224}_{88}Ra + {}^{4}_{2}He + Q}$$

*14 • Write and balance reaction equations for each of the following: (a) ^{240}Pu undergoes spontaneous fission to form two fission fragments and three neutrons. One of the fission fragments is a ^{90}Sr nucleus. (b) A ^{72}Ge nucleus absorbs an alpha particle and ejects a photon. (c) A ^{127}I nucleus absorbs a deuteron and ejects a neutron. (d) A ^{235}U nucleus absorbs a slow neutron and fissions forming a ^{113}Ag nucleus, two neutrons, and another fission fragment. (e) A ^{55}Mn nucleus is struck with a high-energy ^{7}Li nucleus, resulting in a triton, ^{3}H, and a new nucleus. (f) ^{238}U absorbs a slow neutron resulting in a compound nucleus that emits a beta particle. What is the resulting nucleus?

Determine the Concept We can use the information regarding the daughter nuclei to write and balance equations for each of the reactions.

(a) $\boxed{{}^{240}_{94}Pu \rightarrow 3{}^{1}_{0}n + {}^{90}_{38}Sr + {}^{147}_{56}Ba}$

(b) $\boxed{{}^{72}_{32}Ge + {}^{4}_{2}He \rightarrow {}^{1}_{0}n + {}^{75}_{34}Se}$

430   Chapter 40

(c) $\boxed{^{127}_{53}\text{I} + ^{2}_{1}\text{H} \rightarrow ^{1}_{0}\text{n} + ^{128}_{54}\text{Xe}}$

(d) $\boxed{^{235}_{92}\text{U} + ^{1}_{0}\text{n} \rightarrow 2^{1}_{0}\text{n} + ^{113}_{47}\text{Ag} + ^{121}_{45}\text{Rh}}$

(e) $\boxed{^{55}_{25}\text{Mn} + ^{7}_{3}\text{Li} \rightarrow ^{3}_{1}\text{H} + ^{59}_{27}\text{Co}}$

(f) $^{238}_{92}\text{U} + ^{1}_{0}\text{n} \rightarrow ^{239}_{92}\text{U}$; $^{239}_{92}\text{U} \rightarrow ^{0}_{-1}\beta + ^{0}_{0}\nu + ^{239}_{93}\text{Np}$; $^{239}_{93}\text{Np} \rightarrow ^{0}_{-1}\beta + ^{0}_{0}\nu + \boxed{^{239}_{94}\text{Pu}}$

Properties of Nuclei

*17 •   Calculate the binding energy and the binding energy per nucleon from the masses given in Table 40-1 for (a) ^{12}C, (b) ^{56}Fe, and (c) ^{238}U.

Picture the Problem To find the binding energy of a nucleus we add the mass of its neutrons to the mass of its protons and then subtract the mass of the nucleus and multiply by c^2. To convert to MeV we multiply this result by 931.5 MeV/u. The binding energy per nucleon is the ratio of the binding energy to the mass number of the nucleus.

(a) For ^{12}C, Z = 6 and N = 6. Add the mass of the neutrons to that of the protons:

$$6m_p + 6m_n = 6 \times 1.007825\,\text{u} + 6 \times 1.008665\,\text{u} = 12.098940\,\text{u}$$

Subtract the mass of ^{12}C from this result:

$$(6m_p + 6m_n) - m_{^{12}\text{C}} = 12.098940\,\text{u} - 12\,\text{u} = 0.098940\,\text{u}$$

Multiply the mass difference by c^2 and convert to MeV:

$$E_b = (\Delta m)c^2 = (0.098940\,\text{u})c^2 \times \frac{931.5\,\text{MeV}/c^2}{1\,\text{u}} = \boxed{92.2\,\text{MeV}}$$

and the binding energy per nucleon is $\dfrac{E_b}{A} = \dfrac{92.2\,\text{MeV}}{12} = \boxed{7.68\,\text{MeV}}$

(b) For ^{56}Fe, Z = 26 and N = 30. Add the mass of the neutrons to that of the protons:

$$26m_p + 30m_n = 26 \times 1.007825\,\text{u} + 30 \times 1.008665\,\text{u} = 56.463400\,\text{u}$$

Subtract the mass of ^{56}Fe from this result:

$$(26m_p + 30m_n) - m_{^{12}C} = 56.463400\,u - 55.934942\,u = 0.528458\,u$$

Multiply the mass difference by c^2 and convert to MeV:

$$E_b = (\Delta m)c^2 = (0.528458\,u)c^2 \times \frac{931.5\,\text{MeV}/c^2}{1\,u} = \boxed{492\,\text{MeV}}$$

and the binding energy per nucleon is $\dfrac{E_b}{A} = \dfrac{492\,\text{MeV}}{56} = \boxed{8.79\,\text{MeV}}$

(c) For ^{238}U, Z = 92 and N = 146. Add the mass of the neutrons to that of the protons:

$$92m_p + 146m_n = 92 \times 1.007825\,u + 146 \times 1.008665\,u = 239.984990\,u$$

Subtract the mass of ^{238}U from this result:

$$(92m_p + 146m_n) - m_{^{238}U} = 239.984990\,u - 238.050783\,u = 1.934207\,u$$

Multiply the mass difference by c^2 and convert to MeV:

$$E_b = (\Delta m)c^2 = (1.934207\,u)c^2 \times \frac{931.5\,\text{MeV}/c^2}{1\,u} = \boxed{1802\,\text{MeV}}$$

and the binding energy per nucleon is $\dfrac{E_b}{A} = \dfrac{1802\,\text{MeV}}{238} = \boxed{7.57\,\text{MeV}}$

*21 •• The neutron, when isolated from an atomic nucleus, decays into a proton, electron, and an antineutrino as follows: $^1_0 n \rightarrow ^1_1 H + ^{\,\,\,0}_{-1} e + ^0_0 \overline{\nu}$. The thermal energy of a neutron is of the order of kT, where k is the Boltzmann constant. (a) In both joules and electron volts, calculate the energy of a thermal neutron at 25°C. (b) What is the speed of this thermal neutron? (c) A beam of monoenergetic thermal neutrons is produced at 25°C with intensity I. After traveling 1350 km, the beam has an intensity of $I/2$. Using this information, estimate the half-life of the neutron. Express your answer in minutes.

Picture the Problem The speed of the neutrons can be found from their thermal energy. The time taken to reduce the intensity of the beam by one-half, from I to $I/2$, is the half-life of the neutron. Because the beam is monoenergetic, the neutrons all travel at the same speed.

432   Chapter 40

(a) The thermal energy of the neutron is:

$$E_{thermal} = kT$$
$$= (1.38 \times 10^{-23} \text{ J/K})(25 + 273)\text{K}$$
$$= \boxed{4.11 \times 10^{-21} \text{ J}}$$
$$= 4.11 \times 10^{-21} \text{ J} \times \frac{1 \text{eV}}{1.60 \times 10^{-19} \text{ J}}$$
$$= \boxed{25.7 \text{ meV}}$$

(b) Equate $E_{thermal}$ and the kinetic energy of the neutron to obtain:

$$E_{thermal} = \tfrac{1}{2} m_n v^2$$

Solve for v to obtain:

$$v = \sqrt{\frac{2 E_{thermal}}{m_n}}$$

Substitute numerical values and evaluate v:

$$v = \sqrt{\frac{2(4.11 \times 10^{-21} \text{ J})}{1.67 \times 10^{-27} \text{ kg}}} = \boxed{2.22 \text{ km/s}}$$

(c) Relate the half-life, $t_{1/2}$, to the speed of the neutrons in the beam:

$$t_{1/2} = \frac{x}{v}$$

Substitute numerical values and evaluate $t_{1/2}$:

$$t_{1/2} = \frac{1350 \text{ km}}{2.22 \text{ km/s}} = 608 \text{ s} \times \frac{1 \text{ min}}{60 \text{ s}}$$
$$= \boxed{10.1 \text{ min}}$$

*24 •• In 1920, twelve years before the discovery of the neutron, Rutherford argued that proton-electron pairs might exist in the confines of the nucleus in order to explain the mass number, A, being greater than the nuclear charge, Z. He also used this argument to account for the source of beta particles in radioactive decay. Rutherford's scattering experiments in 1910 showed that the nucleus had a diameter of about 10 fm. Using this nuclear diameter, the uncertainty principle, and given that beta particles have an energy range of 0.02 MeV to 3.40 MeV, show why electrons cannot be contained within the nucleus.

Picture the Problem The Heisenberg uncertainty principle relates the uncertainty in position, Δx, to the uncertainty in momentum, Δp, by $\Delta x \Delta p \geq \frac{1}{2} \hbar$.

Solve the Heisenberg equation for Δp:

$$\Delta p \approx \frac{\hbar}{2 \Delta x}$$

Substitute numerical values and evaluate Δp:

$$\Delta p \approx \frac{1.05 \times 10^{-34} \text{ J} \cdot \text{s}}{2(10 \times 10^{-15} \text{ m})}$$
$$= 5.25 \times 10^{-21} \text{ kg} \cdot \text{m/s}$$

The kinetic energy of the electron is given by:	$K = pc$
Substitute numerical values and evaluate K:	$K = (5.25 \times 10^{-21} \text{ kg} \cdot \text{m/s})(3 \times 10^8 \text{ m/s})$ $= 1.58 \times 10^{-12} \text{ J} \times \dfrac{1 \text{eV}}{1.60 \times 10^{-19} \text{ J}}$ $= 9.88 \text{ MeV}$

This result contradicts experimental observations that show that the energy of electrons in unstable atoms is of the order of 1 to 1000 eV.

Radioactivity

*31 •• Plutonium is a highly hazardous and toxic material to the human body. Once it enters the body it collects primarily in the bones, although it can also be found in other organs. Red blood cells are synthesized within the marrow of the bones. The isotope ^{239}Pu is an alpha emitter with a half-life of 24,360 years. Since alpha particles are an ionizing radiation, the blood-making ability of the marrow is, in time, destroyed by the presence of ^{239}Pu. In addition, many kinds of cancers will also be initiated in the surrounding tissues by the ionizing effects of the alpha particles. (*a*) If a person accidentally ingested 2.0 μg of ^{239}Pu and it is absorbed by the bones of the victim, how many alpha particles are produced per second within the skeleton? (*b*) When, in years, will the activity be 1000 alpha particles per second?

Picture the Problem Each ^{239}Pu nucleus emits an alpha particle whose activity, *A*, depends on the decay constant of ^{239}Pu and on the number *N* of nuclei present in the ingested ^{239}Pu. We can find the decay constant from the half-life and the number of nuclei present from the mass ingested and the atomic mass of ^{239}Pu. Finally, we can use the dependence of the activity on time to find the time at which the activity be 1000 alpha particles per second.

(*a*) The activity of the nuclei present in the ingested ^{239}Pu is given by:	$A = \lambda N$ (1)
Find the constant for the decay of ^{239}Pu:	$\lambda = \dfrac{\ln 2}{t_{1/2}} = \dfrac{0.693}{(24360 \text{ y})(31.56 \text{ Ms/y})}$ $= 9.02 \times 10^{-13} \text{ s}^{-1}$
Express the number of nuclei present in the quantity of ^{239}Pu ingested:	$N = m_{Pu} \dfrac{N_A}{M_{Pu}}$ where M_{Pu} is the atomic mass of ^{239}Pu.
Substitute numerical values and evaluate *N*:	$N = (2.0 \, \mu\text{g}) \left(\dfrac{6.02 \times 10^{23} \text{ nuclei/mol}}{239 \text{ g/mol}} \right)$ $= 5.04 \times 10^{15} \text{ nuclei}$

Substitute numerical values in equation (1) and evaluate A:

$$A = (9.02 \times 10^{-13}\,\text{s}^{-1})(5.04 \times 10^{15}\,\alpha)$$
$$= \boxed{4.55 \times 10^3\,\alpha/\text{s}}$$

(b) The activity varies with time according to:

$$A = A_o e^{-\lambda t}$$

Solve for t to obtain:

$$t = \frac{\ln\left(\dfrac{A}{A_o}\right)}{-\lambda}$$

Substitute numerical values and evaluate t:

$$t = \frac{\ln\left(\dfrac{1 \times 10^3\,\alpha/\text{s}}{4.55 \times 10^3\,\alpha/\text{s}}\right)}{-(9.02 \times 10^{-13}\,\text{s}^{-1})\left(\dfrac{31.56\,\text{Ms}}{1\,\text{y}}\right)}$$
$$= \boxed{5.32 \times 10^4\,\text{y}}$$

*33 •• The fissile material ^{239}Pu is an alpha emitter. Write the equation of this reaction. Given that ^{239}Pu, ^{235}U, and an alpha particle have respective masses of 239.052 156 u, 235.043 923 u, and 4.002 603 u, use the equations appearing in Problem 32 to calculate the kinetic energies of the alpha particle and the recoiling daughter nucleus.

Picture the Problem We can write the equation of the decay process by using the fact that the post-decay sum of the Z and A numbers must equal the pre-decay values of the parent nucleus. The Q value in the equations from Problem 32 is given by $Q = -(\Delta m)c^2$.

^{239}Pu undergoes alpha decay according to:

$$\boxed{^{239}_{94}\text{Pu} \rightarrow\, ^{235}_{92}\text{U} + ^{4}_{2}\alpha + Q}$$

The Q value for the decay is given by:

$$Q = [(m_{\text{Pu}}) - (m_{\text{U}} + m_\alpha)]\left(\frac{931.5\,\text{MeV}}{1\,\text{u}}\right)$$

Substitute numerical values and evaluate Q:

$$Q = [(239.052156\,\text{u}) - (235.043923\,\text{u} + 4.002603\,\text{u})]\left(\frac{931.5\,\text{MeV}}{1\,\text{u}}\right) = \boxed{5.24\,\text{MeV}}$$

From Problem 32, the kinetic energy of the alpha particle is given by:

$$K_\alpha = \left(\frac{A-4}{A}\right)Q$$

Substitute numerical values and evaluate K_α:

$$K_\alpha = \left(\frac{239-4}{239}\right)(5.24\,\text{MeV})$$
$$= \boxed{5.15\,\text{MeV}}$$

From Problem 32, the kinetic energy of the ^{239}U is given by:

$$K_U = \frac{4Q}{A}$$

Substitute numerical values and evaluate K_U:

$$K_U = \frac{4(5.24\,\text{MeV})}{239} = \boxed{87.7\,\text{keV}}$$

*36 ••   Radiation has long been used in medical therapy to control the development and growth of cancer cells. Cobalt-60, a gamma emitter of 1.17 and 1.33 MeV energies, is used to irradiate and destroy deep-seated cancers. Small needles made of ^{60}Co of a specified activity are encased in gold and used as body implants in tumors for time periods that are related to tumor size, tumor cell reproductive rate, and the activity of the needle. (*a*) A 1.00 μg sample of ^{60}Co, of half-life of 5.27 y, is prepared in the cyclotron of a medical center to irradiate a small internal tumor with gamma rays. In curies, determine the initial activity of the sample. (*b*) What is the activity of the sample after 1.75 years?

Picture the Problem We can use $R_0 = \lambda N$ to find the initial activity of the sample and $R = R_0 e^{-\lambda t}$ to find the activity of the sample after 1.75 y.

(*a*) The initial activity of the sample is the product of the decay constant λ for ^{60}Co and the number of atoms N of ^{60}Co initially present in the sample:

$$R_0 = \lambda N \quad (1)$$

Express N in terms of the mass m of the sample, the molar mass M of ^{60}Co, and Avogadro's number N_A:

$$N = \frac{m}{M} N_A$$

Substitute numerical values and evaluate N:

$$N = \left(\frac{1.00 \times 10^{-6}\,\text{g}}{60\,\text{g/mol}}\right)(6.02 \times 10^{23}\,\text{nuclei/mol}) = 1.00 \times 10^{16}\,\text{nuclei}$$

The decay constant is given by:

$$\lambda = \frac{\ln 2}{t_{1/2}}$$

Substitute numerical values and evaluate λ:

$$\lambda = \frac{0.693}{(5.27\,\text{y})(31.56\,\text{Ms/y})} = 4.17 \times 10^{-9}\,\text{s}^{-1}$$

Substitute numerical values in equation (1) and evaluate A_0:

$$R_0 = (4.17 \times 10^{-9}\,\text{s}^{-1})(1.00 \times 10^{16}\,\text{nuclei})$$

$$= 4.17 \times 10^{7}\,\text{s}^{-1} \times \frac{1\,\text{Ci}}{3.7 \times 10^{10}\,\text{s}^{-1}}$$

$$= \boxed{1.13\,\text{mCi}}$$

436 Chapter 40

(b) The activity varies with time according to:

$$R = R_o e^{-\lambda t} = R_o e^{-\left(\frac{0.693 t}{5.27 y}\right)}$$

Evaluate R at $t = 1.75$ y:

$$R = (1.13\,\text{mCi})\, e^{-\left(\frac{0.693 \times 1.75 y}{5.27 y}\right)}$$

$$= \boxed{0.898\,\text{mCi}}$$

***41 ••** The counting rate from a radioactive source is measured every minute. The resulting counts per second are 1000, 820, 673, 552, 453, 371, 305, and 250. Plot the counting rate versus time on semilogarithmic graph paper, and use your graph to find the half-life of the source.

Picture the Problem The following graph was plotted using a spreadsheet program. Excel's "Add Trendline" feature was used to determine the equation of the line.

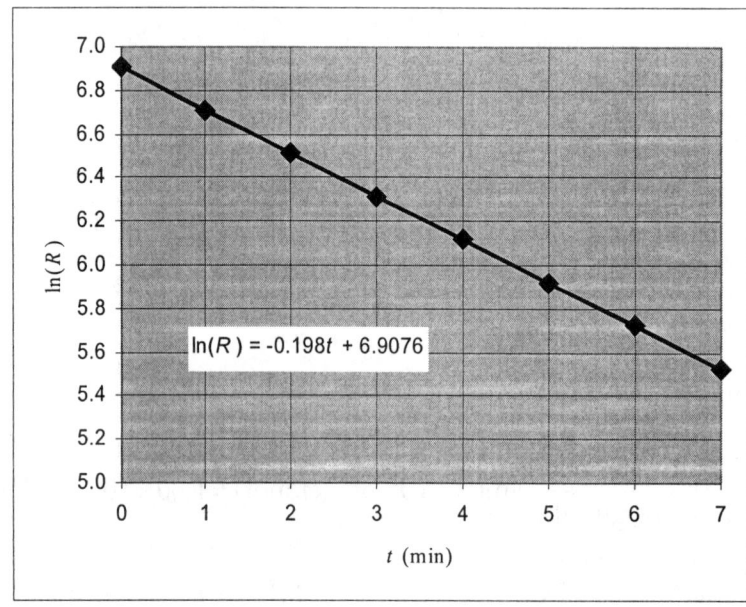

The linearity and negative slope of this graph tells us that it represents an exponential decay.

The decay rate equation is:

$$R = R_o e^{-\lambda t}$$

Take the natural logarithm of both sides of the equation to obtain:

$$\ln R = \ln e^{-\lambda t} + \ln R_0$$
$$= -\lambda t + \ln R_0$$

This equation is of the form:

$$y = mx + b$$

where $y = \ln R$, $x = t$, $m = -\lambda$, and

The half-life of the radioisotope is:
$$t_{1/2} = \frac{\ln 2}{\lambda} = \frac{\ln 2}{0.198\,\text{min}^{-1}} = \boxed{3.50\,\text{min}}$$

$$b = \ln R_0.$$

Nuclear Reactions

*47 •• (a) Use the atomic masses $m = 14.003\,242$ u for $^{14}_{6}\text{C}$ and $m = 14.003\,074$ u for $^{14}_{7}\text{N}$ to calculate the Q value (in MeV) for the β decay

$$^{14}_{6}\text{C} \rightarrow {}^{14}_{7}\text{N} + \beta^- + \overline{\nu}_e$$

(b) Explain why you do not need to add the mass of the β^- to that of atomic N for this calculation.

Picture the Problem We can use $Q = -(\Delta m)c^2$ to find the Q values for this reaction.

(a) The masses of the atoms are:
$$m_{^{14}\text{C}} = 14.003\,242\,\text{u}$$
$$m_{^{14}\text{N}} = 14.003\,074\,\text{u}$$

Calculate the increase in mass:
$$\Delta m = m_f - m_i$$
$$= 14.003\,074\,\text{u} - 14.003\,242\,\text{u}$$
$$= -0.000168\,\text{u}$$

Calculate the Q value:
$$Q = -(\Delta m)c^2$$
$$= -(-0.000168\,\text{u})c^2 \left(931.5\,\frac{\text{MeV}/c^2}{\text{u}}\right)$$
$$= \boxed{0.156\,\text{MeV}}$$

(b)

The masses given are for atoms, not nuclei, so for nuclear masses the masses are too large by the atomic number times the mass of an electron. For the given nuclear reaction, the mass of the carbon atom is too large by $6m_e$ and the mass of the nitrogen atom is too large by $7m_e$. Subtracting $6m_e$ from both sides of the reaction equation leaves an extra electron mass on the right. Not including the mass of the beta particle (electron) is mathematically equivalent to explicitly subtracting $1m_e$ from the right side of the equation.

438   Chapter 40

Fission and Fusion

*49 • Assuming an average energy of 200 MeV per fission, calculate the number of fissions per second needed for a 500-MW reactor.

Picture the Problem The power output of the reactor is the product of the number of fissions per second and energy liberated per fission.

Express the required number N of fissions per second in terms of the power output P and the energy released per fission $E_{\text{per fission}}$:

$$N = \frac{P}{E_{\text{per fission}}}$$

Substitute numerical values and evaluate N:

$$N = \frac{500\,\text{MW}}{200\,\text{MeV}}$$

$$= \frac{5 \times 10^8 \dfrac{\text{J}}{\text{s}} \times \dfrac{1\,\text{eV}}{1.60 \times 10^{-19}\,\text{J}}}{200\,\text{MeV}}$$

$$= \boxed{1.56 \times 10^{19}\,\text{s}^{-1}}$$

*51 •• Consider the following fission process:
$^{235}_{92}\text{U} + ^{1}_{0}\text{n} \rightarrow ^{95}_{42}\text{Mo} + ^{139}_{57}\text{La} + 2\,^{1}_{0}\text{n} + Q$. The masses of the neutron, U, Mo, and La are 1.008 665 u, 235.043 923 u, 94.905 842 u, and 138.906 348 u, respectively. Calculate the Q-value, in MeV, for this fission process. Compare the result to the result obtained in Problem 23.

Picture the Problem We can use $Q = -(\Delta m)c^2$, where $\Delta m = m_f - m_i$, to calculate the Q value.

The Q value is given by:

$$Q = -(\Delta m)c^2 \times \frac{931.5\,\text{MeV}/c^2}{1\,\text{u}}$$

Calculate the change in mass Δm:

$\Delta m = m_f - m_i$
$= 94.905842\,\text{u} + 138.906348\,\text{u} + 2(1.008665\,\text{u}) - (235.043923\,\text{u} + 1.008665\,\text{u})$
$= -0.223068\,\text{u}$

Substitute for Δm and evaluate Q:

$$Q = -(-0.223068\,\text{u}) \times \frac{931.5\,\text{MeV}}{1\,\text{u}}$$

$$= \boxed{208\,\text{MeV}}$$

Nuclear Physics 439

The ratio of Q to U found in Problem 23 is:

$$\frac{Q}{U} = \frac{208 \text{ MeV}}{236 \text{ MeV}} = \boxed{88.1\%}$$

*54 ••• The fusion reaction between ^{2}H and ^{3}H is

$$^3\text{H} + {}^2\text{H} \rightarrow {}^4\text{He} + n + 17.6 \text{ MeV}$$

Using the conservation of momentum and the given Q value, find the final energies of both the ^{4}He nucleus and the neutron, assuming the initial kinetic energy of the system is 1.00 MeV and the initial momentum of the system is zero.

Picture the Problem We can use the conservation of momentum and the given Q value to find the final energies of both the ^{4}He nucleus and the neutron, assuming that the initial momentum of the system is zero.

Apply conservation of energy to obtain:

$$18.6 \text{ MeV} = \tfrac{1}{2} m_{\text{He}} v_{\text{He}}^2 + \tfrac{1}{2} m_n v_n^2 \quad (1)$$
$$= K_{\text{He}} + K_n$$

Apply conservation of momentum to obtain:

$$m_{\text{He}} v_{\text{He}} + m_n v_n = 0 \quad (2)$$

Solve equation (2) for v_{He}:

$$v_{\text{He}} = -\frac{m_n v_n}{m_{\text{He}}} \Rightarrow v_{\text{He}}^2 = \left(\frac{m_n}{m_{\text{He}}}\right)^2 v_n^2$$

Substitute for v_{He}^2 in equation (1):

$$18.6 \text{ MeV} = \tfrac{1}{2} m_{\text{He}} \left(\frac{m_n}{m_{\text{He}}}\right)^2 v_n^2 + \tfrac{1}{2} m_n v_n^2$$

or

$$18.6 \text{ MeV} = \tfrac{1}{2} m_n v_n^2 \left(1 + \frac{m_n}{m_{\text{He}}}\right)$$
$$= K_n \left(1 + \frac{m_n}{m_{\text{He}}}\right)$$

Solve for K_n:

$$K_n = \frac{18.6 \text{ MeV}}{1 + \dfrac{m_n}{m_{\text{He}}}}$$

Substitute numerical values for m_n and m_{He} and evaluate K_n:

$$K_n = \frac{18.6 \text{ MeV}}{1 + \dfrac{1.008665 \text{ u}}{4.002603 \text{ u}}} = \boxed{14.86 \text{ MeV}}$$

Use equation (1) to find K_{He}:

$$K_{He} = 18.6\,\text{MeV} - K_n$$
$$= 18.6\,\text{MeV} - 14.86\,\text{MeV}$$
$$= \boxed{3.74\,\text{MeV}}$$

General Problems

*57 • The counting rate from a radioactive source is 6400 counts/s. The half-life of the source is 10 s. Make a plot of the counting rate as a function of time for times up to 1 min. What is the decay constant for this source?

Picture the Problem We can use the given information regarding the half-life of the source to find its decay constant. We can then plot a graph of the counting rate as a function of time.

The decay constant is related to the half-life of the source:

$$\lambda = \frac{\ln 2}{t_{1/2}} = \frac{\ln 2}{10\,\text{s}} = \boxed{0.0693\,\text{s}^{-1}}$$

The activity of the source is given by:

$$R = R_0 e^{-\lambda t} = (6400\,\text{Bq})e^{-(0.0693\,\text{s}^{-1})t}$$

The following graph of $R = (6400\,\text{Bq})e^{-(0.0693\,\text{s}^{-1})t}$ was plotted using a spreadsheet program.

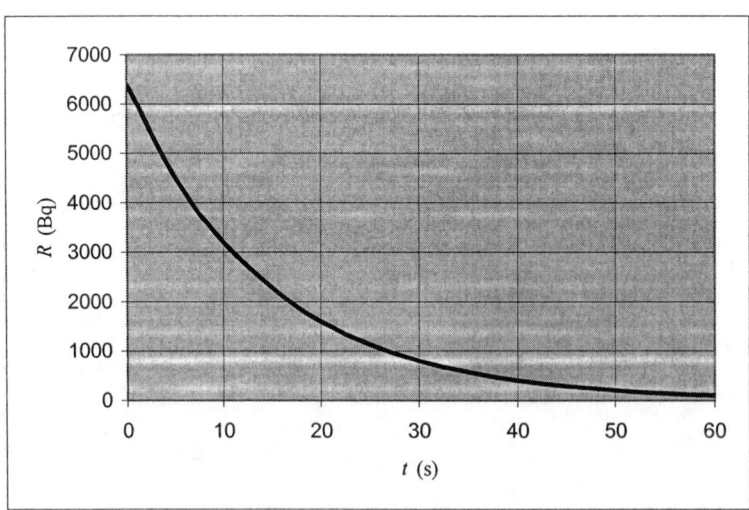

*61 •• Show that the ^{109}Ag nucleus is stable against alpha decay, $^{109}_{47}\text{Ag} \rightarrow \,^4_2\text{He} + \,^{105}_{45}\text{Rh} + Q$. The mass of the ^{109}Ag nucleus is 108.904 756 u, and the products of the decay are 4.002 603 u and 104.905 250 u, respectively.

Picture the Problem We can show that ^{109}Ag is stable against alpha decay by demonstrating that its Q value is negative.

The Q value for this reaction is:

$$Q = -[(m_{Rh} + m_\alpha) - m_{Ag}]c^2 \left(931.5 \frac{MeV/c^2}{u}\right)$$

Substitute numerical values and evaluate Q:

$$Q = -[(4.002603\ u + 104.905250\ u) - 108.904756\ u](931.5 MeV/u)$$
$$= \boxed{-2.88\ MeV}$$

Remarks: Alpha decay occurs spontaneously and the Q value will equal the sum of the kinetic energies of the alpha particle and the recoiling daughter nucleus, $Q = K_\alpha + K_D$. Kinetic energy cannot be negative; hence, alpha decay cannot occur unless the mass of the parent nucleus is greater than the sum of the masses of the alpha particle and daughter nucleus, $m_P > m_\alpha + m_D$. Alpha decay cannot take place unless the total rest mass decreases.

*66 •• (a) Determine the closest distance of approach of an 8-MeV α particle in a head-on collision with a nucleus of ^{197}Au and a nucleus of ^{10}B, neglecting the recoil of the struck nuclei. (b) Repeat the calculation taking into account the recoil of the struck nuclei.

Picture the Problem We can solve this problem in the center of mass reference frame for the general case of an α particle in a head-on collision with a nucleus of atomic mass M u and then substitute data for a nucleus of ^{197}Au and a nucleus of ^{10}B.

In the CM frame, the kinetic energy is:

$$K_{CM} = \frac{K_{lab}}{1 + \frac{m_\alpha}{M}} = \frac{K_{lab}}{1 + \frac{4u}{M}}$$

At the point of closest approach:

$$K_{CM} = \frac{kq_1q_2}{R_{min}} = \frac{k(2e)(Ze)}{R_{min}} = \frac{ke^2 2Z}{R_{min}}$$

or, because $ke^2 = 1.44$ MeV·fm,

$$K_{CM} = \frac{(1.44\ MeV \cdot fm)(2Z)}{R_{min}}$$

Solve for R_{min} to obtain:

$$R_{min} = \frac{(1.44\ MeV \cdot fm)(2Z)}{K_{CM}} \quad (1)$$

(a) Neglecting the recoil of the target nucleus is equivalent to replacing K_{CM} by K_{lab}. Evaluate equation (1) for ^{197}Au:

$$R_{min} = \frac{(1.44\,\text{MeV}\cdot\text{fm})(2\times 79)}{8\,\text{MeV}}$$

$$= \boxed{28.4\,\text{fm}}$$

Evaluate equation (1) for ^{10}B:

$$R_{min} = \frac{(1.44\,\text{MeV}\cdot\text{fm})(2\times 5)}{8\,\text{MeV}}$$

$$= \boxed{1.80\,\text{fm}}$$

(b) Find K_{CM} for the ^{197}Au nucleus:

$$K_{CM} = \frac{8\,\text{MeV}}{1+\dfrac{4\,\text{u}}{197\,\text{u}}} = 7.841\,\text{MeV}$$

Substitute numerical values in equation (1) and evaluate R_{min}:

$$R_{min} = \frac{(1.44\,\text{MeV}\cdot\text{fm})(2\times 79)}{7.841\,\text{MeV}}$$

$$= \boxed{29.0\,\text{fm}}$$

Note that this result is about 2% greater that R_{min} calculated ignoring recoil.

Find K_{CM} for the ^{10}B nucleus:

$$K_{CM} = \frac{8\,\text{MeV}}{1+\dfrac{4\,\text{u}}{10\,\text{u}}} = 5.714\,\text{MeV}$$

Substitute numerical values in equation (1) and evaluate R_{min}:

$$R_{min} = \frac{(1.44\,\text{MeV}\cdot\text{fm})(2\times 5)}{5.714\,\text{MeV}}$$

$$= \boxed{2.52\,\text{fm}}$$

Note that this result is about 40% greater that R_{min} calculated ignoring recoil.

*70 •• The total energy consumed in the United States in 1 y is about 7.0×10^{19} J. How many kilograms of ^{235}U would be needed to provide this amount of energy if we assume that 200 MeV of energy is released by each fissioning uranium nucleus, that all of the uranium atoms undergo fission, and that all of the energy-conversion mechanisms used are 100 percent efficient?

Picture the Problem The mass of ^{235}U required is given by $m_{235} = \dfrac{N}{N_A} M_{235}$, where M_{235} is the molecular mass of ^{235}U and N is the number of fissions required to produce 7.0×10^{19} J.

Relate the mass of ^{235}U required to the number of fissions N required:

$$m_{235} = \frac{N}{N_A} M_{235} \quad (1)$$

where M_{235} is the molecular mass of ^{235}U.

Determine N:

$$N = \frac{E_{annual}}{E_{per\ fission}}$$

Substitute numerical values and evaluate N:

$$N = \frac{7.0 \times 10^{19}\ \text{J}}{200\ \text{MeV} \times \frac{1.60 \times 10^{-19}\ \text{J}}{\text{eV}}}$$

$$= 2.18 \times 10^{30}$$

Substitute numerical values in equation (1) and evaluate m_{235}:

$$m_{235} = \frac{2.18 \times 10^{30}}{6.02 \times 10^{23}\ \text{nuclei/mol}} (235\ \text{g/mol}) = \boxed{8.51 \times 10^5\ \text{kg}}$$

*73 ••• Assume that a neutron decays into a proton plus an electron without the emission of a neutrino. The energy shared by the proton and electron is then 0.782 MeV. In the rest frame of the neutron, the total momentum is zero, so the momentum of the proton must be equal and opposite the momentum of the electron. This determines the relative energies of the two particles, but because the electron is relativistic, the exact calculation of these relative energies is somewhat difficult. (*a*) Assume that the kinetic energy of the electron is 0.782 MeV and calculate the momentum *p* of the electron in units of MeV/*c*. (*Hint*: Use Equation 39-28.) (*b*) Using your result from Part (*a*), calculate the kinetic energy $p^2/2m_p$ of the proton. (*c*) Since the total energy of the electron plus the proton is 0.782 MeV, the calculation in Part (*b*) gives a correction to the assumption that the energy of the electron is 0.782 MeV. What percentage of 0.782 MeV is this correction?

Picture the Problem The momentum of the electron is related to its total energy through $E^2 = p^2 c^2 + E_0^2$ and its total relativistic energy E is the sum of its kinetic and rest energies.

(*a*) Relate the total energy of the electron to its momentum and rest energy:

$$E^2 = p^2 c^2 + E_0^2 \quad (1)$$

The total relativistic energy E of the electron is the sum of its kinetic

$$E = K + E_0$$

444   Chapter 40

energy and its rest energy:

Substitute for E in equation (1) to obtain:
$$(K + E_0)^2 = p^2c^2 + E_0^2$$

Solve for p:
$$p = \frac{\sqrt{K(K+2E_0)}}{c}$$

Substitute numerical values and evaluate p:

$$p = \frac{\sqrt{(0.782\,\text{MeV})(0.782\,\text{MeV} + 2 \times 0.511\,\text{MeV})}}{c} = \boxed{1.188\,\text{MeV}/c}$$

(b) Because $p_p = -p_e$:
$$K_p = \frac{p_p^2}{2m_p}$$

Substitute numerical values (see Table 7-1 for the rest energy of a proton) and evaluate K_p:
$$K_p = \frac{(1.188\,\text{MeV}/c)^2}{2(938.28\,\text{MeV}/c^2)} = \boxed{752\,\text{eV}}$$

(c) The percent correction is:
$$\frac{K_p}{K} = \frac{752\,\text{eV}}{0.782\,\text{MeV}} = \boxed{0.0962\%}$$

*77 •••   Frequently, the daughter of a radioactive parent is itself radioactive. Suppose the parent, designated by A, has a decay constant λ_A; while the daughter, designated B, has a decay constant λ_B. The number of nuclei of B are then given by the solution to the differential equation
$$dN_B/dt = \lambda_A N_A - \lambda_B N_B$$
(a) Justify this differential equation. (b) Show that the solution for this equation is

$$N_B(t) = \frac{N_{A0}\lambda_A}{\lambda_B - \lambda_A}\left(e^{-\lambda_A t} - e^{-\lambda_B t}\right)$$

where N_{A0} is the number of A nuclei present at $t = 0$ when there are no B nuclei.
(c) Show that $N_B(t) > 0$ whether $\lambda_A > \lambda_B$ or $\lambda_B > \lambda_A$. (d) Make a plot of $N_A(t)$ and $N_B(t)$ as a function of time when $\tau_B = 3\tau_A$.

Picture the Problem We can differentiate $N_B(t) = \dfrac{N_{A0}\lambda_A}{\lambda_B - \lambda_A}\left(e^{-\lambda_A t} - e^{-\lambda_B t}\right)$ with respect to t to show that it is the solution to the differential equation

$dN_B/dt = \lambda_A N_A - \lambda_B N_B$.

(a) The rate of change of N_B is the rate of generation of B nuclei minus the rate of decay of B nuclei. The generation rate is equal to the decay rate of A nuclei, which equals $\lambda_A N_A$. The decay rate of B nuclei is $\lambda_B N_B$.

(b) We're given that:
$$\frac{dN_B}{dt} = \lambda_A N_A - \lambda_B N_B \quad (1)$$

$$N_B(t) = \frac{N_{A0} \lambda_A}{\lambda_B - \lambda_A} \left(e^{-\lambda_A t} - e^{-\lambda_B t}\right) \quad (2)$$

$$N_A = N_{A0} e^{-\lambda_A t} \quad (3)$$

Differentiate equation (2) with respect to t to obtain:

$$\frac{d}{dt}[N_B(t)] = \frac{N_{A0} \lambda_A}{\lambda_B - \lambda_A} \frac{d}{dt}\left[\left(e^{-\lambda_A t} - e^{-\lambda_B t}\right)\right] = \frac{N_{A0} \lambda_A}{\lambda_B - \lambda_A}\left[-\lambda_A e^{-\lambda_A t} + \lambda_B e^{-\lambda_B t}\right]$$

Substitute this derivative in equation (1) to get:

$$\frac{N_{A0} \lambda_A}{\lambda_B - \lambda_A}\left[-\lambda_A e^{-\lambda_A t} + \lambda_B e^{-\lambda_B t}\right] = \lambda_A N_{A0} e^{-\lambda_A t} - \lambda_B \left[\frac{N_{A0} \lambda_A}{\lambda_B - \lambda_A}\left(e^{-\lambda_A t} - e^{-\lambda_B t}\right)\right]$$

Multiply both sides by $\dfrac{\lambda_B - \lambda_A}{\lambda_B \lambda_A}$ and simplify to obtain:

$$\frac{N_{A0}}{\lambda_B}\left[-\lambda_A e^{-\lambda_A t} + \lambda_B e^{-\lambda_B t}\right] = \frac{\lambda_B - \lambda_A}{\lambda_B} N_{A0} e^{-\lambda_A t} - N_{A0}\left(e^{-\lambda_A t} - e^{-\lambda_B t}\right)$$

$$= N_{A0} e^{-\lambda_A t} - \frac{N_{A0} \lambda_A}{\lambda_B} e^{-\lambda_A t} - N_{A0} e^{-\lambda_A t} + N_{A0} e^{-\lambda_B t}$$

$$= -\frac{N_{A0} \lambda_A}{\lambda_B} e^{-\lambda_A t} + N_{A0} e^{-\lambda_B t}$$

$$= \frac{N_{A0}}{\lambda_B}\left[-\lambda_A e^{-\lambda_A t} + \lambda_B e^{-\lambda_B t}\right]$$

which is an identity and confirms that equation (2) is the solution to equation (1).

(c) If $\lambda_A > \lambda_B$ the denominator and the expression in the parentheses are both negative for $t > 0$. If $\lambda_A < \lambda_B$ the denominator and the expression in the parentheses are both positive for $t > 0$.

(*d*) The following graph was plotted using a spreadsheet program.

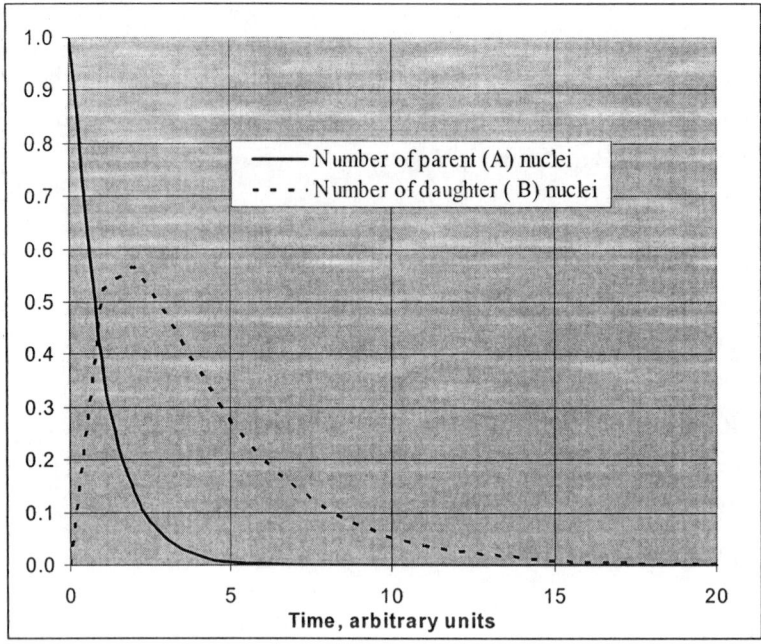

Chapter 41
Elementary Particles and the Beginning of the Universe

Conceptual Problems

*3 • How can you tell whether a decay proceeds via the strong interaction or the weak interaction?

Determine the Concept A decay process involving the strong interaction has a very short lifetime ($\sim 10^{-23}$ s), whereas decay processes that proceed via the weak interaction have lifetimes of order 10^{-10} s.

*9 • Based on the assumption that a pion, π^+, interacts with an antiproton, $\bar{p}$, is it possible that a proton, p, could be produced by such an interaction?

Determine the Concept No. Such a reaction is impossible. A proton requires three quarks. Three quarks are not available because a pion is made of a quark and an antiquark and the antiproton consists of three antiquarks.

Spin and Antiparticles

*12 • Two pions at rest annihilate according to the reaction $\pi^+ + \pi^- \rightarrow \gamma + \gamma$. (a) Why must the energies of the two γ-rays be equal? (b) Find the energy of each γ-ray. (c) Find the wavelength of each γ-ray.

Picture the Problem We can use both conservation of energy and momentum to explain why the energies of the two γ-rays must be equal. We can find the energy of each γ-ray in Table 41-1 and find their wavelengths using $\lambda = hc/E$.

(a) The initial momentum is zero; therefore, the final momentum must be zero. The momentum of a photon is E/c. To conserve both momentum and energy the two photons must have the same momentum magnitude. Hence, they have the same energy.

(b) From Table 41-1:

$$E_\gamma = \boxed{139.6 \text{ MeV}}$$

(c) The wavelength of each γ ray is given by:

$$\lambda = \frac{hc}{E} = \frac{1240 \text{ MeV} \cdot \text{fm}}{E}$$

Substitute numerical values and evaluate λ:

$$\lambda = \frac{1240 \text{ MeV} \cdot \text{fm}}{139.6 \text{ MeV}} = \boxed{8.88 \text{ fm}}$$

The Conservation Laws

*20 •• Test the following decays for violation of the conservation of energy, electric charge, baryon number, and lepton number: (*a*) n $\to \pi^+ + \pi^- + \mu^+ + \mu^-$; (*b*) $\pi^0 \to e^+ + e^- + \gamma$. Assume that linear momentum and angular momentum are conserved. State which conservation laws (if any) are violated in each decay.

Picture the Problem A decay process is allowed if energy, charge, baryon number, and lepton number are conserved.

(*a*) Energy conservation: | Because $m_n > 2m_\pi + 2m_\mu$, energy conservation is not violated.

Charge conservation:
$0 \to +1 + (-1) + 0 + 0 = 0$ | Because the total charge is 0 before and after the decay, charge is conserved.

Baryon number:
$1 \to 0 + 0 + 0 + 0 = 0$ | Because baryon number changes from +1 to 0, conservation of baryon number is violated.

Lepton number:
$0 \to 0 + 0 + 1 + (-1) = 0$ | Because $L_\mu = 0$ before and after the decay, the lepton number for muons is conserved.

The process is not allowed because it violates conservation of baryon number.

(*b*) Energy conservation: | Because $m_\pi > 2m_e$, energy conservation is not violated.

Charge conservation:
$0 \to +1 + (-1) + 0 + 0 = 0$ | Because the total charge is 0 before and after the decay, charge is conserved.

Baryon number:
$0 \to 0 + 0 + 0 + 0 = 0$ | Because $B = 0$ before and after the decay, the baryon number is conserved.

Elementary Particles and the Beginning of the Universe 449

Lepton number:

$0 \to 0 + 0 + 0 + 0 = 0$

Because $L_e = 0$ before and after the decay, the lepton number is conserved.

The decay satisfies all conservation laws and is allowed.

Quarks

*27 •• Find a possible quark combination for the following particles: (a) Λ^0, (b) $\bar{p}^-$, and (c) Σ^-.

Picture the Problem Because Λ^0, $\bar{p}^-$, and Σ^- are baryons, they are made up of three quarks. We can use Tables 41-1 and 41-2 to find combinations of quarks with the correct values for electric charge, baryon number, and strangeness for these particles.

(a) For Λ^0 we need:
$Q = 0$
$B = +1$
$S = -1$

The quark combination that satisfies these conditions is $\boxed{uds}$.

(b) For $\bar{p}^-$ we need:
$Q = -1$
$B = -1$
$S = +1$

The quark combination that satisfies these conditions is $\boxed{\bar{u}\bar{u}\bar{d}}$.

(c) For Σ^- we need:
$Q = -1$
$B = +1$
$S = -1$

The quark combination that satisfies these conditions is $\boxed{dds}$.

The Evolution of the Universe

*31 • A galaxy is receding from the earth at 2.5 percent the speed of light. Estimate the distance from the earth to this galaxy.

Picture the Problem We can use Hubble's law to find the distance from the earth to this galaxy.

The recessional velocity of galaxy is $\quad v = Hr$

related to its distance by Hubble's law:

Solve for r:
$$r = \frac{v}{H}$$

Substitute numerical values and evaluate r:
$$r = \frac{(0.025)c}{\frac{23 \text{ km/s}}{10^6 c \cdot y}} = \frac{(0.025)(3 \times 10^5 \text{ km/s})}{\frac{23 \text{ km/s}}{10^6 c \cdot y}}$$
$$= \boxed{3.26 \times 10^8 \, c \cdot y}$$

*34 •• The red line in the spectrum of atomic hydrogen is frequently referred to as the Hα line, and it has a wavelength of 656.3 nm. Using Hubble's law and the relativistic Doppler equation from Problem 33, determine the wavelength of the Hα line in the spectrum emitted from galaxies at distances of (a) 5×10^6 c·y, (b) 50×10^6 c·y, (c) 500×10^6 c·y, and (d) 5×10^9 c·y from the earth.

Picture the Problem Using Hubble's law, we can rewrite the equation from Problem 31 as $\lambda' = \lambda_0 \sqrt{\frac{1 + Hr/c}{1 - Hr/c}}$.

From Problem 33 we have:
$$\lambda' = \lambda_0 \sqrt{\frac{1 + v/c}{1 - v/c}}$$

Use Hubble's law to relate v to r:
$$v = Hr$$

Substitute for v to obtain:
$$\lambda' = \lambda_0 \sqrt{\frac{1 + Hr/c}{1 - Hr/c}}$$

(a) For $r = 5 \times 10^6$ c·y:

$$\lambda' = 656.3 \text{ nm} \sqrt{\frac{1 + \left(\frac{23 \text{ km/s}}{10^6 c \cdot y}\right)\frac{(5 \times 10^6 c \cdot y)}{(3 \times 10^5 \text{ km/s})}}{1 - \left(\frac{23 \text{ km/s}}{10^6 c \cdot y}\right)\frac{(5 \times 10^6 c \cdot y)}{(3 \times 10^5 \text{ km/s})}}} = \boxed{656.6 \text{ nm}}$$

(b) For $r = 50 \times 10^6$ c·y:

$$\lambda' = 656.3 \text{ nm} \sqrt{\frac{1 + \left(\frac{23 \text{ km/s}}{10^6 c \cdot y}\right)\frac{(50 \times 10^6 c \cdot y)}{(3 \times 10^5 \text{ km/s})}}{1 - \left(\frac{23 \text{ km/s}}{10^6 c \cdot y}\right)\frac{(50 \times 10^6 c \cdot y)}{(3 \times 10^5 \text{ km/s})}}} = \boxed{658.8 \text{ nm}}$$

(c) For $r = 500 \times 10^6$ c·y:

$$\lambda' = 656.3\,\text{nm}\sqrt{\dfrac{1+\left(\dfrac{23\,\text{km/s}}{10^6\,\text{c}\cdot\text{y}}\right)\dfrac{\left(500\times10^6\,\text{c}\cdot\text{y}\right)}{\left(3\times10^5\,\text{km/s}\right)}}{1-\left(\dfrac{23\,\text{km/s}}{10^6\,\text{c}\cdot\text{y}}\right)\dfrac{\left(500\times10^6\,\text{c}\cdot\text{y}\right)}{\left(3\times10^5\,\text{km/s}\right)}}} = \boxed{682.0\,\text{nm}}$$

(d) For $r = 5 \times 10^9$ c·y:

$$\lambda' = 656.3\,\text{nm}\sqrt{\dfrac{1+\left(\dfrac{23\,\text{km/s}}{10^6\,\text{c}\cdot\text{y}}\right)\dfrac{\left(5\times10^9\,\text{c}\cdot\text{y}\right)}{\left(3\times10^5\,\text{km/s}\right)}}{1-\left(\dfrac{23\,\text{km/s}}{10^6\,\text{c}\cdot\text{y}}\right)\dfrac{\left(5\times10^9\,\text{c}\cdot\text{y}\right)}{\left(3\times10^5\,\text{km/s}\right)}}} = \boxed{983.0\,\text{nm}}$$

General Problems

*37 •• In Problem 36, one of the reactions is $\pi^0 \to \gamma + \gamma$. (a) In terms of the quark model, show how this reaction can take place. (b) Why is it that the number of photons produced must be at least two?

Picture the Problem The π^0 particle is composed of two quarks, $u\bar{u}$. Hence, the reaction $\pi^0 \to \gamma + \gamma$ is equivalent to $u\bar{u} \to \gamma + \gamma$.

(a) $\boxed{\text{The u and } \bar{u} \text{ annihilate resulting in the photons.}}$

(b) $\boxed{\text{Two or more photons are required to conserve linear momentum.}}$

*39 •• Using Figure 41-2 and the laws of conservation of charge number, baryon number, strangeness, and spin, identify the unknown particle in each of the following strong reactions: (a) $p + \pi^- \to \Sigma^0 + ?$, (b) $p + p \to \pi^+ + n + K^+ + ?$, and (c) $p + \overline{K}^- \to \Xi^- + ?$.

Picture the Problem We can systematically determine Q, B, S, and s for each reaction and then use these values to identify the unknown particles.

(a) For the strong reaction: $\qquad\qquad\qquad p + \pi^- \to \Sigma^0 + ?$

Charge number: $\qquad\qquad\qquad\qquad\qquad +1 - 1 = 0 + Q \Rightarrow Q = 0$

Baryon number: $\qquad\qquad\qquad\qquad\qquad +1 + 0 = +1 + B \Rightarrow B = 0$

Strangeness: $\qquad\qquad\qquad\qquad\qquad\quad 0 + 0 = -1 + S \Rightarrow S = +1$

Spin: $+\frac{1}{2} + 0 = +\frac{1}{2} + s \Rightarrow s = 0$

> These properties indicate that the particle is the kaon, K^0.

(b) For the strong reaction: $p + p \rightarrow \pi^+ + n + K^+ + ?$

Charge number: $+1 + 1 = +1 + 0 + 1 + Q \Rightarrow Q = 0$

Baryon number: $+1 + 1 = 0 + 1 + 0 + B \Rightarrow B = +1$

Strangeness: $0 + 0 = 0 + 0 + 1 + S \Rightarrow S = -1$

Spin: $+\frac{1}{2} + \frac{1}{2} = 0 + \frac{1}{2} + 0 + s \Rightarrow s = +\frac{1}{2}$

> These properties indicate that the particle is either the Σ^0 or the Λ^0 baryon.

(c) For the strong reaction: $p + \overline{K}^- \rightarrow \Xi^- + ?$

Charge number: $+1 - 1 = -1 + Q \Rightarrow Q = +1$

Baryon number: $+1 + 0 = +1 + B \Rightarrow B = 0$

Strangeness: $0 - 1 = -2 + S \Rightarrow S = -1$

Spin: $+\frac{1}{2} + 0 = +\frac{1}{2} + s \Rightarrow s = 0$

> These properties indicate that the particle is the kaon, K^+.

*43 ••• A Σ^0 particle at rest decays into a Λ^0 plus a photon. (a) What is the total energy of the decay products? (b) Assuming that the kinetic energy of the Λ^0 is negligible compared with the energy of the photon, calculate the approximate momentum of the photon. (c) Use your result from Part (b) to calculate the kinetic energy of the Λ^0. (d) Use your result from Part (c) to obtain a better estimate of the momentum and the energy of the photon.

Picture the Problem The total kinetic energy of the decay products is the rest energy of the Σ^0 particle. We can find the momentum of the photon from its energy and use the conservation of momentum to calculate the kinetic energy of the Λ^0.

(a) The total kinetic energy of the decay products is given by:

$$K_{tot} = (m_\Sigma)c^2$$

Substitute numerical values (see Table 41-1) and evaluate K_{tot}:

$$K_{tot} = \left(1193 \frac{\text{MeV}}{c^2}\right)c^2 = \boxed{1193\,\text{MeV}}$$

(b) The momentum of the photon is given by:

$$p_\gamma = \frac{E_\gamma}{c} = \frac{E - m_\Lambda c^2}{c}$$

Substitute numerical values and evaluate p_γ:

$$p_\gamma = \frac{1193\,\text{MeV} - \left(1116\,\frac{\text{MeV}}{c^2}\right)c^2}{c}$$

$$= \boxed{77.0\,\frac{\text{MeV}}{c}}$$

(c) The kinetic energy of the Λ^0 is given by:

$$K_\Lambda = \frac{p_\Lambda^2}{2m_\Lambda}$$

or, because $p_\Lambda = p_\gamma$,

$$K_\Lambda = \frac{p_\gamma^2}{2m_\Lambda}$$

Substitute numerical values and evaluate K_Λ:

$$K_\Lambda = \frac{\left(77.0\,\frac{\text{MeV}}{c}\right)^2}{2\left(1116\,\frac{\text{MeV}}{c^2}\right)} = \boxed{2.66\,\text{MeV}}$$

(d) A better estimate of the energy of the photon is:

$$E_\gamma = E - m_\Lambda c^2 - K_\Lambda$$

Substitute numerical values and evaluate E_γ:

$$E_\gamma = 1193\,\text{MeV} - \left(1116\,\frac{\text{MeV}}{c^2}\right)c^2 - 2.66\,\text{MeV} = \boxed{74.3\,\text{MeV}}$$

The improved estimate of the momentum of the photon is:

$$p_\gamma = \frac{E_\lambda}{c} = \frac{74.3\,\text{MeV}}{c} = \boxed{74.3\,\frac{\text{MeV}}{c}}$$